BIODEGRADATION OF CELLULOSE

HOW TO ORDER THIS BOOK

BY PHONE: 800-233-9936 or 717-291-5609, 8AM–5PM Eastern Time

BY FAX: 717-295-4538

BY MAIL: Order Department
Technomic Publishing Company, Inc.
851 New Holland Avenue, Box 3535
Lancaster, PA 17604, U.S.A.

BY CREDIT CARD: American Express, VISA, MasterCard

BY WWW SITE: http://www.techpub.com

BIODEGRADATION OF CELLULOSE
ENZYMOLOGY AND BIOTECHNOLOGY

Anthony J. Clarke, Ph.D.
University of Guelph

Biodegradation of Cellulose
a**TECHNOMIC**® publication

Published in the Western Hemisphere by
Technomic Publishing Company, Inc.
851 New Holland Avenue, Box 3535
Lancaster, Pennsylvania 17604 U.S.A.

Distributed in the Rest of the World by
Technomic Publishing AG
Missionsstrasse 44
CH-4055 Basel, Switzerland

Printed in the United States of America
10 9 8 7 6 5 4 3 2 1

Main entry under title:
Biodegradation of Cellulose: Enzymology and Biotechnology

A Technomic Publishing Company book
Bibliography: p.
Includes index p. 261

Library of Congress Catalog Card No. 96-60383
ISBN No. 1-56676-414-9

To
Lee Anne
and
Michael, Andrew and Chelsea

Table of Contents

Preface

THE first scientific reports concerning the activity of cellulolytic enzymes were published over a century ago, and the term "cellulase" was first coined by Pringsheim in 1912 [1]. In spite of this early interest in the biodegradation of cellulose, little progress was made during the first half of this century regarding an understanding of the enzymology of cellulose hydrolysis. Thus, before 1945 only two cellulases had been studied in any detail; a cellulase from the snail *Helix pomatia*, and an enzyme from the fungus *Aspergillus oryzae*. These early investigations were obviously hampered by the technical limitations of the day, and the slow progress is reflected in the questions that are posed in the introduction of a seminal paper published by Reese and co-workers in 1950 [2]. The three questions that constituted the purpose of that particular study have, in fact, formed and directed most subsequent research into the mode of action of cellulolytic enzymes. Indeed, one of these questions, "How do the number, nature, and location of the substituents affect the availability of the modified cellulose?" could easily comprise an introduction to an investigation published today!

Such was the state of affairs until the so-called oil crisis of the 1970s shifted much attention on the recovery of energy from renewable resources. This era also marked the dawning of biotechnology, and hence great emphasis was placed on the production of inexpensive products on a large scale. Although enthusiastically embraced by biotechnologists, venture capitalists, and government funding agencies, the success of these pursuits was thwarted by lack of a basic understanding of the structure and function relationship of the enzymes involved in the biodegradation of cellulosic and heteroxylosic materials. The experimental approaches in the 1970s and 1980s concerning the enzymology of these enzymes, like most enzymes at that time, was restricted to chemical modification and amino acid sequence homology studies and only one X-ray crystal structure of a cellulolytic enzyme was solved prior to 1990. This situation has radically changed in the past five years with the design and synthesis of new mechanism-based inhibitors and the application of molecular biological techniques which has allowed the production of forms of the enzymes suitable for crystallization. Hence, a total of twelve three-dimensional structures of cellulases, cellobiohydrolases and xylanases are now known, with half of these appearing in the literature during the past year alone. In addition, a number of others have been crystallized and their structures are currently being solved.

The plethora of literature now available on the biodegradation of cellulosic and heteroxylosic materials is truly impressive and stands as a testament to the power, vitality and potential of the modern scientific approach. As a consequence of the imagination, ingenuity and skill of the many researchers who have contributed to our current understanding of the mechanism of action of the cellulolytic and heteroxylanolytic enzymes, the task of providing a comprehensive and complete review of this literature is now well beyond the capability of a single author and I have certainly made no attempt to achieve this. Instead, I hope that this book reflects these features as attention is focused on the structure and function relationship of the four major hydrolytic enzymes, cellulases, cellobiohydrolases, β-glucosidases, and xylanases as it relates to their mechanism of action. It should be of interest to biotechnologists and industrial researchers interested in manipulating these enzymes to their full potential as catalysts for various current and new applications. The book should also be of interest to individuals working in the areas of biochemistry, botany, crop science, ecology, microbiology and mycology, in addition to those in the forestry and forest product industries. It begins with an overview of the nature of cellulose and heteroxylan, followed by a description of the enzymes involved in its hydrolysis, their general structure, characteristics, and classification.

Chapter 3 discusses how these various enzymes are integrated and associated for the efficient solubilization of cellulose and heteroxylan. This includes a review of the literature concerning the cellulosomes and other cellulolytic complexes. The chapter on production and purification provides an overview of this subject matter. This is by no means comprehensive, as the volume of literature pertaining to the induction, repression, production, assay, isolation, and purification of the respective cellulolytic and xylanolytic enzymes is immense. Rather than attempt to provide a complete catalogue citing the contribution of all workers in these fields, which would certainly have led to the inadvertent exclusion of many, I have tried to present the salient features. These first four chapters thus set the stage for the more comprehensive discussion of the mechanism of action of each of the primary hydrolases which follows. As noted above, very rapid advances have been made in this area of study over the past five years, and I have attempted to present most of the observations relating to the structure and function relationship of the enzymes as found in the literature up until August 1995.

It was my intent to convince the reader that we now, after over forty years of research, have sufficient evidence to substantiate the hypotheses presented by Koshland in 1953 for the mechanism of hydrolysis by carbohydrases [3], at least for the β-glycosidases. Thus, the data pertaining to the catalytic mechanism of both the retaining and inverting forms of these enzymes since such studies were initiated on cellulases by Whitaker and co-workers in 1954 [4] is reviewed in Chapter 5. The book concludes with an overview of the mode of action of the enzymes and a discussion, citing a few examples, of how the modern methods of molecular biology, enzymology, and X-ray crystallography are being used to manipulate selected enzymes for a variety of biotechnological and industrial purposes.

On a technical note, the figures of the three-dimensional structures of the enzymes and domains discussed in the text (Chapters 2, 5 and 6) have been reproduced in color, as black and white reproductions simply do not do justice to both the structures themselves and the researchers who have devoted months, and in some cases years, elucidating them.

REFERENCES

1 Pringsheim, H. 1912. "Über der fermentativen Abbau der Zellulose," *Z. Physiol. Chem.* 78:266–291.

2 Reese, E. T., R. G. H. Siu, and H. S. Levinson. 1950. "The biological degradation of soluble cellulose derivatives and its relationship to the mechanism of cellulose hydrolysis," *J. Bacteriol.* 59:485–497.

3 Koshland, D. E. 1953. "Stereochemistry and the mechanism of enzymatic reactions," *Biol. Rev.* 28:416–436.

4 Whitaker, D. R. 1954. "Mutarotation after hydrolysis of cellopentaose by *Myrothecium verrucaria* cellulase," *Arch. Biochem. Biophys.* 53:439–449.

Acknowledgements

MANY people have been indirectly involved in this production. I wish to acknowledge the encouragement and stimulation provided by present and past graduate students who have pursued research in this area of cellulase and xylanase enzymology under my direction: Mark Bray, Anastasia (Taz) Jager, Kelly McAllister, Teresa Sanelli, and Henri Strating. To the rest of my current research group, Sarah Broughton, Alan Carriere, Karen Lee, Zusheng Li, Ken Payie, Gaynor Watson, and Steve Watt, who persevered in my absence from the laboratory during the summer of '95, I offer my thanks for their patience and understanding. I would also like to express my appreciation to my mentors, whom I still regard with the utmost respect and highly value their friendship: Thammiah Viswanatha for both stimulating my interest in enzymology and taking the risk; Birte Svensson for switching my attention to carbohydrases and convincing me to tackle this project; and Makoto Yaguchi for providing me with the opportunity to initiate research on cellulases. Most of all, I deeply thank my family: my parents for providing home-offices away from home; my children, Michael, Andrew and Chelsea for their patience, understanding, and countless pots of tea; and my wife Lee Anne, for her unending love, encouragement, and support.

Studies in my laboratory concerning the enzymology of cellulolytic and heteroxylanolytic enzymes have continuously been supported by operating grants from the Natural Sciences and Engineering Research Council of Canada.

CHAPTER 1

Chemistry and Structure of Cellulose and Heteroxylan

THE plant cell wall is a complex matrix comprised of two main layers, the primary and secondary walls, which are situated between the plasma membrane and middle lamella (Figure 1.1). Details of the primary wall were mainly derived from studies on cultured sycamore cells that revealed it to be a gel-like matrix of cellulose, heteroxylans, and pectins (Figure 1.2). The pectins are often found cross-linked to the heteroxylans, which, in turn, are hydrogen bonded to the surface of cellulose microfibrils. Pectins are also abundant in the middle lamella, the region between the cell walls of adjacent cells. The secondary wall is a multilayered complex of predominantly cellulose microfibrils with successive layers being deposited in different orientations. Lignin, a complex, insoluble polymer of phenolic residues, is particularly abundant in wood where it associates with the primary wall and in the secondary walls of the xylem. The following provides an overview of the salient features of the structure and composition of the major plant cell wall polymers. This cursory introduction to the voluminous literature published on the cell wall polymers is simply intended to provide a background for what follows concerning the enzymology of its biodegradation, and the reader is directed to the reviews of Young and Rowell [1] and Carpita and Gibeaut [2] for a comprehensive treatment of this subject.

The carbohydrate composition of the lignocellulosic materials isolated from hardwood, softwood, wheat straw, and sugar cane are summarized in Table 1.1. With glucose being present in the highest concentration of the monosaccharides, followed by xylose, it is not surprising to find that the glucan and heteroxylan polysaccharides constitute over 50% of the dry weight material present in the different cell wall types (Table 1.2). The primary polysaccharide of both hardwood and softwood is cellulose, which constitutes, on average, over 42% of the dry weight. Associated with the cellulose of hardwoods are partially acetylated, acidic heteroxylans and, to a lesser extent, some mannans. In contrast, the predominant polymers

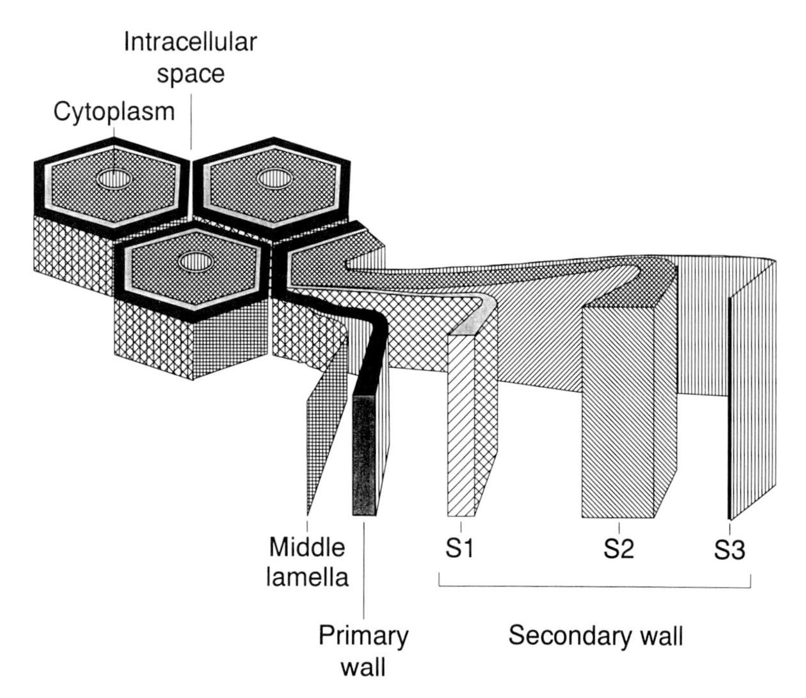

Figure 1.1 Schematic representation of the structure of a typical wood cell. The middle lamella that is composed predominantly of lignin forms the outer layer around the primary wall. The secondary wall is comprised of three layers: outer layer (S1), middle layer (S2), and inner layer (S3). In most cases, the cellulose content is highest in the S1 and S2 layers. These two layers also contain the highest levels of uronic acid-containing heteroxylans, whereas S3 is rich in L-arabino-D-glucurono-D-xylans. Each layer is built up of a series of cellulose sheets. The cellulose fibers of individual sheets are aligned in a more or less parallel orientation, but the direction varies within different sheets (adapted from Reference [95]).

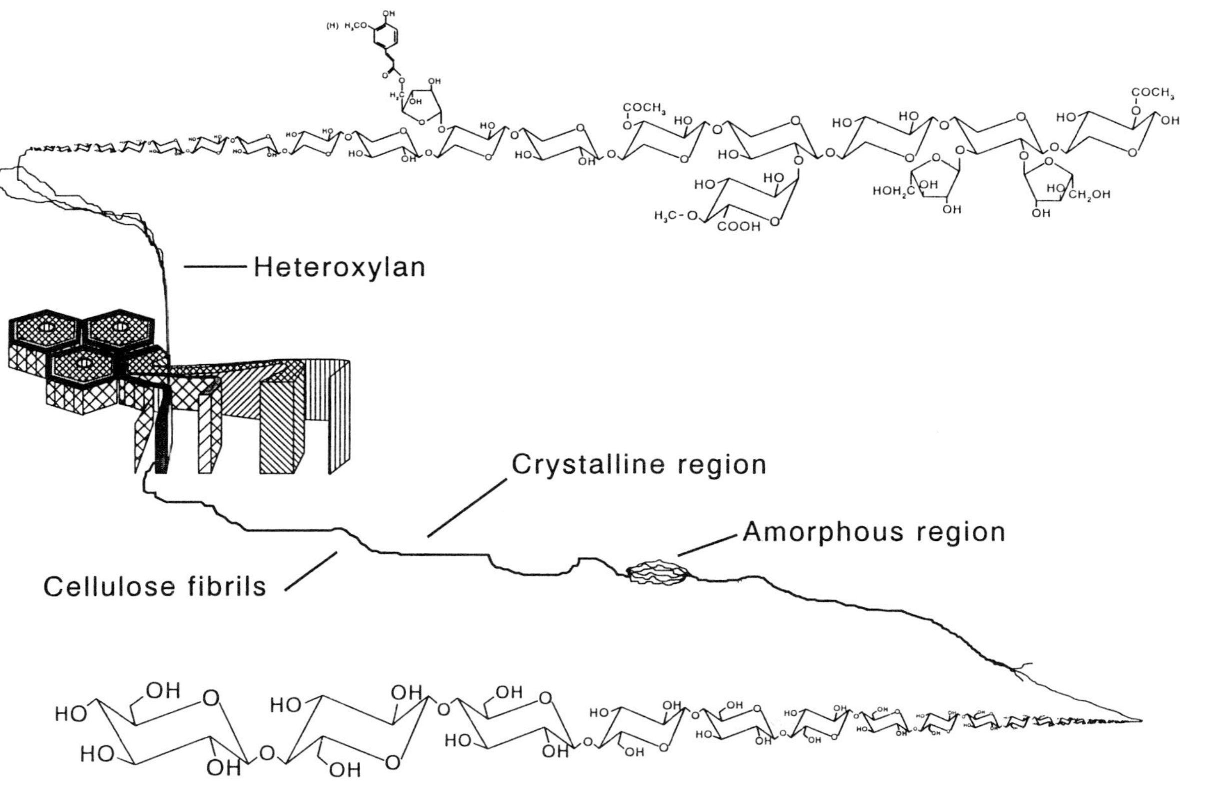

Figure 1.2 Schematic representation of the polysaccharides comprising the primary cell wall of a typical wood cell.

TABLE 1.1. Compositional Analysis[a] of Carbohydrates in the Hydrolysates of Four Different Representative Lignocellulosic Materials.

Source	Arabinose	Galactose	Glucose	Lignin	Mannose	Xylose
Cottonwood	0.27	0.34	44.1	22.7	2.15	13.7
Pine	1.29	2.33	44.1	27.1	11.4	6.19
Straw	2.46	0.66	38.1	22.8	0.17	22.6
Bagasse	1.47	0.35	44.5	22.7	0.21	22.5

[a]Values presented as percentages based on dry weight of raw material.
Source: Compiled from Reference [41].

TABLE 1.2. Polysaccharides[a] Associated with Four Different Representative Lignocellulosic Materials.

Polysaccharide	Cottonwood	Pine	Straw	Bagasse
Glucan	42.4	42.4	31.9	38.8
Xylan	13.0	5.94	18.9	21.4
Mannan	2.05	11.0	0.15	0.20
Arabinan	0.25	1.28	2.08	1.40
Galactan	0.33	2.29	0.56	0.34
Total	58.3	62.9	53.8	62.5

[a]Values presented as percentages based on dry weight of raw material.
Source: Compiled from Reference [41].

associated with softwood cellulose are galactoglucomannans, whereas a heteroxylan similar to that of hardwoods constitutes only a small fraction, typically approximately 3%, of the total polysaccharides present. Arabinogalactans, comprising up to 20% of the dry weight, have also been observed in some softwoods, e.g., larchwood.

1.1 CELLULOSE

Cellulose is the major carbohydrate synthesized by plants and thus is the most abundant organic polymer produced in nature. It constitutes the fibrillar component of all higher plant walls, comprising approximately 2–4% in cereal endosperm walls [3] to approximately 94% in secondary walls of cotton seed hairs [4]. The French chemist Anseleme Payen (1795–1871) is credited as the first to attempt the separation of wood into its constituents [5]. The fibrous substance extracted from wood after treatment with nitric acid was coined cellulose, and its formula $C_6O_{10}H_5$, which is isomeric with starch, was determined in 1839. Cellulose was later found to be associated with lignin by Schulze, who reserved the term cellulose for the fraction resistant to aqueous acid hydrolysis [6]. Most subsequent studies on the physicochemical properties of cellulose have been conducted on the residual cellulose fraction remaining after extraction of plant cell walls with aqueous chelating agents and strong alkali. This fraction, termed α-cellulose, is usually associated with the hydroxyproline-rich glycoprotein extensin and some nonglucosyl carbohydrates (for example, References [7,8]). It is still unclear as to the nature of the association between the nonglucosyl component and cellulose, but it is likely to involve only noncovalent interactions.

Cellulose is a linear polymer of β-(1→4)-linked anhydrous glucose residues that, depending upon the source, may extend to 15,000 residues in length. The glucose moieties exist in the lowest energy conformation of β-D-glucopyranose, the chair configuration. Because of the nature of the β-linkage, the linear polymers form an extended ribbon structure with a twofold screw axis and each glucopyranose residue is oriented 180° relative to its neighbours. As a result, the basic repeating unit of cellulose is not glucose but instead cellobiose (Figure 1.3). Intramolecular hydrogen bonding between adjacent residues confers rigidity, and two stable ordered conformations of cellulose have been observed [9]. In one form, termed k_I, the oxygen of the C-6 hydroxyl and the ring oxygen of the same glucosyl residue are donors in a bifurcated hydrogen bond with the proton of the C-3 hydroxyl from an adjacent residue serving as the acceptor [Figure 1.3(a)]. Thus, in this form of cellulose, the C-6 hydroxyl groups on adjacent glucosyl residues are nonequivalent as are the glycosidic linkages. The intramolecular hydrogen bonding observed in k_{II} cellulose involves the simple interaction

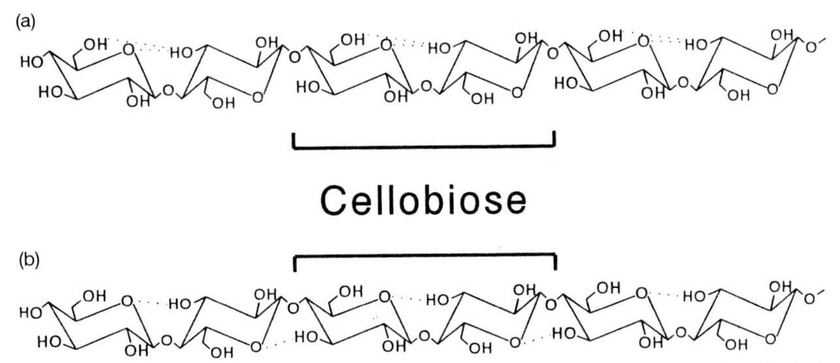

Figure 1.3 Composition and hydrogen-bonding network of cellulose (a) k_{I} and (b) k_{II}. The basic recurring unit of a cellulose chain is cellobiose giving rise to the nonequivalence centered at the glucosidic linkages.

between the ring oxygen of one residue and the C-3 hydroxyl of an adjacent residue [Figure 1.3(b)]. In this second form, nonequivalence occurs only at the glycosidic linkages.

In nature, cellulose chains pack into ordered arrays to form insoluble microfibrils that are stabilized by cross-links involving intermolecular hydrogen bonds. Microfibrils involving k_{I} cellulose chains may form highly ordered crystalline regions in which neighbouring strands are staggered to ensure the highest degree of hydrogen bonding, or they can be arranged in less ordered amorphous regions. These less ordered regions may be comprised of both k_{I} and k_{II} celluloses. The microfibrils range in lateral dimension from 3 to 4 nm in higher plants up to 20 nm for those of the alga *Valonia macrophysa.* Although each cellulose chain may be several thousand units long [10,11], they begin and end at different locations within a microfibril. This structure can be likened to a length of cotton thread that is composed of many thousands of individual cotton fibers each only several centimeters long.

The crystallinity of cellulose was first proposed in 1858 by Carl von Nägli using polarizing microscopy [12], but it took eighty years to confirm these observations by X-ray crystallography. Naturally occurring cellulose was found to have a monoclinic unit cell, and on the basis of further X-ray data, early workers proposed an antiparallel orientation of the cellulose chains within a microfibril [13]. This view persisted until as recently as 1974 with the report on the crystalline structure of the cellulose from the green alga *Valonia ventricosa* [14]. These early studies were conducted on regenerated forms of cellulose obtained from the natural form. It is now recognized that the extraction process alters the structure of cellulose [15] and on the basis of X-ray and electron diffraction patterns, the dominant orientation of the chains in the natural celluloses isolated from the bacteria *Acetobacter xylinum* and *Phaseolus aureus*, the green alga *V. ventricosa*, cotton (*Gossypium* spp.) seed hairs, and ramie (*Boehmeria nivea*) bast fibers [9,16–18] is generally believed to be parallel. It has been reasoned that this parallel configuration, designated cellulose I, would form the strongest arrangement of intramolecular hydrogen bonds while ensuring low intermolecular attractive forces, thereby combining strength with flexibility [19]. Although the limiting size of crystals used for study by diffraction techniques prevents unambiguous assignment [16,17,20–22], other direct and indirect evidence does further support the parallel assignment and precludes both chain folding and antiparallel packing. Biosynthetic considerations would suggest that this parallel orientation exists for all native celluloses [23]. On the other hand, the antiparallel configuration of cellulose II, which is comprised of k_{II} cellulose, is exclusively obtained upon recrystallization of cellulose I, suggesting the former represents the more thermodynamically stable configuration [19]. Hence, it is possible that eventually both forms of cellulose will be found to exist in nature [9].

Solid state carbon-13 NMR studies have indicated that two slightly different forms of type I cellulose, named I_α and I_β, occur in nature which differ in intermolecular hydrogen bonding patterns. Observations made with celluloses from *A. xylinum* and *V. ventricosa*, and two fibrous celluloses, cotton linters and ramie, indicated that both Type I forms exist in the various celluloses but to different extents. Crude estimates suggested that *Acetobacter* cellulose is 60–70% I_α, whereas cotton is approximately 60–70% I_β. This led to the proposal that all native celluloses are composites of the two Type I forms with the structural cellulose of higher plant forms being richer in Type I_β [9]. These and other results obtained by the characterization of the solid-state structure of cellulose are reviewed in Reference [24].

The degree of crystallinity, degree of polymerization, and width of cellulose microfibrils will depend on its source, age, and pretreatment and can vary from 0% for amorphous, acid-swollen cellulose to nearly 100% for the cellulose isolated from *V. macrophysa* [25]. Cellulose from cotton is approximately 70% crystalline [26], whereas the degree of crystallinity of most commercial celluloses varies between 30 and 70%. Cotton seed hairs are the best characterized in terms of degree of polymerization. The cellulose of the primary cell wall is relatively low and heterogeneous (DP 2,000–6,000), whereas secondary wall cellulose is characterized by higher degrees of polymerization and is more homogeneous (DP 14,000) [27,28].

1.2 HETEROXYLANS AND ASSOCIATED POLYMERS

1.2.1 COMPOSITION AND STRUCTURE

The major noncellulosic polysaccharides of primary walls of gymnosperms (grasses) and secondary walls of all angiosperms have been collec-

TABLE 1.3. Nature of the Side Chain Substitutions Associated with Heteroxylans.

Substitution	Typical Linkage to Main Chain Xylose	Heteroxylan	Reference
Terminal single units			
α-D-Glucuronic acid		Angiosperms	[30]
α-4-*O*-Methyl-D-glucuronic acid	1→2	Gymnosperms	[31]
α-L-Arabinofuranose	1→3	Gymnosperms	[31]
Complex oligosaccharides			
β-D-Xyl*p*-(1→2)-α-L-Ara*f*	1→3	Corn cob	[88]
β-D-Gal*p*-(1→5)-α-L-Ara*f*	1→3	Bamboo leaves	[89,90]
β-D-Gal*p*-(1→4)-D-Xyl*p*-(1→2)-α-L-Ara*f*	1→3	Bamboo leaves	[89,90]
4-Me-α-D-Glc*p*A-(1→4)-D-Xyl*p*-(1→4)-D-Gal*p*	1→3	Bamboo leaves	[89,90]
Arabinose side chains			
Oligo-(1→2)-L-arabinofuranose	1→3	Wheat kernel	[91]
Oligo-(1→3)-L-arabinofuranose	1→3	Wheat kernel	[91]
Oligo-(1→5)-L-arabinofuranose	1→3	Wheat kernel	[91]
Oligo-(1→2,3)-L-arabinofuranose	1→3	Wheat kernel	[91]
Nonsaccharide substitutions			
Acetic acid	→2, →3, →2,3	Angiosperms	[92,93]
	L-Arabinofuranose	Angiosperms	[38]
Ferulic acid	(→5)-L-Arabinofuranose	Monocots	[94]

Source: Compiled from Reference [58].

tively labelled as hemicellulose. Historically, the term hemicellulose was used to identify the polysaccharides extracted from plant cell walls by alkali. However, this term has also been widely used as a label for specific xylosidic components of plant walls. Because chemically related material is often soluble in both alkali and water, whereas others are alkali insoluble, Bacic and colleagues [29] made an attempt to define the chemical composition of the wall polysaccharides based on structure rather than reflect the nature of its isolation. This form of chemical classification based on structure rather than solubility is more consistent with regulations adopted by I.U.P.A.C. Moreover, as noted by Bacic et al. [29], such nomenclature parallels the International Union of Biochemistry regulations for the naming of the associated hydrolases (and indeed, all enzymes) that are based on the nature of both the substrate and the bonds cleaved. The persistent use of the term hemicellulose has led to the coining of the *phenomenase* term hemicellulase for enzymes that hydrolyse hemicellulose, which are primarily xylanases. Unfortunately, even a cursory survey of the current literature reveals that *hemicellulase* is still widely used to define xylanase.

Heteroxylans comprise a family of heteropolysaccharides having a backbone consisting of β-(1→4)-linked D-xylosyl residues. The β-(1→4) xylans are often seen to be highly branched with both the composition and structure of the polymers being dependent on both source and manner of extraction. The most commonly encountered substituents associated with the heteroxylans of various origins are listed in Table 1.3. Those from woody plants contain a variety of different carbohydrate residues, the most common of which are D-mannose, D-glucose, D-galactose, L-arabinose, D-galacturonic acid, D-glucuronic acid, and 4-*O*-methyl-D-glucuronic acid [29]. In contrast, the heteroxylans from graminaceous plants are associated with fewer types of sugars, predominantly D-glucose, D-galactose, and L-arabinose, but both the variety and abundance of linkages are greater. Several excellent and exhaustive reviews have been written on the composition and structure of the heteroxylans and associated polymers from both wood and grasses, the most notable of which include those by Timell [30,31] and Wilkie [32].

1.2.1.1 Hardwoods and Grasses

In general, *O*-acetyl-(4-*O*-methylglucurono)xylan (Figure 1.4) is the predominant form of heteroxylan in hardwoods [33] (Table 1.4). This structure has a linear framework of, on average, 200 β-(1→4)-linked D-xylopyranose residues with single 4-*O*-methyl-α-D-glucuronic substituents occurring at the C-2 position of approximately 10% of its xylosyl residues [30]. Although few in number, examples of unbranched linear xylan homopolymers have been isolated including those from tobacco stalks [34], guar seed husk [35], and esparto grass [36]. The best characterized is the xylan from esparto grass, which, because of its relative simplicity, has served as a model xylan for chemical and physical studies [36]. Acetyl substituents are present at the C-2 and/or C-3 positions on up to approximately 70% of the xylose residues associated with hardwood heteroxylan [37]. The distribution of both the acetyl moieties and the *O*-methylglucuronic acid residues along the main-chain xylose residues may not necessarily be even. Analysis of a carefully extracted sample of the heteroxylan from *Mimosa scabrella* (Bracitanga), which is comprised of xylose, 4-*O*-methylglucuronic acid and *O*-acetyl groups in the molar ratio 60:7:33, revealed an uneven distribution with clusters of up to six consecutive xylose residues modified by both acetyl and *O*-methylglucuronic acid substituents, whereas five consecutive xylose residues were observed to be *O*-acetylated alone [38]. Minor amounts of rhamnose and galacturonic acid also were observed to be associated with the main chain of various hardwood heteroxylans [39].

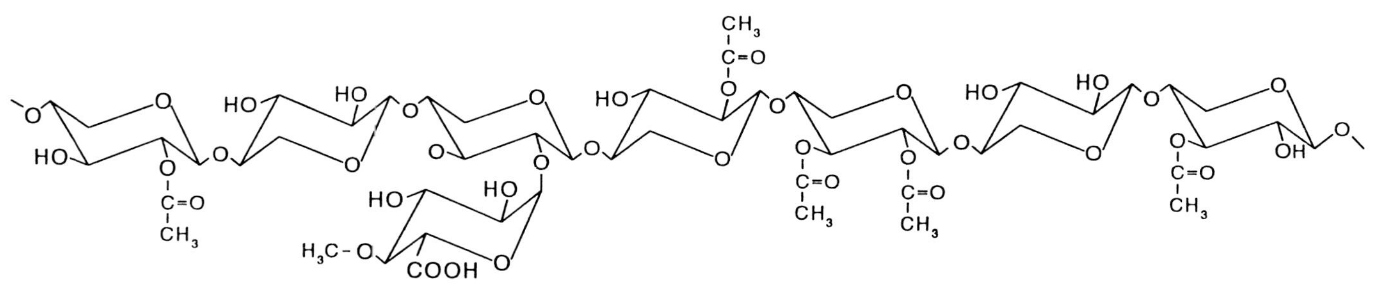

Figure 1.4 General structure of *O*-acetyl-(4-*O*-methylglucurono)xylan from hardwoods.

TABLE 1.4. Polysaccharides Comprising the Cell Walls of Hardwoods and Grasses.

Polysaccharide	Occurrence	Percentage of Material[a]	Carbohydrate Composition	Molar Ratio	Linkages	Degree of Polymerization
O-Acetyl-4-*O*-methylglucuronoxylan	Hardwoods	10–35	β-D-Xyl*p*	10	1→4	200
			4-*O*-Me-α-D-Glu*p*A	1	1→2	
			O-Acetic acid	7		
Glucomannan	Hardwoods	3–5	β-D-Man*p*	1–2	1→4	>70
			β-D-Glu*p*	1	1→4	
Arabinoxylan	Grasses	20–40	β-D-Xyl*p*	Varying	1→4	>70
			L-Ara*f*		1→3,1→2,3	
			4-*O*-Me-α-D-Glu*p*A		1→2	
			Ferulic acid			
			p-Coumaric acid			
			O-Acetic acid			

[a]Based on dry weight.
Source: Adapted from Reference [41].

A small proportion (3–5%) of the lignified secondary walls of hardwoods consists of glucomannans, a family of polysaccharides with linear extended backbones of β-(1→4)-linked residues of randomly arranged β-D-mannopyranose and β-D-glucopyranose (Figure 1.5). The glucose:mannose ratio is approximately 1:2, and the degrees of polymerization range from less than 100 to several thousand [31,40]. Little substitution with either acetyl groups or other carbohydrate residues has been observed on the glucomannans from hardwoods.

A wide variety of heteroxylans may be found among both the various species of the grasses and the different tissue cells, but in general, they are distinguished from wood heteroxylans by containing smaller proportions of uronic acids while being more highly branched and containing larger amounts of L-arabinofuranosyl residues. For example, every seventh xylose residue of the heteroxylan from the internodes of wheat was found to be substituted by L-arabinofuranose [41], whereas the xylose:arabinose ratio observed in trisaccharide units isolated from the heteroxylan from wheat leaves was 1:2 [32]. The single L-arabinofuranose residues are linked predominantly to the C-3 hydroxyls of the heteroxylan backbone residues (Figure 1.6), but binding to the C-2 position also was observed [42]. When present, 4-*O*-methylglucuronic acid substitutions occur at the C-2 hydroxyl of xylose backbone residues.

The arabinoxylans from graminaceous plants are known to be extensively esterified with both acetyl and phenolic acid groups (Figure 1.6). Approximately 1–2% of the dry weight of cell walls of grasses may be comprised of *O*-acetyl groups that esterify the C-2 or C-3 of xylose residues. A similar weight ratio of phenolic acids (viz., 1–2%) is also associated with these arabinoxylans where their linkage is to the C-6 hydroxyl of the arabinose residues. For barley straw heteroxylan, approximately one in every fifteen arabinose residues is esterified with ferulic acid, whereas one in every thirty-one is esterified with *p*-coumaric acid [43].

1.2.1.2 Softwoods

Although the linear backbone of the major softwood heteroxylans is also comprised of β-(1→4)-linked D-xylopyranose residues, the composition and pattern of the various substituents are quite different from the other heteroxylans. Thus, approximately seven of ten xylopyranose residues are substituted at the C-2 position with 4-*O*-methyl-α-D-glucuronic acid (Figure 1.7). In place of the acetyl residues found on hardwood heteroxylans, approximately 12% of the xylose residues of softwood heteroxylans are directly linked through C-3 to α-L-arabinofuranosyl groups. Other sugar decorations that have been observed to be associated with softwood heteroxylans include D-glucose, D-mannose, and D-galactose.

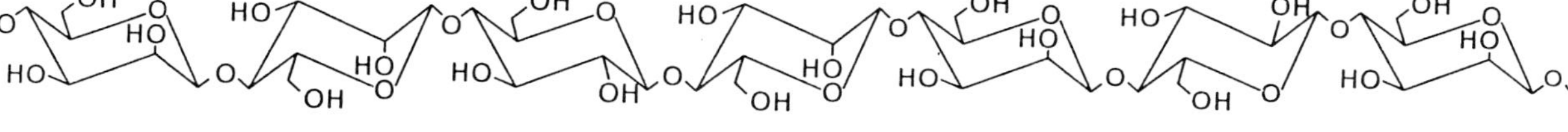

Figure 1.5 General structure of glucomannan from hardwood.

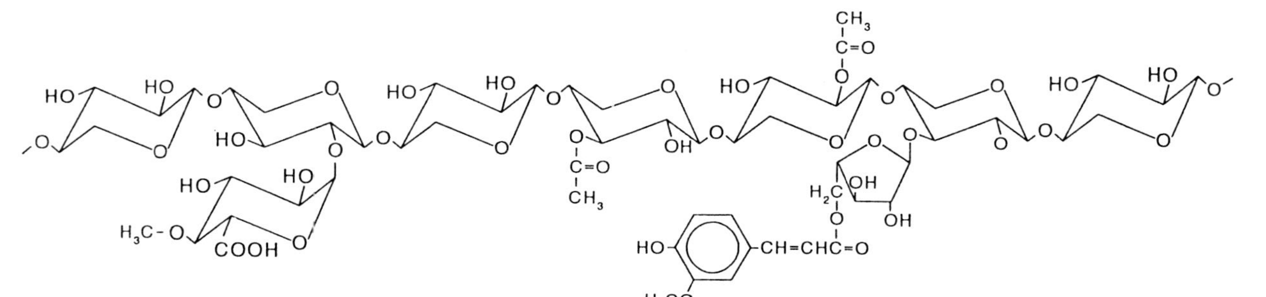

Figure 1.6 Structure of a hypothetical arabinoxylan from grasses, containing acetyl, feruloyl, and 4-*O*-methylglucuronyl substitutions.

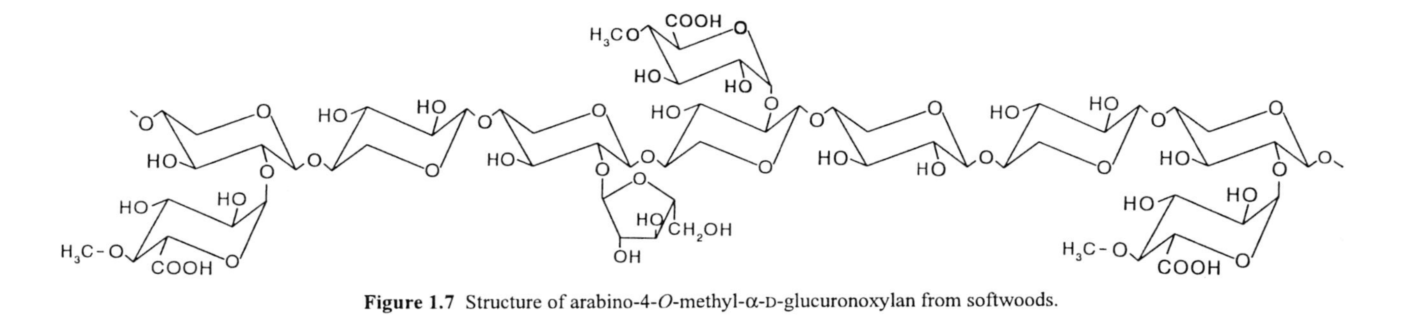

Figure 1.7 Structure of arabino-4-*O*-methyl-α-D-glucuronoxylan from softwoods.

At least two types of galactoglucomannans are known to be associated with heteroxylans, but they occur in relatively larger proportions in the lignified secondary walls of softwoods (12–15%) compared with hardwoods (3–5%). The predominant galactoglucomannans typically have a galactose:glucose:mannose ratio of 1:1:3, whereas the ratio of the second, less soluble form is 0.1:1:3. Because of its low galactose content, the latter polymer is often referred to as glucomannan. Both are comprised of randomly distributed β-(1→4)-linked D-glucopyranose and D-mannopyranose residues. Degrees of polymerization range from less than 100 to several thousand [31,40]. The single α-D-galactopyranose side chains are randomly linked to the C-6 of either backbone hexose (Figure 1.8). In addition, some mannopyranosyl residues may be substituted at C-6 with β-D-galactopyranosyl-(1→2)-α-D-galactopyranosyl side chains. The galactoglucomannans are also acetylated, with acetyl contents ranging from 5.9 [44] to 8.8% [37]. Most acetylation occurs equally distributed between C-2 and C-3 of the mannopyranosyl residues [37,44], but there is at least one report [45] of both mannose and glucose acetylation occurring at only the C-3 position.

One other major class of polymers associated with cell walls of softwoods are the pectins, a group of complex acidic and neutral oligosaccharides and polysaccharides. Type II arabinogalactan, the predominant softwood pectin, consists of β-(1→3)-linked D-galactopyranose each substituted at C-6. The substituents can be the following: two β-(1→6)-linked arabinopyranose residues; a trisaccharide of 3-*O*-β-L-arabinosyl-α-L-arabinofuranosyl-3-β-D-galactopyranose; or a single α-L-arabinofuranose, β-D-galactopyranose, or β-D-glucuronic acid residue (Figure 1.9) [46,47]. In some instances (e.g., ryegrass endosperm cells [48]), Type II arabinogalactans have been observed to be associated with the plasma membrane or constituents of the extracellular space.

1.2.2 CONFORMATION AND INTERACTIONS OF HETEROXYLAN

Much of what is known about the chemical structure of heteroxylans has been acquired through classical chemical, enzymatic, and spectroscopic analyses of the alkali-soluble fraction. This approach has provided the detailed information discussed above on the general characteristics of only the true hemicellulose fraction of woods, and, moreover, such data represent only average structures. Until recently, little or no attention was paid to the finer details of heteroxylan and associated polymer content in complete samples where structural features previously considered to be insignificant or due to impurities may profoundly affect the mode of action of the various enzymes involved with biomass saccharification. Such enzymatic activity is also affected by both the relative distribution of side chains and the physical conformation of the polysaccharides (for example, see Reference

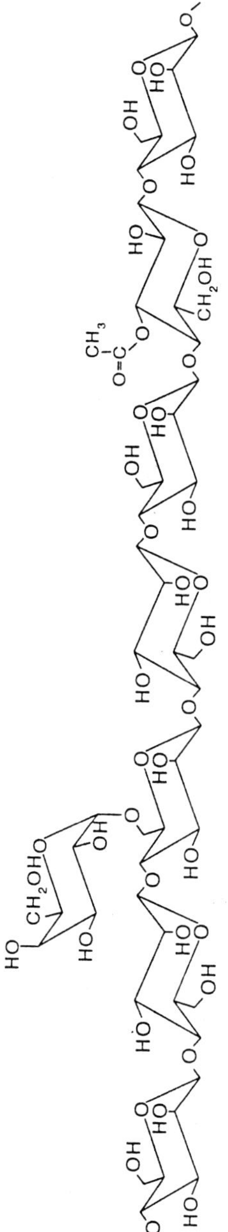

Figure 1.8 Structure of acetyl-galactoglucomannan from softwoods.

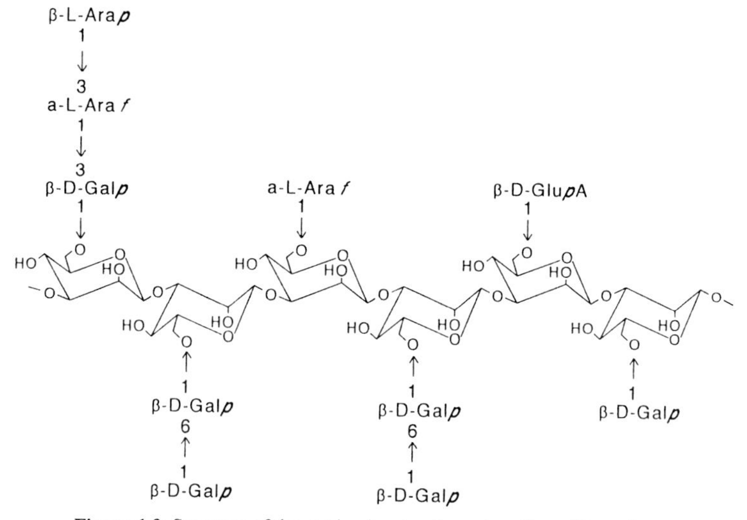

Figure 1.9 Structure of the predominant softwood pectin arabinogalactan.

[49]). Investigations involving specific degradation by chemical reagents [50] or enzymes in conjunction with the modern techniques available for the analysis of oligosaccharides, such as high performance liquid chromatography, two-dimensional NMR, and fast-atom bombardment (FAB)-mass spectrometry, are now providing details of the true primary structures of heteroxylans [51–54].

Exhaustive acid hydrolysis of the heteroxylan from the softwood larch, followed by GC-MS analysis of the released oligosaccharides and undegraded limit polymer, suggested that the uronic acid substituents are irregularly distributed along the heteroxylan backbone [55]. Similar results have been found for the distribution of uronic acids in the heteroxylan from another softwood, *Sequoia sempervirens* [56] and suggest that open sections of unsubstituted heteroxylan exist that are more accessible and readily hydrolysed by endoxylanases with the subsequent production of highly decorated limit heteroxylans. Details of the fine structure of these limit heteroxylans are now also emerging. For example, treatment of wheat arabinoxylan with an endo-(1→4)-β-D-xylanase from a species of *Aspergillus* yielded a mixture of heteroxylans that was resolved by high-performance liquid chromatography separation and characterized by both monosaccharide and methylation analysis, FAB-mass spectrometry and high-resolution one- and two-dimensional-[^{1}H]NMR spectrometry techniques. These analyses identified twenty-seven different components involving ten specific branching patterns [54].

X-Ray crystallography has aided the elucidation of the three-dimensional structure, interactions, and properties of heteroxylans. In general, β-(1→4)-linked xylan chains form twisted ribbon-like helical strands with a threefold, left-handed symmetry. In contrast, mannans tend to be more untwisted, ribbon-like structures, formally as twofold helices [57]. The critical difference between the heteroxylans and mannans (and, for that matter, cellulose) is the ability of the latter to use the hydroxymethyl group at position 5 of the ring to form cooperative networks of inter- and intrachain hydrogen bonds. Change of this hydroxymethyl to a single hydrogen in the xylan backbone totally suppresses this hydrogen-bonding pattern. Because cooperative hydrogen bonds involving hydroxymethyl groups (lattice forces) hold the cellulose chain into the twofold conformation within a microfibril, the chains at the surface will tend to be perturbed by twisting because they are not so firmly held to this conformation. Therefore, the capacity of heteroxylans to associate with cellulose depends on their conformation in situ in the cell wall relative to cellulose. The current scenario of interaction is that heteroxylan chains complex by molecular recognition with the surfaces of cellulose or cellulose-like fibrils and form a hydrated, more compliant matrix [57]. In contrast to proteins, where the side groups of constituent amino acids influence the three-dimensional shape of the polymer chain, structural analyses indicate that adding bulky side chains such as arabinose does not perturb the basic conformation of the glycosidic linkage geometry of the xylose chains. However, the negative charges brought about by the glucuronosyl substituents of heteroxylans may be an influencing factor on chain conformation.

On the basis of the drastic alkaline conditions needed to extract heteroxylans from secondary walls, it has long been thought that strong bonds between heteroxylans and wall polymers must exist. Xylanases have been utilized to isolate lignin-carbohydrate complexes for determination of the linkages associating carbohydrate and lignin [58]. The reagent 2,3-dichloro-5,6-dicyano-1,4-benzoquinone, which specifically effects an oxidative cleavage of benzyl ether linkages, has also been used for the analysis of lignin-carbohydrate complexes. The best-documented covalent-bond linkages between heteroxylans and other polymers include glycosidic, ester, and ether linkages that exist between the main-chain or branch sugars and various noncarbohydrate moieties such as ferulic acid and lignin [58–61]. Of these various interactions, the best characterized is the benzyl ester linkage between lignin and the carboxylate of 4-*O*-methylglucuronic acid residues of glucuronoxylans [62–64]. It is thought that this ester bond is formed between the D-glucuronic acid residues and quinonmethide intermediates during the polymerization of lignin [65]. Another commonly encountered covalent interaction between lignin and heteroxylan is the benzyl ether linkages to either xylose residues [60,66] or L-arabinose side chains [59]. Other covalent interactions between lignin and heteroxylan may involve the oxygen-containing bonds of acetals and glycosidic linkages. Glycosidically linked arabinose residues have been observed in enzymatic digests of lignin-carbohydrate complexes [67,68], but although acetal bonds have been proposed [69], only limited evidence for their existence has been documented [70].

Esters of *p*-hydroxycinnamic acids may interconnect separate heteroxylan chains or link them to other cell wall polymers [71–75]. Treatment of gramaceous cell walls with peroxidases led to the formation of cross-links between arabinoxylan chains involving an alkali-stable dimerization of hydroxycinnamic acids [75]. Analogous studies on the cell walls of corn (*Zea mays*) employing feraxanases, enzymes that catalyze the hydrolysis of feruloylated arabinoxylans (Feraxans), suggested that in addition to the existence of linkages between galacturonans and arabinoxylans, cross-links between arabinoxylan chains involving ferulic acid may exist [76]. Finally, several other types of covalent interactions between heteroxylans and both other cell wall polysaccharides, such as pectins, and structural proteins and polyphenols have been proposed [77].

1.3 LOCALIZATION OF CELLULOSE AND HETEROXYLANS

Electron microscopy in combination with immunogold labelling has provided some details on the ultrastructural localization of cellulose and heteroxylans within plant cell walls. Early studies involved a difference (or subtractive) microscopy technique where electron micrographs of plant tissues were obtained before and after treatment with either cellulase or xylanases. Results obtained with the holocelluloses of the secondary walls of beech wood and spruce wood indicated an uneven distribution of both heteroxylans and mannans but application of xylanases with mannases revealed some synergy, suggesting an intimate association of the two polysaccharides. Application of this technique to investigations on the elongating internode of a gramaceous plant confirmed the role of heteroxylans in cementing cellulose microfibrils [78].

Immunogold labelling techniques have been more recently employed to visualize the intact and native plant cell wall morphology. Using immunogold-labelled xylanase and cellobiohydrolase, the glucuronoxylans of linden wood [79] and quince seed (*Cydonia oblonga*) [80], respectively, were shown to be in tight association with cellulose and to form a charged coat around each cellulose microfibril, thereby contributing to the observed helical pattern of the microfibrils. Using colloidal gold of a smaller diameter, the distribution of heteroxylan in *Arundo donax* was observed to be slightly different, but confirmation of a high concentration of heteroxylan in primary cell walls was obtained [81]. Similar studies have been extended to the analysis of kraft pulps in which the heteroxylans were removed by treatment with chaotropic reagents, followed by reassociation and labelling with the xylanase-immunogold complex. These studies demonstrated the strong interaction, sorption, and retention of heteroxylans with cellulose microfibril surfaces [82]. A refinement of this technique has involved the generation of antibodies directed against a variety of cell wall monosaccharides and oligosaccharides [83]. For example, a (1→3,1→4)-β-glucan-specific monoclonal antibody was found to bind strongly to the walls of the aleurone in thin sections of immature wheat (*Triticum aestivum*) grains but not to the middle lamella region [84], whereas an anti-(1→3)-β-glucan antibody bound specifically to discrete patches on the aleurone walls [85]. Immunocytology using the electron microscope has localized the various saccharides in dividing and differentiating cells of bean (*Phaseolus vulgaris*) root and callus tissue. The method has since been applied to the study of two varieties of corn, differing in their lignin content [86].

In the future, the development of techniques that permit the observation and analysis of polymers in their natural location in the cell wall, such as high-resolution electron microscopy, solid-state NMR, and Fourier transform infrared spectroscopy (FTIR), should give more useful detailed information about native conformation and the various interactions of cellulose and heteroxylans in the plant cell wall. Indeed, neutron diffraction and proton nuclear magnetic resonance studies have been employed as complementary probes of in situ cellulose dimensions and primary cell wall structure [87].

CHAPTER 2

Enzymology of Biodegradation of Cellulose and Heteroxylans

IN view of the complicated nature and interactions of the various polysaccharides comprising plant cell walls, it is not too surprising that their biodegradation requires a complex system of secreted enzymes with different specificities and modes of action. These β-(1→4) glycoside hydrolases, like all enzymes, are grouped and classified according to their specificities and action patterns. Thus, the microbial degradation of cellulosic biomass requires the action of at least the following three distinct enzymes: cellulase (EC 3.2.1.4; endoglucanase), cellobiohydrolase (EC 3.2.1.91), and β-glucosidase (EC 4.2.1.21) which act synergistically to hydrolyse the β-1,4 bonds of cellulose to glucose. Figure 2.1 presents a simplistic view of this concerted action of the three enzymes. Another exoacting enzyme, an exoglucohydrolase (EC 3.2.1.74) has also been described [96,97], but its role in the saccharification of cellulose has not been delineated.

Cellulose is exposed to the cellulolytic complex of enzymes by the action of the endo-1,4-β-xylanases (EC 3.2.1.8; 1,4-β-D-xylan xylanohydrolase, xylanase), which catalyze the hydrolysis of the xylose backbone of heteroxylan chains layered on cellulose microfibrils. The xylanases act at random locations on the heteroxylan chains that have themselves become exposed by the action of the various debranching enzymes [33] (Figure 2.2). These latter enzymes include α-*N*-arabinofuranosidases (EC 3.2.1.55), α-glucuronidases (EC 3.2.1.x), β-xylosidases (EC 3.2.1.37), acetylesterases (EC 3.1.1.6), and feruloyl (*p*-coumaroyl) esterases (EC 3.1.1.x). A complete list of enzymes currently known to participate in the biodegradation of celluloses, heteroxylans, and associated polysaccharides is presented in Table 2.1. Some of the listed enzymes have been only recently discovered, and, hence, very little information on their properties is currently available. With a very narrow knowledge base, few generalizations can presently be drawn. However, because cellulose and the heteroxylans constitute the bulk of the plant cell matter, the enzymes directly involved in their hydrolysis have

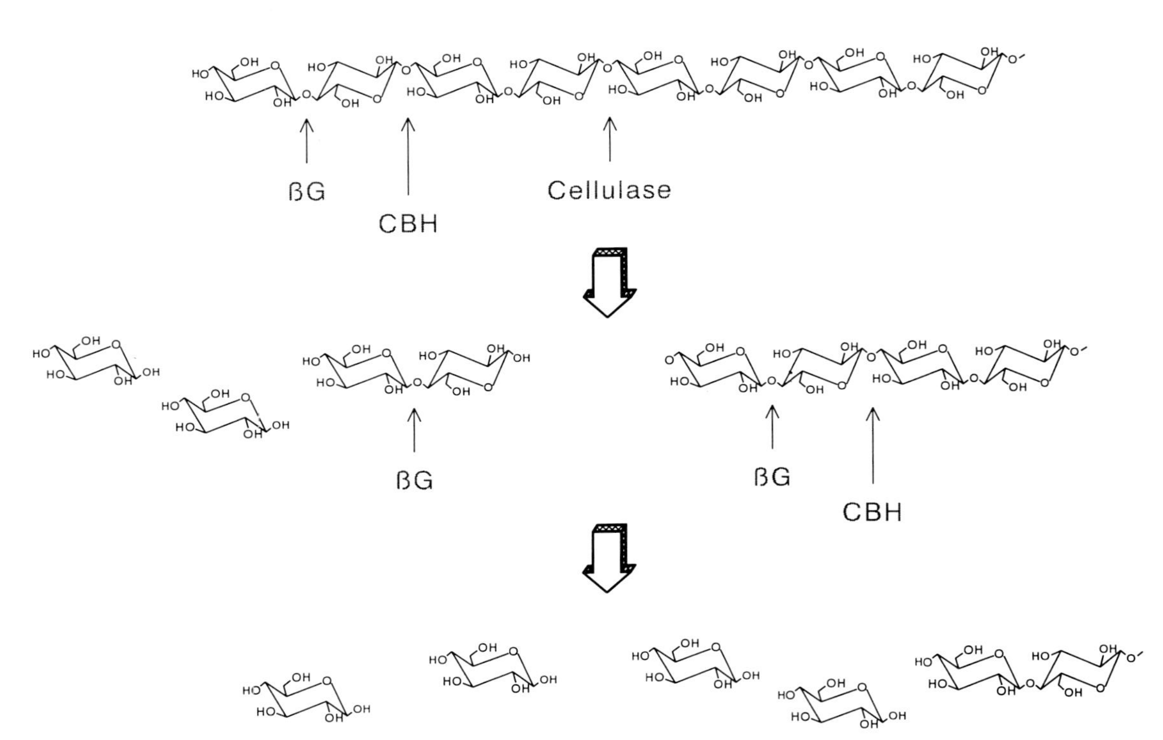

Figure 2.1 The enzymatic hydrolysis of cellulose. The sites of attack on the β-1,4-glucosidic linkages of cellulose by cellulase, cellobiohydrolase (CBH) and β-glucosidase (βG) are indicated by the thin arrows.

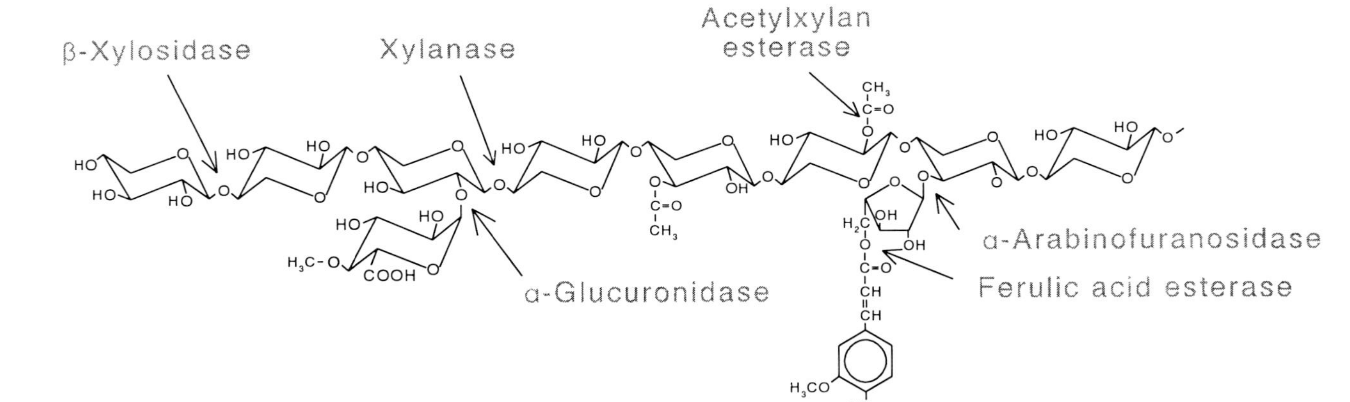

Figure 2.2 Enzymes involved in the hydrolysis of a hypothetical heteroxylan.

TABLE 2.1. Enzymes Known to Facilitate the Hydrolysis of Cellulose, Heteroxylans, and Associated Polymers.

Enzyme Name				
Common	Systematic	EC Number	Description	Reference
Cellulose				
Cellulase (endoglucanase)	1,4-(1,3:1,4)- β-D-Glucan 4-glucanohydrolase	3.2.1.4	Random cleavage of β-1,4 linkages of cellulose with preference for soluble and amorphous forms of the substrate. Affinity decreases with decreasing degree of polymerization with no activity on cellobiose.	[242]
Cellobiohydrolase	1,4-β-D-Glucan cellobiohydrolase	3.2.1.91	Release of cellobiose from the nonreducing ends of cellulose with preference for crystalline forms of the substrate.	[242]
β-Glucosidase	β-D-Glucoside glucohydrolase	3.2.1.21	Release of β-D-glucose from the nonreducing ends of a wide variety of cellulose, cello-oligosaccharides, and a wide variety of β-1,4-glucosides.	[242]
Glucan 1,4-β-glucosidase	1,4-β-D-Glucan glucohydrolase	3.2.1.74	Release of β-D-glucose from 1,4-β-D-glucans, but not cellobiose.	[242]
Xylan backbone				
Endoxylanase	1,4-β-D-Xylan xylanohydrolase	3.2.1.8	Random cleavage of β-1,4 linkages of xylans with preference for unsubstituted regions. Affinity decreases with decreasing degree of polymerization with no activity on xylobiose.	[242]
	1,4-β-D-Arabinoxylan xylanohydrolase	3.2.1.x	Cleavage of arabinoxylans with preference at linkages in vicinity of arabinosyl substituents.	[243]

TABLE 2.1. (continued).

Enzyme Name				
Common	Systematic	EC Number	Description	Reference
Xylan backbone (continued)				
	1,4-β-D-Glucuronoxylan xylanohydrolase	3.2.1.x	Cleavage of glucuronoxylans with preference at linkages in vicinity of 4-*O*-methyl-glucuronic acid substituents.	[164]
β-D-Xylosidase	β-D-Xyloside xylohydrolase	3.2.1.37	Release of xylose from nonreducing ends of β-1,4-linked heteroxylans and xylo-oligosaccharides, including xylobiose.	[242]
Exoxylanase	β-D-Xylan xylohydrolase	3.1.2.x	Release of xylose from nonreducing ends of β-1,4-linked heteroxylans and xylo-oligosaccharide, excluding xylobiose.	[169]
Xylan debranching hydrolases				
Acetylxylan esterase		3.1.1.x	Release of acetyl groups from heteroxylans.	[244]
α-L-Arabinosidase	1,4-α-Arabinoxylan arabinofuranohydrolase	3.2.1.x	Release of 1,3- or 1,2-α-L-arabinosyl substituents from arabinoxylans.	[245]
	α-L-Arabinofuranoside	3.2.1.55	Release of 1,3- or 1,2-α-L-arabinosyl arabinohydrolase substituents from artificial arabinosides and arabinosyl-substituted pectins in addition to arabinoxylans.	[242]
Coumaric acid esterase		3.1.1.x	Release of arabinose-linked coumaric acid from cereal arabinoxylans.	[246,247]
Ferulic acid esterase		3.1.1.x	Release of arabinose-linked ferulic acid from cereal arabinoxylans.	[246,247]

(continued)

TABLE 2.1. (continued).

Enzyme Name		EC Number	Description	Reference
Common	Systematic			
Xylan debranching hydrolases (continued)				
α-1,2-L-Fucosidase	2-*O*-α-L-Fucopyranosyl-β-D-galactoside galactohydrolase	3.2.1.63	Release of L-fucose α-1,2 linked to D-galactosyl substituents of xyloglucans.	[242]
α-D-Glucuronidase		3.2.1.x	Release of α-1,2-linked D-glucuronic acid or 4-*O*-methylglucuronic acid residues from substituted xylo-oligomers released by endoacting enzymes.	[248]
Associated hydrolases				
β-L-Arabinosidase	β-L-Arabinopyranoside arabinohydrolase	3.2.1.x	Release of β-linked L-arabinosyl side chains of arabinogalactans.	[166]
Endoarabinase	Endo-(1,5)-α-L-arabinase	3.2.1.99	Random cleavage of α-1,5-linked backbone of arabinan.	[242]
Endogalactanase	1,4-β-D-Galactan galactanohydrolase	3.2.1.x	Random cleavage of β-1,4-linked backbone of Type I arabinogalactans.	[166]
Endomannase	1,4-β-D-Mannan mannanohydrolase	3.2.1.78	Random cleavage of β-1,4-linked backbone of galactoglucomannans.	[242]
α-D-Galactosidase	α-D-Galactoside galactohydrolase	3.2.1.22	Release of α-1,6-linked D-galactopyranosyl side chains from galactoglucomannans.	[242]
β-D-Galactosidase	β-D-Galactoside galactohydrolase	3.2.1.23	Release of β-1,6-linked D-galactopyranosyl side chains from galactoglucomannans.	[242]

Adapted from Reference [166].

received much attention, and a number of excellent comprehensive reviews have been published [98–102]. The following briefly summarizes some of the more salient features of the four major hydrolases.

2.1 PROPERTIES OF CELLULOLYTIC AND HETEROXYLANOLYTIC ENZYMES

2.1.1 CELLULASE

The cellulases are a class of secreted enzymes that randomly attack and hydrolyse the β-(1→4) bonds of cellulose to produce cello-oligosaccharides. According to I.U.B. recommendations, cellulase is the recommended name for this specific enzyme with endoglucanase serving as an alternative. Unfortunately, however, the term *cellulase* is very often applied to the complete complex of cellulolytic enzymes, which leads to some confusion in the literature. To comply with the current I.U.B. regulations then, the use of *cellulase* should be restricted to the specific enzyme, whereas *cellulase complex* and, still better, *cellulolytic complex* could be used to identify the complement of three different enzymes (viz., cellulase, cellobiohydrolase, and β-glucosidase).

Cellulases rapidly decrease the viscosity of cellulose solutions, thus providing a relatively specific means for assaying their activity. This activity appears, in general, to be confined to amorphous regions of cellulose because very few isolated cellulases have been shown to hydrolyse crystalline cellulose. The isolated cellulases from the fungi *Trichoderma viride* [103,104] and *T. koningii* [105] represent exceptions to this general rule because they decrease the viscosity of both amorphous Avicel and untreated cotton linters. The size of the limit cello-oligosaccharides resulting from cellulase activity alone will vary depending on the source of the enzyme, but very few are capable of efficiently hydrolysing substrates with a degree of polymerization less than 4. However, their binding sites appear to be flexible enough to accommodate xylan and thereby effect its slow hydrolysis. For example, the rate of hydrolysis of a cellulase (EGI) from the white-rot fungus (basidiomycete) *Schizophyllum commune* toward xylan is approximately 1% of that for carboxymethyl cellulose (unpublished observations). Intrinsic xylanase activity has also been noted for the *Irpex lacteus* cellulase [106]. In this case, the K_M of the enzyme for xylan was greater than that for carboxymethyl cellulose by a factor of 1.6, whereas the V_{max} for xylan was approximately 18% of the value for the cellulose substrate. Likewise, the *Clostridium thermocellum* cellulase E (EGE) has high activity against xylans, with its ratio of xylanase:carboxymethyl cellulose activity approximately 1:10 [107]. The physicochemical properties of some isolated and purified cellulases are presented in Table 2.2. The molecular weights of the fungal enzymes range between 11 and over 100 kDa, but the majority of the

TABLE 2.2. Properties of Some Fungal and Bacterial Cellulases.

Organism	M_r[a] (k)	p*I*	Glycosylation (% by wt.)[b]	pH_{opt}	$Temp_{opt}$ (°C)	Stability		Reference
						pH	Temp (°C)	
Fungi								
Aspergillus aculeatus	25	4.8	0	4.5	50	2–9	45	[249]
	38	3.4	7.0	4.0	65	3.5–9	65	
	66	4.0	27.0	5.0	70	3–8	70	
	68	3.5	8.3	2.5	60	3.5–6	50	
Aspergillus fumigatus		7.1						[121]
Aspergillus niger	23.6	4.47	0	4.8				[250]
	26		0	3.8–4.0	45	1–9		[251]
	31	3.67	0	4.0		5.0–8.0	70	[252]
	46	3.3						[253]
Coriolus versicolor	29.5			5.0	55	4.0–6.0	55	[254]
Dichomitus squalens	42	4.8		4.8–5.0	55–60	4–8	65	[255]
	47	4.1		4.6–4.8	55–60	4–8	65	
	56	4.3		4.8	55–60	4–8	65	
Eupenicillium javanicum	21			5.0–5.5	55			[256]
	30			5.0–5.5	55			
	41			5–5.5	55			
	47	n.d.		5–5.5	55			
	65			5–5.5	55			
Fusarium solani	37	4.75–5.5						[128,257]
Geotrichum condidum	11	4.0						[258]
	36	4.0						
	67	3.3						
	82	3.75						
	145	4.0						

TABLE 2.2. (continued).

Organism	M_r[a] (k)	p*I*	Glycosylation (% by wt.)[b]	pH_{opt}	$Temp_{opt}$ (°C)	Stability		Reference
						pH	Temp (°C)	
Fungi (continued)								
Humicola insolens	57			5.0	50	3.5–9.5	65	[259]
Humicola grisea	63		39.0	5.0	50	3.5–9.5	65	[259]
Irpex lacteus	35.6		10.0	4.0–5.0	40	3.0–6.0	50	[106,260]
Meruliporia (Serpula) incrassata	25.2	<3.6	+	2.5–4.0		2.0–5.0		[261]
	48.5	<3.6	+	2.5–4.0		2.0–5.0		
	57.1	<3.6	+	2.5–4.0		2.0–5.0		
Phanerochaete chrysosporium	28.3	4.40	7.8					[262]
	32.3	5.32	10.5					
	36.7	4.72	0					
	37	4.20	2.2					
	37.5	4.65	4.7					
Polyporus versicolor	11.4	4.50	33					[263]
Schizophyllum commune	38	3.5	+	5.5	40	4–8	30–70	[264,265]
	40	3.7	+	5.5	40	4–8	30–70	
Sclerotium rolfsii	27.5	4.20	+	2.8–3	50	3–7		[266]
	50 (2)	4.55	+	4.0	74	4–7		
	77.6	4.51	+	4.0	50	4–7		
Sporotrichum pulverulentum	28.3	4.4	7.8					[267]
	32.3	5.32	10.5					
	36.7	4.72	0					
	37	4.2	2.2					
	37.5	4.65	0					
Thermoascus aurantiacus	34	—	1.8	4.5–5	65	5–9	65	[268]

(continued)

TABLE 2.2. (continued).

Organism	M_r[a] (k)	p*I*	Glycosylation (% by wt.)[b]	pH_{opt}	$Temp_{opt}$ (°C)	Stability pH	Stability Temp (°C)	Reference
Fungi (continued)								
	49		2.6	4.5–5	68	2–12	65	
	78		5.5	5	75	5–9	65	
Trichoderma koningii	13	4.72						[105]
	31	5.09						
	48	4.32						
	48	4.32						
Trichoderma reesei	20	7.5	0					[117–119]
	43	4.0						
	48	5.5	6.0					
	55	4.5	10.0					
	56	5.0	27					
	67	6.5						
Trichoderma viride								
Onozuka SS	12.5	4.6	21.0					[103]
	50	3.39	12.0					
Pancellase SS	37		4.5					[269]
	49		15.0					
	52		15.0					
QM 9414	23.5	7.7		5.5				[116]
	45	6.9	+	4.0				
	50	5.3	+	5.1				
	52	3.5	+	4.5				
	57	4.4	+	4.5				
	58.5	6.5	+	5.0				

TABLE 2.2. (continued).

Organism	M_r[a] (k)	pI	Glycosylation (% by wt.)[b]	pH_{opt}	$Temp_{opt}$ (°C)	Stability		Reference
						pH	Temp (°C)	
Bacteria								
Bacillus sp. strainN-4 (ATCC 21833)	45.7	6 –10.5						[270]
	54.3							
strain PDV	32.5			4.5–7.0	60	4–10	70	[271]
Cellulomonas uda	48		+					[272]
	48		+					
	55		+					
	55.5		+					
	56		+					
	81	4.4	+	5.5–6.5	45–50	5.5–8		
Clostridium cellulolyticum	48			6.0	48			[109]
Clostridium thermocellum	39	6.2		6.0–6.6	60		70	[273]
	56	6.2		6.0	60	5–7	65	[274]
	64			5.9	60	5–7		[275]
	91–99		2	6.4	80	5–7	85	[276]
	94	6.72	11.2	5.2	62			[277,278]
Fibrobacter succinogenes S85	50.1	4.25		5.9	39–44			[279]
	58			6.4	30		<50	[156]
	65	4.75–4.9	0	6.4	39	5.9–7.1	<45	[123]
	94	9.18	0	5.8	39	5.4–6.2	<45	[123]
	118	9.4	0					[123]

(continued)

TABLE 2.2. (continued).

Organism	M_r[a] (k)	p*I*	Glycosylation (% by wt.)[b]	pH_{opt}	$Temp_{opt}$ (°C)	Stability		Reference
						pH	Temp (°C)	
Bacteria (continued)								
Pseudomonas fluorescens	40		13.0	8.0		7–8		[280]
	100		36.0	8.0		7–8		
			33.0	7.0		7–8		
Ruminococcus albus 8	30	6.0–6.1						[281]
AR67								
SY3								
F-40								
Sporocytophaga myxococcoides	46	7.5	0	7.0				[120]
	52	4.75	0	5.5–7.5				
Streptomyces lividans	36	4.2						[282]
Thermomonospora fusca	84 (2)	4.7	+					[283,284]
	45	4.5	0					[283]
	71	3.1	+					[283]
	106	3.6	0					[283]
	108	3.2	0					[283]
Thermotoga maritima	27			6.0–7.5	95		80	[285]

[a]Values in parentheses indicate the number of subunits in the holoenzyme.
[b]+ denotes the presence of an unknown amount of glycosylation.

characterized enzymes are 30–55 kDa in size. The bacterial cellulases tend to be a little larger with many having molecular weights greater than 65 kDa. Most of the fungal enzymes are glycoproteins with the carbohydrate chains being linked via both asparagine and serine and threonine residues. Some of the bacterial enzymes are also known to be glycosylated [108–110]. As with other glycoproteins, the glycan chains are thought to provide both protection from proteolytic attack [111] and thermostability [112,113]. In addition, glycosylation has also been implicated in the adsorption of the cellulases to insoluble substrates [114].

The majority of the fungal cellulases studied to date are characterized by acidic p*I* values ranging from as low as 2.86 for the cellulase IV of *Talaromyces emersonii* [115] to neutrality (Table 2.2), but there are a few exceptions. Although barely alkaline, the p*I* values of cellulases from *Trichoderma viride* QM9414 (EG IV) [116], *T. reesei* [117–119], *Sporocytophaga myxococcoides* [120], and *Aspergillus fumigatus* [121] are 7.7, 7.5, 7.5, and 7.1, respectively. Continued investigation of the cellulases produced by plants may reveal the production of a number of truly basic enzymes because studies on the bean plant (*Phaseolus vulgaris* L. cv. Red Kidney) has led to the discovery of a cellulase with a p*I* of 9.5 [122]. Bacterial cellulases are also generally acidic in nature, but the exceptions are extreme. Although most bacterial enzymes have p*I* values between pH 3.1 and 6.1, two cellulases from *Fibrobacter succinogenes* are relatively alkaline, with p*I* values of 9.2 and 9.4 [123].

The pH optima of most cellulases also tends to be acidic (pH 2.5–7.0) but again there are a few exceptions among the bacteria, especially species of alkalothermophilic *Bacillus*. The pH activity profile of the *Bacillus* sp. NK1 cellulase is very broad, ranging from 5 to 10.9 [124,125]. This broad pH specificity is apparently dependent on only two specific amino acid residues because their replacement (Ser287Asn and Ala296Ser) results in a decrease in enzymatic activity in the alkaline range [126]. Moreover, the enzymatic activity of the wild-type NK1 cellulase in the alkaline range varies with the size of the oligomeric substrate and, thus, appears to require substrates spanning distal substrate-binding subsites [127].

As secreted enzymes, the cellulases are relatively resilient to extremes of temperature and pH, and this is reflected in their optimum temperatures for activity. These temperature optima are quite broad, ranging from 30 to 75°C, but the majority of the enzymes are most active at temperatures equal to or greater than 50°C. These values, as with all of the most established kinetic and enzymatic properties have been obtained in vitro with isolated and purified enzymes, and it is quite possible that the cellulases together with the other cellulolytic and heteroxylanolytic enzymes are even more stable when stabilized by the environmental conditions in situ. Indeed, temperatures of up to 60–65°C have been recorded within compost piles.

2.1.2 CELLOBIOHYDROLASE

Cellobiohydrolases release cellobiose from the nonreducing ends of a cellulosic substrate. The exoaction of this enzyme has been conclusively shown from studies in which the decrease in viscosity or degree of polymerization of a cellulose substrate treated with the enzyme from a variety of sources does not parallel the increase in reducing sugar concentrations [104,128]. Cellobiose is released from various substrates with either chemically reduced [128] or blocked [129] reducing ends, confirming that the enzyme attacks the nonreducing ends of its substrates. Unlike cellulases, fungal cellobiohydrolases display a broad specificity, hydrolysing both crystalline and amorphous celluloses (Table 2.3), but they are generally found to be inactive toward substituted celluloses, such as carboxymethyl cellulose. The cellobiohydrolases are slightly larger than cellulases (molecular weights of 41–85 kDa), but they also have acidic p*I* values (pH 3.6–6.3). All of the characterized enzymes from fungi are known to be glycosylated. The pH activity profiles of most cellobiohydrolases show a single sharp optimum between pH 4 and 5, but interestingly, those of the enzymes from both *Penicillium funiculosum* and *Trichoderma koningii* are bimodal, with optima at pH 2.5 and pH 4–5.5 [130].

Cellobiohydrolases are often the most abundant protein in the filtrates of cultured white-rot fungi (for example, *Trichoderma reesei* [131]). In direct contrast, brown-rot fungi express little or no cellobiohydrolase and consequently, they typically do not enzymatically catalyze the hydrolysis of crystalline cellulose [132]. Instead, these fungi are presumed to effect the degradation of complex carbohydrates by oxidative reactions catalyzed by released metabolic substrates, such as hydrogen peroxide [133]. Most of the cellobiohydrolase produced by fungi is secreted but species of *Trichoderma* appear to also produce a form of the enzyme (CBH II) that remains bound to conidia [134]. CBH II may play a role in the induction of the cellulolytic enzymes by cellulose (discussed further in Chapter 3).

Until recently, it was thought that bacteria do not produce cellobiohydrolases. It is likely that these enzymes escaped prior detection because of both their weaker activity on the plate assays used to screen recombinant clones for cellulolytic enzymes and the presence of highly active cellulases in enzyme isolates. The list of detected cellobiohydrolases from bacteria is thus relatively short (Table 2.3). Few have been unambiguously identified as true cellobiohydrolases [135–138], although all hydrolyse insoluble cellulose but not soluble carboxymethyl cellulose. Nevertheless, further study has revealed that as predicted [139] some bacteria, especially the fungal-like Actinomycetes and related Corynebacteria, degrade crystalline cellulose in a manner analogous to the fungi, involving a variety of physically nonassociated cellulolytic enzymes. Indeed, anaerobic bacteria (for example, the

TABLE 2.3. Properties of Some Fungal and Bacterial Cellobiohydrolases.

Organism		M_r (k)	p*I*	Glycosylation (% by wt.)[a]	pH_{opt}	$Temp_{opt}$ (°C)	Substrates[b]	Reference
Fungi								
Coniophora puteana		50	3.55	2	5.0		A cel, Avicel, *p*NPC, *p*NPL, cellodextrins	[286]
		52	3.6	4	5.0		A cel, Avicel, *p*NPC, *p*NPL, cellodextrins	
Fusarium solani		41	4.75	21			Avicel, cotton, G_3-G_6	[128]
			4.9	12				
Humicola insolens		72		26.1	5.0	50	Avicel	[259]
Irpex lacteus		56		12.2	4.0–5.0	55		[287]
		65		2.4	5.0	50	Cellulose, CMC, G_3-G_6, *p*NPC	[288]
Penicillium pinophilum		46	4.36	9	2.5		A-S cel, G_3	[289]
		50.7	5.0	19	4.5		A-S cel	
Penicillium funiculosum		46.3	4.36	9	2.5; 4–5.5	60	A cel, Avicel	[130]
Penicillium occitanis		60	5.2	20	4–5	60	A-S cel, Avicel, FP, *p*NPC, *p*NPL	[290]
		55	5.9	40	4–5	65	A-S cel, Avicel, FP, *p*NPC	
Sclerotium rolfsii		41.7	4.32	7	4.5	50	A-S cel, Avicel, G_3-G_7	[291]
Sporotrichum pulverulentum		48.6	4.3	0			Avicel, cotton, cellodextrins, Walseth cellulose	[267]
Sporotrichum thermophile		63.8	4.52	7.5	3.5	80	A cel	[292]
Trichoderma koningii		62	3.8	33	2.5; 5.0		Avicel, G_4, G_6	[130,293]
Trichoderma reesei		64	3.9	5.6				[117]
		53	5.9	18			Avicel	
CBHI		65	3.6–4.2	10			A cel, Avicel, cotton, spruce & birch cellulose	[294]
CBHII		85.3	6.3	8			A cel, Avicel, CMC, G_6, HEC, β-glucan, GlcMan	
Trichoderma viride	Onozuka SS	41.8	3.8	9.2			Avicel	[104]
	Meicelase P	53		1.4–10.4				[295]

(continued)

TABLE 2.3. (continued).

Organism	M_r (k)	p*I*	Glycosylation (% by wt.)[a]	pH_{opt}	$Temp_{opt}$ (°C)	Substrates[b]	Reference
Bacteria							
Bacillus circulans	72			4.5	40	CMC, G_3-G_6, *p*NPC, xylan	[144]
	82	5.4		4.5	50	Avicel, CMC,FP, G_4-G_6, *p*NPC, xylan	[145]
Fibrobacter succinogenes	75	6.7	8–16	6.2	39–45	CMC, G_3-G_6, *p*NPG, *p*NPC	[143]
	40	4.9		5.9–6.2	45–50	*p*NPC	[141,142]
Cellulomonas fimi	56		+			CMC, *p*NPC, xylan	[162]
Clostridium thermocellum	75				<50	A cel, Avicel, CMC, G_4-G_6, xylan	[296]
Clostridium stercorarium	30						[297]
Ruminococcus albus	200 (2)			6.8	7.4	Avicel, CMC, G_3-G_5, *p*NPC	[298]
Ruminococcus flavefaciens FD-1	230 (2)		0	5.0	39–45	*p*NPC	[137]
Streptomyces flavogriseus ATCC 33331	45	4.15	+			A-S cel	[135]
Thermotoga maritima	29			6.0–7.5	95	Avicel, CMC, G_3, β-glucan, *p*NPC	[285]

[a] + denotes the presence of an unknown amount of glycosylation.

[b] A cel, amorphous cellulose; A-S cel, acid-swollen cellulose; CMC, carboxymethyl cellulose; FP, filter paper; G_3-G_6, cellotriose-cellohexaose; HEC, *O*-hydroxymethyl cellulose; *p*NPC, *p*-nitrophenyl-β-D-cellobioside; *p*NPL, *p*-nitrophenyl-β-D-lactoside.

rumen bacteria such as *Clostridium thermocellum* and *Fibrobacter succinogenes,* formerly *Bacteroides succinogenes*) have long been known to efficiently hydrolyse crystalline cellulose to cellobiose [140].

The first reported cellobiohydrolase of bacterial origin was that from *Clostridium stercorarium*, which was shown to release cellobiose from acid-swollen cellulose and aid cellulase in the efficient saccharification of ordered substrates, such as filter paper [136]. Since then, investigations have revealed that the bacterial cellobiohydrolases closely resemble their fungal counterparts in both their production and function. Thus, the cellobiohydrolase from *C. thermocellum* ATCC 27405 is a major component of the cellulolytic system of this bacterial thermophile [133], whereas the cellobiohydrolase from *Cellulomonas fimi* has been shown to be very similar to cellobiohydrolase II from the fungus *T. reesei* in both its physical structure, as predicted by amino acid sequence analysis, and activity profile [138].

The cellobiohydrolases produced by some bacteria compared with their fungal counterparts are quite distinct in their substrate profiles (Table 2.4). For example, an enzyme produced by the rumen commensal *F. succinogenes* readily hydrolyses soluble cello-oligosaccharides and analogous chromogenic substrates such as *p*-nitrophenyl-β-D-cellobioside but is not at all active toward Avicel or amorphous cellulose despite being able to weakly bind these insoluble substrates [141,142]. Interestingly, this enzyme also does not hydrolyse carboxymethyl cellulose, but a distinct chloride-stimulated form of the enzyme is weakly active toward this soluble form of cellulose [143]. The substrate profile of this latter enzyme thus resembles that of two recently characterized cellobiohydrolases from *Bacillus circulans* [144,145]. In an effort to distinguish these enzymes from the fungal cellobiohydrolases, which are highly active toward insoluble substrates, the terms *cellobiosidase* and *cellodextrinase* have been adopted (for example, References [141–143]), but this terminology has not yet been officially recognized by the Nomenclature Committee of the International Union of Biochemistry and Molecular Biology.

2.1.3 β-GLUCOSIDASE

As a class of glycosidases, the β-glucosidases display very broad specificity with respect to both the aglycon and glycon moieties of substrates. Natural substrates for the β-glucosidases in addition to cellulose include, for example, the steroid β-glucosides and β-glucosylceramides of mammals, and the cyanogenic β-glucosides of plant secondary metabolism. The fungal and bacterial enzymes expressed to hydrolyse cellulose bind cellobiose and soluble cello-oligosaccharides in one productive mode and release glucosyl residues sequentially from the nonreducing end. These enzymes not only provide the source of energy and carbon in the form of glucose to the host

TABLE 2.4. Properties of Some Fungal and Bacterial β-Glucosidases.

Organism	M_r[a] (k)	pI	Glycosylation (% by wt.)[b]	pH_{opt}	$Temp_{opt}$ (°C)	Substrates[c]	Reference
Fungi							
Aspergillus fumigatus	340 (4)	4.5		4.5	65	G_2, gentiobiose, *p*NPG	[299,300]
	41		8	5.0		G_2, *p*NPG	[301]
Aspergillus niger	150 (2)		0.03				
	137	3.8	12.5			G_2-G_7, *p*NPC, *p*NPG, sophorose, laminarbiose, gentiobose	[302]
Novozyme	118	4.0		4.0	70–75	G_2-G_5	[303]
Aspergillus oryzae	218 (2)	4.3	10	4.0–5.0		G_2, *p*NPG	[304]
Aspergillus phoenicis				4.3		G_2, *p*NPG	[305]
Aspergillus wentii	170		22	1.5–5.0		G_2, *p*NPG	[306]
Aureobasidium pullulans	340 (2)		+	4.5	75	G_2-G_7, *p*NPG	[307]
Botryodiplodia theobromae	350 (8)					G_2, *p*NPG	[308,309]
Candida guilliermondii	48			6.8		*p*NPG	[310]
Humicola insolens	250	4.23	2.5	5.0	50	G_2, *p*NPG, salicin	[259]
Neocallimastix frontalis	153	3.9		6.0	50	A-S cel, G_2, laminaribiose, *p*NPG, sophorose	[311]
Penicillium funiculosum	—	4.65					[130]
Saccharomyces cerevisiae	313			6.8		G_2, *p*NPG	[312]
	300			6.4–6.8	45	G_2, *p*NPG	[313]
Saccharomyces fragilis				5.7–6.2		G_2, *p*NPG	[314]

TABLE 2.4. (continued).

Organism	M_r[a] (k)	p*I*	Glycosylation (% by wt.)[b]	pH_{opt}	$Temp_{opt}$ (°C)	Substrates[c]	Reference
Fungi (continued)							
Schizophyllum commune	96		+	5.4	52	G_2-G_6, *p*NPG	[151,315,316]
	110		+	5.4	52		
Sclerotium rolfsii	95		+	4.2–4.5	68	G_2-G_5, *p*NPC, *p*NPG	[317]
	95.5		+	4.2–4.5	68		
	95.5		+	4.2–4.5	68		
Sporotrichum pulverulentum	165	4.8					[318]
	165	5.15					
	172	4.52					
	175	4.56					
	182	4.87					
Taloromyces emersonii	45.7	3.6		5.1	70	G_2-G_6, *p*NPG	[150]
	57.6	4.41–4.50		5.7	35	*p*NPG	
	100					*p*NPG	
	135	3.4–4.17	50	4.1	70	G_2, *p*NPG	
Thermoascus aurantiacus	87		33.0	4.5–5.0	70	*p*NPG	[268]
Torulopsis wickerhamii	143.6 (2)	3.2	12	4.5	50	Avicel, G_2-G_6, *p*NPG	[319]
Trichoderma koningii	39.8	5.53	0			G_2-G_5, *p*NPG	[320]
	39.8	5.85	2.0				

(continued)

TABLE 2.4. (continued).

Organism	M_r[a] (k)	p*I*	Glycosylation (% by wt.)[b]	pH_{opt}	$Temp_{opt}$ (°C)	Substrates[c]	Reference
Fungi (continued)							
Trichoderma reesei	70	8.2	+	6.0		G_2, *p*NPG, sophorose	[149]
	73		10			G_2, *p*NPG	[321]
	81.6	8.5	1.3	4.5–5		G_2-G_5, MeUmbGlc, *p*NPG	[148]
	98			6.5		G_2	[322]
Trichoderma viride Onozuka SS	47	5.74	0			G_2, G_4, *p*NPG	[323]
	76			5.0		G_2	[324]
Bacteria							
Acetivibrio cellulolyticus	81						[325]
Agrobacterium faecalis	51.5					*d*NPG, G_2-G_5, *p*NPG, salicin	[326]
Alcaligenes faecalis	122			6–7		G_2-G_4, laminaribiose, *p*NPG	[327]
Bacteroides fibrisolvens						G_2-G_5	
Clostridium thermocellum	50			6.0	65	G_2, *p*NPG	[328]
Erwinia herbicola	122			6.0–7.5			[329]
Ruminococcus albus	82		12	6.5	30–35	G_2-G_5, *p*NPC, *p*NPG	[153]
Streptomyces lividans	66						
Streptomyces sp. QM-B814	52.6	4.4		6.5	50	G_2-G_4, lactose, *p*NPC, *p*NPG	[330]

[a]Values in parentheses indicate the number of subunits in the holoenzyme.
[b]+ denotes the presence of an unknown amount of glycosylation.
[c]G_2-G_6, cellobiose-cellohexaose; MeUmbGlc, 4-*O*-methylumbelliferyl-β-D-glucoside; *p*NPC, *p*-nitrophenyl-β-D-cellobioside; *p*NPG, *p*-nitrophenyl-β-D-glucoside.

microorganism but also facilitate the further efficient hydrolysis of cellulose by clearing the end product and competitive inhibitor of cellobiohydrolase and cellulase, cellobiose [146].

The β-glucosidases represent the largest of the cellulolytic enzymes, and quite often they are di- or multimeric. The sizes of the monomeric enzymes range between 41 and 170 kDa with holoenzymes achieving molecular weights of up to 350 kDa for the octameric *Botryodiplodia theobromae* β-glucosidase [147]. The bacterial β-glucosidases tend to remain monomeric but are still relatively large with molecular weights of 50–122 kDa. Although the biochemical properties of relatively few of the enzymes have been thoroughly characterized, all those investigated have acidic p*I* values (pH 3.2–5.9), with the exception of a *Trichoderma reesei* 81.5 kDa β-glucosidase [148]. As with the cellulases and cellobiohydrolases, most of the β-glucosidases are glycosylated; the exceptions being the 39.8 and 47 kDa enzymes from *Trichoderma koningii* and *T. viride*, respectively. Likewise, the optima pH for activity ranges within the acidic region, 1.5–6.8, with the majority of characterized enzymes being most active at pH values between 4 and 5.

β-Glucosidases produced by various microorganisms may be released extracellularly, retained by the cell, or both. Both intra and extracellular β-glucosidases have been detected in a number of fungi, including *T. reesei* [149], *Talaromyces emersonii* [150], and *Schizophyllum commune* [151]. Early studies with the latter white-rot fungus suggested that the extracellular enzyme participates in the hydrolysis of cellulosic matter for nutrient requirements, whereas the intracellular enzyme performs a physiological role in morphogenetic events [151]. However, more recent evidence indicates that the major role of the enzyme is to provide glucose for nutritional purposes. Considerable amounts of β-glucosidase are produced constitutively and secreted by *T. reesei* QM 9414, but the majority remains associated with the cell wall [149]. The cellobiose produced by the concerted action of the extracellular activities of the cellulases and cellobiohydrolases must diffuse to the fungal cell whereupon it is cleaved by the wallbound β-glucosidase to produce glucose. This glucose produced at the fungal cell wall would then be immediately transported into the cell. The retention of the enzyme at the cell wall thus provides the fungus with a competitive edge over other microorganisms in the immediate environment that would otherwise benefit from the extracellular catabolism of the fungus.

Although fungal β-glucosidases may be either secreted or retained by the cell, the bacterial enzymes seem to predominantly remain with the producing cell. Thus, the enzymes from the rumen bacteria such as *F. succinogenes* [152], *Ruminococcus albus* [153], *R. flavefaciens* [154], and *C. thermocellulum* [155] are all found to be cell associated. In fact, the association of each of the cellulolytic enzymes, either as multimeric complexes (discussed in

Chapter 3) or alone, seems to be a common feature among the bacteria. For example, *F. succinogenes* produces periplasmic and extracellular forms of a cellobiohydrolase (cellodextrinase) [141,142] and cellulase [156] in addition to the cell wall-associated β-glucosidase. With no possible role in bacterial cell wall biosynthesis, the periplasmic and cell wall enzymes must be produced strictly for the degradation of cellulose.

2.1.4 XYLANASES

Xylanases catalyze the hydrolysis of the β-(1→4) bonds between D-xylose residues of heteroxylans and xylo-oligosaccharides, with the exception of xylobiose. As typical for an endo-acting enzyme, the affinity of xylanases for xylo-oligosaccharides and hence the efficiency of their hydrolysis decreases with decreasing degrees of polymerization. Also typical of endoglycosidases, most xylanases are capable of catalyzing transglycosylation reactions, especially when in the presence of high concentrations of soluble xylo-oligomers.

The specificity of xylanases for various substrates can differ widely. Some enzymes are highly specific for the xylan substrates while others show considerable flexibility, being able to hydrolyse other β-(1→4) linked natural polymers such as cellulose or artificial chromophoric substrates. For example, an *Aspergillus niger* xylanase (p*I* 3.65) catalyzes the hydrolysis of carboxymethyl cellulose at a rate of approximately 25% of that for soluble xylan [157], whereas the specific activities of the *Thermoascus aurantiacus* xylanase for carboxymethyl cellulose, filter paper, and nitrophenyl-β-D-xylopyranoside relative to oat-spelt xylan are 5, 0.2, and 2%, respectively [158]. In contrast, the specific activity of XynZ from *C. thermocellum* for carboxymethyl cellulose is 0.5% of that with xylan as substrate [159], and the *Pseudomonas fluorescens* xylanase (XynA) does not act at all on carboxymethyl cellulose even though this enzyme is capable of binding to cellulose [160,161]. The Cex protein from *Cellulomonas fimi* has high xylanase activity, but even higher activity against carboxymethyl cellulose, resulting in its classification as a cellulase [162].

Unlike cellulose, which is a straight-chain homopolymer of glucose, heteroxylans may be heavily substituted (see Chapter 1) leaving few regions of uninterrupted β-1,4-linked xylosyl sequences for xylanase activity. One of the first studies to note the effect of side chains on xylanase-catalyzed hydrolysis of xylan involved the action of a *T. koningii* enzyme with the heavily *O*-acetylated ryegrass xylan [49]. Chemical deacetylation of this substrate caused a marked increase in the rate of hydrolysis when compared with the native xylan. In the absence of debranching enzymes, therefore, the products of xylanase activity are typically larger, substituted xylo-oligosaccharides. However, some xylanases only act at substituted sites within heteroxylans. Enzymes that hydrolyse heteroxylans in the immediate vicinity of substitutions include two xylanases (p*I* 8.0 and 9.2, respectively) from *A. niger* [163]. These enzymes hydrolyse soluble larchwood xylan more rapidly than insoluble substituted larchwood xylan but are inactive on insoluble xylan freed of substitutions and only weakly active on linear xylo-oligosaccharides. Similarly, xylanases from *Bacillus subtilis* (feraxanase) [164] and *Gloeophyllum trabeum* [165] hydrolyse heteroxylans preferentially at xylosyl residues vicinal to residues substituted with 4-*O*-methyl-glucuronic acid. The obvious differences in substrate requirements for these latter enzymes compared with typical xylanases has prompted the suggestion that they be referred to as *appendage-dependent glycanases* [164] and led to a call, championed by the late Michael Coughlan, for a review of the classification of xylanases as solely EC 3.2.1.8 [166].

Although some xylanases reflect relaxed flexibility toward the nature of the carbohydrate moiety of their substrates, most are highly specific for the β-(1→4) intersugar linkage. There is no known example of a xylanase that can hydrolyse the main-chain β-1,3 linkages of natural substrates, such as 1,3-β-xylan, although a *Cryptococcus albidus* xylanase has been reported to synthesize via transglycosylation reactions and subsequently hydrolyse short xylo-oligomers of mixed 1,3 and 1,4 xylosidic linkages [167]. The mixed (1,3:1,4) linkage xylan from the alga (seaweed) *Rhodymenia palmatum* does serve as a substrate for the xylanases from both *Penicillium capsulatum* and *Talaromyces*, but it appears that only the β-1,4 linkages are hydrolysed, resulting in the production of limited xylo-oligomers retaining all β-1,3 linkages intact [168]. Nevertheless, many xylanases have been classified as debranching enzymes, because they possess the ability to liberate free arabinose from heteroxylans in addition to cleaving the main-chain linkages [99,169]. Xylanase I from *F. succinogenes* represents an extreme example of a debranching enzyme in that this enzyme is reported to release free arabinose from the heteroxylans from rye flour and oat spelts before it hydrolyses the β-1,4 linkages of the backbone to generate soluble xylo-oligomers [170]. The activity of debranching xylanases thus represents a dichotomy because it appears that the ability to release α-1,3 linked arabinofuranosyl residues is an integral activity of these enzymes. However, it is quite possible that the homogeneous preparations of the enzymes used in these studies may in fact be contaminated with trace amounts of arabinosidase, and it has been suggested that this issue may be resolved only with the characterization of cloned xylanases expressed in hosts lacking arabinosidases [171].

Fungal and bacterial xylanases have been grouped into two main classes based on their physicochemical properties [172]. Enzymes with molecular weights below 30 kDa are typically basic proteins, whereas those with molecular weights above 30 kDa are usually acidic. Although this trend is not universal, it is particularly apparent among the enzymes from bacteria and the various species of *Trichoderma* (Table 2.5). However, almost

TABLE 2.5. Properties of Some Fungal and Bacterial Xylanases.

Organism	M_r[a] (k)	p*I*	Glycosylation (% by wt.)[b]	pH_{opt}	$Temp_{opt}$ (°C)	Stability pH	Stability Temp (°C)	Reference
Fungi								
Aspergillus ochraceus	48		0	6.0	50	5–10	45–50	[331]
Aspergillus niger Rhozyme I	13	8.6		6.0	45	5–6	60	[163]
II	13	9.0		5.5	45	5–6	60	[163]
	14	4.5		4.9	45	5.6	48	[332]
III	28	3.65		5.0	40–45			[157]
str. 11	31			4	50	4–8	50–60	[333]
	31					6.0–6.5	50	
	50			5.5–6.0	65–80			
	50			4.0–4.5	65–80			
str. 14	27–33	4.2	20	4.0	50	3–8		[334]
str. 15 II	36	7.3		5.0	50			[166]
I	73	4.05		4.5	45–50			
Aspergillus sydowii	30			5.5	60	5	70	[335]
Aureobasidium pullulans	20	2.0		4.5	45			[336]
Caldocellum saccharolyticum	40.5			5.5–6	70		<70	[337]
Ceratocystis paradoxa II		4.5		5.1	80	5–10	<70	[338]
I		9.1–9.6		5.5	40		<50	
Cryptococcus albidus	48	5.7	+					[339]
	48	5.3	+					
Chaetomium thermophile I	26			4.8–6.4	70			[340]
II	7			5.4–6.0	60			[340]
Gloeophyllum trabeum	40.5	5.0		4.0	80	7	50	[165]

TABLE 2.5. (continued).

Organism	M_r[a] (k)	p*I*	Glycosylation (% by wt.)[b]	pH_{opt}	$Temp_{opt}$ (°C)	Stability pH	Stability Temp (°C)	Reference
Fungi (continued)								
Humicola lanuginosa I	21	4.1		6.0	65	5–8	<55	[341]
II	26		1.2	6.0	65	6–9	30–60	[342]
Irpex lacteus	38	7.6–8.0	23	4.6–5.2	60		<60	[343]
Lentinula edodes	41	3.6		4.5–5.0	60		45	[344]
Malbranchea pulchella		8.6		6.0–6.5	70		50	[345]
Myrothecium verrucaria	15.9	4.35	11.0	5.5	45		<50	[346]
Penicillium capsulatum	28.5	5–5.2		4	48			[166]
	29.5	5–5.2		4	48			
Penicillium purpurogenum	70 (3)	5.9		5.0	50		50	[347]
Postia placenta	44	3.8			50–60		60	[348]
Schizophyllum commune	21	4.5	0	5.0	55	4.5–7	<55	[349,350]
Schizophyllum radiatum	25.7			4.9	55	5–7.5	45	[351]
Sporotrichum dimorphosporum	19	7.35						[352]
	20	7.2						
	26	7.1						
	26	5.5		4.5–5	50			
	32	3.9–4.7		4.5–5	65–70			
Talaromyces byssochlamydoides	45		14.2	5.0	70		70	[353]
	54		31.5	4.5	70		70	
	76		36.6	5.5	75		70	

(continued)

TABLE 2.5. (continued).

Organism	M_r[a] (k)	p*I*	Glycosylation (% by wt.)[b]	pH_{opt}	$Temp_{opt}$ (°C)	Stability		Reference
						pH	Temp (°C)	
Fungi (continued)								
Talaromyces emersonii	30.1	3.98		4.2	73			[166]
	35.7	4.25		4.7	75			
	36.6	4.02		4.4	79			
	45	3.87		4.2	80			
	47.9	4.38		4.4	75			
	48.6	4.06		4.4	78			
	52.8	4.42		4.1	73			
	54	4.2		3.5	67			
	54.3	4.32		4.7	78			
	58.5	4.05		4.3	77			
	59	4.2		4.3	77			
	62.2	4.25		4.0	79			
	74.9	5.3		4.2	78			
Thermoascus aurantiacus	32	7.1		5.1	80		<80	[158]
Thielaviopsis basicola				5	60	4–8	<50	[354]
Trametes hirsuta	23			5.0–5.5	50	4–8	<45	[355]
Trichoderma harzianum E58 I	20	9.4		5.5	50		50	[356]
II	22	8.5		4.5–5.0	45–50			[357]
III	29	9.5		5.0	60		40	[356]
Trichoderma koningii IMI 73022 II	17.7	7.3	0		50			[358]
I	29	7.24	0		60–65			
G-39	21.5	8.9		5.5	60			[359]

TABLE 2.5. (continued).

Organism	M_r[a] (k)	p*I*	Glycosylation (% by wt.)[b]	pH_{opt}	$Temp_{opt}$ (°C)	Stability		Reference
						pH	Temp (°C)	
Fungi (continued)								
Trichoderma lignorum B	20	8.7		6.5	45			[360]
A	21	5.1		3.5	45			
Trichoderma pseudokoningi	15	9.6		5.0				[361]
Trichoderma reesei QM 9419	8.5	10.3						[33]
	11.7	8.6						
	14.4	6.6						
	20	9.0		5.3				[362]
	40.7	8.5						[33]
	55	4.5						[117,363]
VTT-D-80133	23	6.4		4–5				[364]
	32	4.2		4–5				
Trichoderma viride Onozuka SS	16			5.5–6.0				[365]
	17.8	9.2		4.8	59			[366-368]
	53							[369]
K-10-34	19	9.1		5.0	55			[370]
Bioxylanase	19.9	8.4		4.5	50			[371]
	16							
	13.1							
Cellulysin		21.5						[372,373]
	14.5							
	5.4							

(continued)

TABLE 2.5. (continued).

Organism	M_r[a] (k)	p*I*	Glycosylation (% by wt.)[b]	pH_{opt}	$Temp_{opt}$ (°C)	Stability pH	Stability Temp (°C)	Reference
Fungi (continued)								
ATCC 52438	21.5							[372,373]
	42							
Sigma V	22	9.3		5.0	53			[374]
Maxazyme CL	23.5	7.7						[375,376]
	53	5.3						
	57	4.4						
Meicellase				3.5	50			[377]
Tyromyces palustris	56	3.6			76		70	[378]
Bacteria								
Aeromonas caviae ME-1	41							[379]
	46	5.4		6.8	30–37	5.0–8.6	30–37	[380]
Bacillus sp. str. W1	21.5	8.5		6.0	65			[174]
	49	3.6		7–9	70			
str. W2	22.5	8.3		6.0	65			[174]
	50.5	3.7		7–9.5	70			
str. C-125	16			6–7	70			[173]
	43					6–10	70	
str. 11-1S	56			4.0	80			[176]
str. C-59-2		6.3		6–8	60			[175]
str. XE	22	7.7		6.0	75		60	[381]
str. 41M-1	36	5.5		9.0	50	7–11	55	[382]
Bacillus circulans WL-12	15	9.1		5.5–7				[383]

TABLE 2.5. (continued).

Organism	M_r[a] (k)	p*I*	Glycosylation (% by wt.)[b]	pH_{opt}	$Temp_{opt}$ (°C)	Stability pH	Stability Temp (°C)	Reference
Bacteria (continued)								
	85	4.5		5.5–7				
Bacillus coagulans 26	24	10.0		6.0	37			[384]
Bacillus polymxya	61	4.7		6.5	50			[385]
Bacillus pumilus IPO	22.4		0	6.5	45–50	5–8	40	[336,386]
Bacillus subtilis PAP115	32			5.0	50			[387]
Novo Ban	9.5			5.5	50			[388]
Bacteroides fibrisolvens H17c	49							
Bipolaris sorokiniana Shoemaker	30	9.5		5.5	70	5–8	50–55	[389]
Caldocellum saccharolyticum				5.5–6	70			[337]
Clostridium acetobutylicum ATCC 824	29	8.5		5.5–6	60			[390]
	64	4.4		5.0	50			
P262	28	10		6.0	37–43			[391]
Clostridium stercorarium	44	4.5		6.5	75			[392]
	62	4.4		6.0	75			
	72	4.4		6–7	75			
Clostridium thermolacticum TC21	39	4.45		6.5	80			[393]
	55	4.55		6.5	80			
	65	4.65		6.5	80			
Fibrobacter succinogenes S85	53.7	8.9		7.0	40			[170]
Streptomyces sp. str. B-12-2	23.8	8.3						[394]
	26.4	6.5						

(continued)

TABLE 2.5. (continued).

Organism	M_r[a] (k)	p*I*	Glycosylation (% by wt.)[b]	pH_{opt}	$Temp_{opt}$ (°C)	Stability		Reference
						pH	Temp (°C)	
Bacteria (continued)								
	36.2	5.0						
	36.2	5.4						
	40.5	4.8						
str. KT-23	44	6.9		5.5	55			[395]
str. E-86	40.5	7.3		5.5–6	55–60			[396]
str. 3137	25	10.1		5–6	60–65			[397]
	25	10.3		5–6	60–65			
	50	7.1		5.5–6.5	60–65			
Streptomyces exfoliatus MC1				5.5	50			[398]
				7.0	55			
				5.5	55			
Streptomyces flavogriseus CD-45	24			5.5				[399]
	42			5–7.5				
Streptomyces halstedii	45			6.3	60	4–10	4–50	[400]
Streptomyces lividans 1326	43	5.2		6.0	60			[401]
Streptomyces thermoviolaceus	33	8.0		7.0	60	5–9	<60	[402]
	54	4.2		7.0	70	5–9	<50	
Streptomyces xylophagus				6.2	55–60			[403]
Thermoanaerobacterium sp.	350 (?)	4.37	6.0	6.2	80		70	[404]
Thermomonospora fusca	32			7.0	75	5.0–9.0	75	[405]
Thermotoga sp.	33			5.0–5.5	105	4.5–5.5	100	[178]

[a]Values in parentheses indicate the number of subunits in the holoenzyme.
[b]+ denotes the presence of an unknown amount of glycosylation.

without exception, both fungal and bacterial xylanases are single-subunit proteins. Molecular weights range from below 10 to 85 kDa, with the majority of the characterized enzymes being relatively small (15–35 kDa). The molecular weights of less than 15 kDa listed in Table 2.5 are suspect in view of the inherent ability of many xylanases to bind the matrices of gel filtration columns.

The pH optimum of xylanase activity is generally acidic (pH 3.5–7) with species of the thermoalkalophilic *Bacillus* providing the exceptions [173,174]. Many xylanases, in addition to those of the thermoalkalophilic *Bacillus* sp. (temperature optimum, 60–80°C [173–176]), are quite resilient to high temperature and possess relatively high temperature optima for activity. Indeed, extremely thermostable xylanases from *Dictyoglomus* sp. [177] and *Thermatoga* sp. [178] have recently been characterized as having optimal activity at 90 and 105°C, respectively.

2.1.5 β-XYLOSIDASES

The β-D-xylosidases hydrolyse xylo-oligosaccharides to xylose and thus resemble the β-glucosidases in their mode of action. Despite the fact that these enzymes are both essential for the complete digestion of heteroxylans and produced by most heteroxylanolytic microorganisms, they have received very little attention, especially in terms of their enzymology. As far as I am aware, only one study has been reported concerning the mechanism of action of β-xylosidase, but, presumably, it would be similar to that of the xylanases and cellulolytic enzymes.

Most of the characterized β-xylosidases are not active toward xylan but hydrolyse soluble xylo-oligosaccharides. The enzymatic rate increases with decreasing degrees of polymerization of the oligosaccharides, and xylobiose is the preferred substrate [179–181]. These enzymes are capable of catalyzing transferase reactions, especially at high concentrations of substrate [179,180,182–184], but under in situ conditions, they greatly improve the yield of xylose from xylans [33,182]. As evident from Table 2.6, most of the β-xylosidases are acidic and relatively large enzymes with molecular weights ranging between 100 and 200 kDa. Although all of these characterized enzymes are most active under acidic conditions with temperature optima approximately 50°C, it is likely that more alkali and thermo-tolerant enzymes exist in nature.

2.2 GLYCOSYLATION OF CELLULOLYTIC AND XYLANOLYTIC ENZYMES

Many celluloytic and xylanolytic enzymes of both fungal and bacterial origin are known to be glycoproteins, involving both *O*-linked and *N*-linked

TABLE 2.6. Properties of Some Fungal and Bacterial β-Xylosidases.

Organism	M_r (k)	p*I*	Glycosylation (% by wt.)	pH_{opt}	$Temp_{opt}$ (°C)	Stability		Reference
						pH	Temp (°C)	
Fungi								
Aspergillus niger	<200	4.6		3.0–4.5				[180,406]
Aspergillus fumigatus	360 (4)	5.4		4.5	75	2–8	65	[300]
Chaetomium trilaterale	240 (2)	4.86	20.7	4.2–4.5	60	4–11	60	[407,408]
Cryptococcus albidus				5.0–5.4	55		<60	[409]
Emericella nidulans	240 (2)	3.25	4.0	4.5–5.0	55	4.5–5.0	55	[179,410]
Malbranchea pulchella	26	4.8	0	6.2–6.8	50	6.3–6.7	50	[410]
Penicillium wortmanni	90	5.0	16	3.0–4.0		5.0–6.0		[183]
Sclerotium rolfsii	175	6.8		4.5	50	4.0–5.0	50	[411]
Trichoderma reesei	100	4.7	7.5	4.0	60	3–6	55	[412]
Trichoderma viride	101.5	4.45	4.5	4.5	55	3–4	<55	[184,410]
Bacteria								
Bacillus coagulans	200 (8)	4.9		6.5	30			[384]
Bacillus pumilus	110 (2)	4.4	0	7.2	40	6.5–9.0	<40	[413,414]

carbohydrate chains. For example, the cellobiohydrolase from *Trichoderma viride* contains glucosamine and sixteen to seventeen *O*-glycosidic glycans of one to five hexose residues composed of mannose, glucose, and galactose [185]. The *T. reesei* cellobiohydrolase I also contains both *N* and *O*-linked glycans. The *N*-glycans are of the high-mannose type having the structures $(Man)_5(GlcNAc)_2$ and $(Man)_9(GlcNAc)_2$. They are located at three of the four putative *N*-glycosylation sites (Asn-X-Ser/Thr). The *O*-chains vary from one to four hexoses, mostly mannose, and there are on average eight chains per enzyme molecule [186] located within a short serine and threonine-rich sequence close to the C-terminus of the protein [187]. However, some of the *O*-linked chains, especially those associated with enzymes associated with the bacterial cellulolytic complexes, are quite novel in terms of their composition and structure. For example, the unique oligosaccharide chains comprising both furanose and pyranose forms of galactose were isolated from a 230-kDa subunit of the cellulolytic complex of *Bacteroides cellulosolvens* [188] and share some common elements with the oligosaccharide moieties of the *C. thermocellum* enzymes [189].

Glycosylation of the cellulolytic and xylanolytic enzymes, and indeed of most enzymes, has been implicated in protection against environmental fluctuations (e.g., temperature), protection against proteolysis, secretion, and facilitating adsorption to insoluble substrates. The occurrence of Thr/Pro peptide regions between protein domains (discussed later) in which the majority of the Thr residues are glycosylated has been reported for a number of bacterial cellulases, cellobiohydrolases, and xylanases, including those from *Cellulomonas fimi* [190–192], *Fibrobacter succinogenes* [193], *Thermomonospora fusca* [194] and *Pseudomonas fluorescens* ssp. *cellulosa* [195]. The similar catalytic activities and substrate specificity profiles of a *C. fimi* cellulase and cellobiohydrolase obtained by gene expression in *E. coli*, which resulted in the production of nonglycosylated forms of the enzymes, compared with the wild types, indicated a lack of dependence on glycosylation. However, the nonglycosylated derivatives of these enzymes were susceptible to cleavage by a *C. fimi* protease between protein domains, strongly implicating the protective role of *O*-glycosylation to the linker regions [111]. Treatment of *T. reesei* with tunicamycin, thereby inhibiting *N*-glycosylation, resulted in the production of cellulases with increased sensitivity to proteases while not affecting activity [112]. Hence, one of the major roles of glycosylation in general is protection from proteolytic degradation.

The role of glycosylation in secretion is more controversial. Tunicamycin-induced inhibition of *N*-glycosylation of the *S. commune* has been reported to marginally inhibit secretion of the β-glucosidase and cellulase [196]. With the cellulases from *T. reesei*, *O*-linked but not *N*-linked glycosylation is thought to be necessary for secretion [197]. In direct contrast, similar inhibition studies with a *Cryptococcus albidus* xylanase revealed that glycosylation in fact retards secretion [198].

Early studies monitoring the binding kinetics to cellulose led to the hypothesis that the adsorption properties of cellulolytic enzymes may be determined by their degree of glycosylation [199]. Mannan polymers had previously been shown to have affinity for crystalline cellulose [40], and evidence was obtained with a cellobiohydrolase to suggest that mannose-rich glycan chains may have a functional role in binding the enzyme to insoluble substrates [114]. However, the more recent discovery of cellulose-binding domains within many cellulolytic and xylanolytic enzymes (discussed below) somewhat diminishes the importance that generic glycosylation may have in these interactions.

2.3 CLASSIFICATION OF β-GLYCOSIDASES

With the recent emergence and development of techniques in both molecular biology and the sequencing of DNA, a plethora of deduced amino acid sequences of a broad variety of enzymes has accumulated in the literature. Indeed, the complete or partial amino acid sequences of over 500 glycosyl hydrolases are now known, of which over 150 are the cellulolytic and heteroxylanolytic β-glycosidases. Comparison and alignment of the amino acid sequences of these enzymes together with hydrophobic cluster analysis have revealed considerable homology between the fungal and bacterial enzymes and enabled the establishment of a classification scheme [200–206] that currently comprises forty-six distinct families [205,206]. As indicated in Table 2.7, eighteen of these families involve the cellulolytic and heteroxylanolytic hydrolases (a complete list of the enzymes comprising these families may be found in Appendix I).

In some cases, the levels of similarity between the various members of a given family are somewhat low, being confined to putative catalytic domains. This has given rise to the further division of some families into subtypes [200]. Members of these families are typically characterized by regions of high amino acid sequence similarity interspersed with regions where similarities are either low or totally devoid; for example, all members of Family 9 Subtype 1 contain an N-terminal extension of eighty to ninety amino acid residues that is absent in other Family 9 enzymes [200,207]. With the availability of X-ray crystallographic data for two of the Family 6 enzymes, a detailed model to account for the difference between subtypes has been proposed. The putative catalytic residues of a Subtype 2 cellobiohydrolase are conserved in all other Family 6 cellulases, but a set of amino acid sequences that form extended surface loops in the three-dimensional structure of the cellobiohydrolase [208] are absent in the sequences of the

TABLE 2.7. Classification of Glycosyl Hydrolases.

Family[a]	β-Glycosidase[b]
1	**β-Glucosidases** β-Galactosidases 6-Phospho-β-galactosidases Lactase/phlorizin hydrolase
3	**β-Glucosidases**
5 (A)	**Cellulases** **β-Glucosidases** β-Mannanase Exo-1,3-β-glucanases
6 (B)	**Cellulases** **Cellobiohydrolases**
7 (C)	**Cellulase** **Cellobiohydrolases**
8 (D)	**Cellulases** Lichenase
9 (E)	**Cellulases** Spore-germination specific protein
10 (F)	**Xylanases** **Cellobiohydrolase**
11 (G)	**Xylanases**
12 (H)	**Cellulases**
26 (I)	**Cellulases**
39	**β-Xylosidases** α-L-Iduronase
40	**β-Glucosidases**
41	**β-Glucosidases**
43	**β-Xylosidases**
44 (J)	**Cellulases**
45 (K)	**Cellulases**
46 (L)	**Cellulases** **Cellobiohydrolase**

[a]Letters in parentheses denote the Family according to the scheme of Gilkes et al. [203].
[b]Cellulolytic and xylanolytic enzymes are indicated by boldface type.

Subtype 1 cellulases [209]. These surface loops of the cellobiohydrolase cover the active site cleft of the enzyme and thereby restrict enzymatic activity to the nonreducing ends of celluloses [208]. By lacking the surface loop sequences, the Subtype 1 *Thermomonospora fusca* cellulase A has an open active-site cleft [210] that permits the endoactivity of the enzyme.

2.4 DOMAIN STRUCTURE

2.4.1 CELLULOSE BINDING DOMAINS

On the basis of the disulfide bonding pattern observed over a decade ago in a cellobiohydrolase from *Trichoderma reesei*, the enzyme was postulated to be comprised of at least two protein domains [211], and studies initiated soon after involving limited proteolysis showed this to be the case [212]. Subsequent analogous investigations on a second cellobiohydrolase from this fungus [213] and two cellulases from *Cellulomonas fimi* [111,191] suggested that these structural features may be common among the cellulolytic and xylanolytic enzymes. It is now known that over forty different cellulolytic and xylanolytic enzymes are comprised of two functional domains, a catalytic domain and a noncatalytic cellulose-binding domain (CBD) (for recent reviews, see [98,200,203,214]). The two are usually joined by characteristic linker sequences.

Enzymes that contain a putative CBD are listed in Table 2.8, and a schematic representation of the structural arrangement of the domains in some *Pseudomonas fluorescens* enzymes is presented in Figure 2.3. It should be understood that cellulose-binding properties of isolated domains have been demonstrated for only a fraction of the listed enzymes; the others have been identified solely by sequence alignments. Nevertheless, it is highly likely that the uncharacterized CBDs are functional. Indeed, there is a strong correlation between the capacity of cellulolytic enzymes to degrade crystalline cellulose and either the availability or possession of a CBD [98]. Truncation of the domain by genetic manipulation of the gene encoding a *B. subtilis* cellulase resulted in the expression of an enzyme with a fivefold reduction in activity on an insoluble substrate while, in fact, increasing its activity on carboxymethyl cellulose [215]. Similar trends were observed with limited proteolysis studies of the *T. reesei* cellobiohydrolases CbhI and CbhII [212,213], and *C. fimi* CenA [191]. However, observations made with chimeric enzymes in which CBDs were grafted onto unidomain enzymes are more equivocal. The addition of the *T. fusca* cellulase CBD to a *Prevotella ruminicola* cellulase did indeed yield an enzyme with approximately tenfold higher activity on insoluble cellulose [216]. On the other

TABLE 2.8. Cellulose-Binding Domains of β-Glycanases.

Organism	Enzyme	Function	GenBank Accession Number/Reference
Type 1			
Butyrivibrio fibrisolvens	EgII	Cellulase	[415]
Cellulomonas fimi	CenA	Cellulase	[416]
Cellulomonas fimi	CenB	Cellulase	[417]
Cellulomonas fimi	CenD	Cellulase	L02544
Cellulomonas fimi	Cex	CBH/xylanase	[195]
Cellulomonas flavigena	ORF X	—	[418]
Clostridium cellulovorans	EngD	Cellulase	M37434
Clostridium longisporum	CelA	Cellulase	L02868
Microbispora bispora	CelA	Cellulase	[419]
Pseudomonas fluorescens subsp. *cellulosa*	CelA	Cellulase	[420]
Pseudomonas fluorescens subsp. *cellulosa*	CelB	Cellulase	[190]
Pseudomonas fluorescens subsp. *cellulosa*	CelC	β-Glucosidase	S64954
Pseudomonas fluorescens subsp. *cellulosa*	XynA	Xylanase	[160]
Pseudomonas fluorescens subsp. *cellulosa*	XynB/XynC	Xylanase	[421]
Streptomyces plicatus	ChtA	Chitinase	M82804
Thermomonospora fusca	E_2	Cellulase	M73321
Thermomonospora fusca	E_5	Cellulase	L01577
Type 2			
Agaricus bisporus	Cel1	—	M86356
Humicola grisea	CbhI	Cellobiohydrolase	M64588
Penicillium janthinellum	CbhI	Cellobiohydrolase	X59054
Phanerochaete chrysosporium	CbhI-1	Cellobiohydrolase	S109508
Phanerochaete chrysosporium	CbhI-2	Cellobiohydrolase	S109508
Trichoderma longibrachiatum	EgI	Cellulase	X60652
Trichoderma reesei	CbhI	Cellobiohydrolase	[187]
Trichoderma reesei	CbhII	Cellobiohydrolase	[422]
Trichoderma reesei	EgI	Cellulase	[423]
Trichoderma reesei	EgII	Cellulase	[424]
Trichoderma reesei	EgV	Cellulase	[425]

(continued)

TABLE 2.8. (continued).

Organism	Enzyme	Function	GenBank Accession Number/Reference
Type 3			
Bacillus lautus	CelA	Cellulase	M76588
Bacillus lautus	CelC	Cellulase	[426]
Bacillus subtilis	—	Cellulase	X67044, M28332, M16185, D01057
Caldocellum saccharolyticum	ManA	Mannanase/ cellulase	L01257
Caldocellum saccharolyticum	CelA	Cellulase	[427]
Caldocellum saccharolyticum	CelB	CBH/Cellulase	X13602
Cellulomonas fimi	CenB	Cellulase	M64644
Clostridium cellulolyticum	CelCCG	Cellulase	S114528
Clostridium cellulovorans	CbpA	CBD	M73817
Clostridium sterocarium	CelZ	Cellulase	P23659
Clostridium thermocellum	CipA	Scaffoldin	L08665
Clostridium thermocellum	CipB	Scaffoldin	X68233
Clostridium thermocellum	CelF	Cellulase	X60545
Clostridium thermocellum	CelI	Cellulase	L04735
Type 4			
Cellulomonas fimi	CenC	Cellulase	X57858
Clostridium cellulolyticum	CelCCC	Cellulase	S48579
Streptomyces reticuli	Cel1	Cellulase (Avicelase)	L04735
Unclassified			
Clostridium thermocellum	CelE	Cellulase	M22759
Erwinia chrysanthemi	CelZ	Cellulase	Y00540

Adapted from Reference [98].

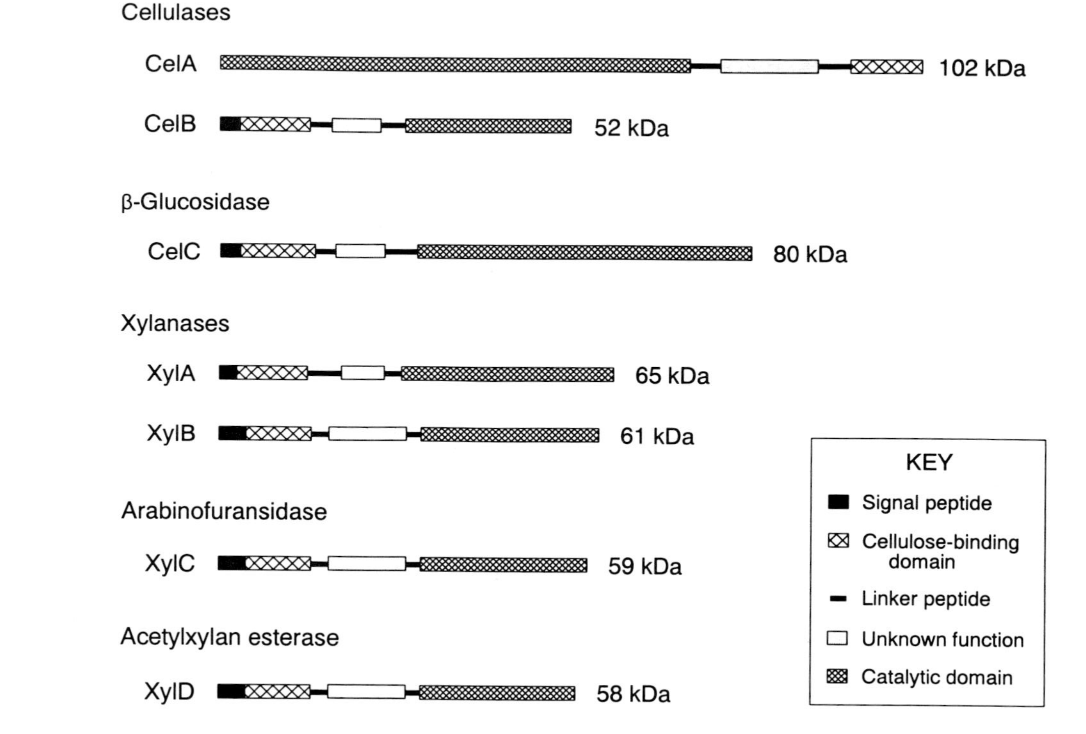

Figure 2.3 Schematic representation of the modular structure of cellulolytic and xylanolytic enzymes from *Pseudomonas fluorescens* subspecies *cellulosa* (adapted from References [171] and [431]).

hand, grafting these domains onto the *Ruminococcus albus* CelA and *Clostridium thermocellum* CelE cellulases did not alter their activities toward either soluble or insoluble substrates [217]. A number of suggestions have been made to explain these observations [98], and the most probable concerns the precise fusion of the two domains that allows for the retention of their active conformations. As alluded to above, adjustments to the length of the linker between domains does impart modulation of catalytic activity [218,219] and conformational states [219]. That the CBD solely confers the ability of a complete enzyme to bind cellulose was demonstrated with *T. reesei* CbhI [220]. A catalytically inactivated enzyme prepared by chemical modification with a carbodiimide, which is known to modify the catalytic carboxyl groups, still possessed the ability to adsorb onto microcrystalline cellulose.

Most of the identified CBDs are classified into one of four families based on sequence similarities, with those associated with *Clostridium thermocellum* CelE and *Erwinia chrysanthemi* CelZ remaining unclassified [98]. It is likely that as more enzymes are characterized and shown to possess CBDs, the number of families will increase. Type 2 domains, which are apparently confined to fungal cellulases and cellobiohydrolases, have received considerable attention and represent the best characterized of the four. Type 2 CBDs share a highly conserved sequence of approximately thirty residues (Figure 2.4) rich in glycyl, cysteinyl, and aromatic residues [203]. In an intriguing study, the CBD of *T. reesei* cellobiohydrolase I was chemically

```
                                                     ** *
Tre CbhI    146     TQSHYGQCGGIGYSGPTVCASGTTCQVLNPYYSQCL
Tre CbhII         QACSSVWGQCGGQNWSGPTCCASGSTCVYSNDYYDQCL
Tre EGI     421   SCTQTHWGQCGGIGYSGCKTCTSGTTCQYSNDYYSCQL
Tre EGII            QQTVWGGQCGIGWSGPTNCAPGSACSTLNPYYAQCI
Tre EGV     205     QQTLYGQCGGAGWTGPTTCQAPGTCKVQNQWYSQCL
Tvi CbhI    460     TQTHYGQCGGIGYIGPTVCASGSTCQVLNPYYSQCL
Pja CbhI    502     GARDWAGQCGNGWTGPTTCVSPYTCTKQNDWYSQCL
Pcr CbhI    480     TVPQWGQCGGIGYTGSTTCASPYTCHVLNPYYSQCY
Hgr CbhI    490     KAGRWQQCGGIGFTGPTQCEEPYICTKLNDWYSQCL
Abi Cel1    284     TIPQYGQCGGIGWTGGTGCVAPYQCKVINDYYSQCL
Abi Cel3            QSPVWGQCGGNGWTGPTTCASGSTCVKQNDFYSQCL
Hin EGV     246   GCTAGRWAQCGGNGWSGCTTCVAGSTCTKINDWYHQCL
```

Figure 2.4 Comparison of deduced amino acid sequences of Type II CBDs. The CBDs (or putative CBDs) are from the following *Trichoderma reesei* cellobiohydrolases I (Tre CbhI) [432] and II (Tre CbhII) [422], and cellulases I (Tre EGI) [423], II (Tre EGII) [424], and V (Tre EGV) [425]; the cellobiohydrolases I from *T. viride* (Tvi CbhI) [433], *Penicillium janthinellum* (Pja CbhI) [434], *Phanerochaete crysosporium* (Pcr CbhI) [435], and *Humicola grisea* (Hgr CbhI) [436]; *Agaricus bisporus* cellulases 1 (Abi Cel1) [437] and 3 (Abi Cel3) [438]; *Humicola insolens* cellulase V (Hin EGV) [439]. Amino acids in bold type denote identity among at least seven of the twelve sequences, whereas shaded residues denote complete identity. The putative cellulose-binding residues are denoted by asterisks (adapted from Reference [221]).

synthesized, and its three-dimensional structure was determined by NMR spectroscopy [221]. Three antiparallel β-strands form a wedge-shaped irregular β-sheet (Figure 2.5) that is held together by two cystines [222]. The CBD is amphipathic with a flat hydrophilic side exposing a number of conserved amino acid residues, including three aromatics that are thought to make contacts with cellulose substrates. One of these conserved aromatic residues, Tyr31, is located at the tip of the wedge, and this residue, together with a second tyrosyl residue (Tyr32) and a conserved glutamine (Gln34), are thought to be responsible for the tight binding of the CBD to cellulose [221]. This postulate was supported by mutagenesis studies involving the replacement of the tyrosyl residues [223].

Type 1 CBDs are associated with a variety of bacterial enzymes that not only include cellulases and cellobiohydrolase, but interestingly, also xylanases, an arabinofuranosidase, and a chitinase. These domains of approximately 100 amino acid residues (Figure 2.5) are comprised of conserved tryptophan, asparagine, and glycyl residues, a high content of hydroxyamino acids, and a low content of charged amino acids [224]. The three-dimensional structure of the Type 1 CBD from *Cellulomonas fimi* exoglucanase or xylanase (Cex) has also been determined by NMR spectrometry [225]. Its general morphology is quite different from that of the Type 2 CBD from *T. reesei* cellobiohydrolase I, but three conserved tryptophan residues are present at the surface of the domain (Figure 2.6), and they likely contribute greatly to binding interactions [224,226]. The precise role and function of these bacterial CBDs is controversial. The isolated CBDs do indeed specifically bind cellulose with varying affinities [190], but its removal from a *Pseudomonas* xylanase did not drastically alter the catalytic properties of the enzyme [227]. However, substitution of the CBD comprising Cex from *C. fimi* with a Type 2 CBD from *T. reesei* cellobiohydrolase resulted in a hybrid enzyme that was not as efficient as the wild type in hydrolysing insoluble substrates [228]; the activity of the hybrid enzyme was two to three times less than the wild type on bacterial microcrystalline cellulose.

The CBDs of Type 3 are larger still, 130 residues, and appear to be associated with the cellulolytic enzyme complexes common among anaerobic bacteria (discussed in Chapter 3).

2.4.2 LINKER SEQUENCES

The majority of the linker sequences that connect the catalytic domain to the CBD of a given β-glycosidase are glycosylated peptides rich in either serine residues or a combination of proline and threonine residues [203]. This glycosylated linker varies in length, typically six to fifty-nine amino acid residues, but the linkers between the domains of the *Ruminococcus flavefaciens* XynA [229] and *Neocallimastix patriciarum* XynA [230] xylanases are unique, being composed of 374 and 455 residues, respectively. The

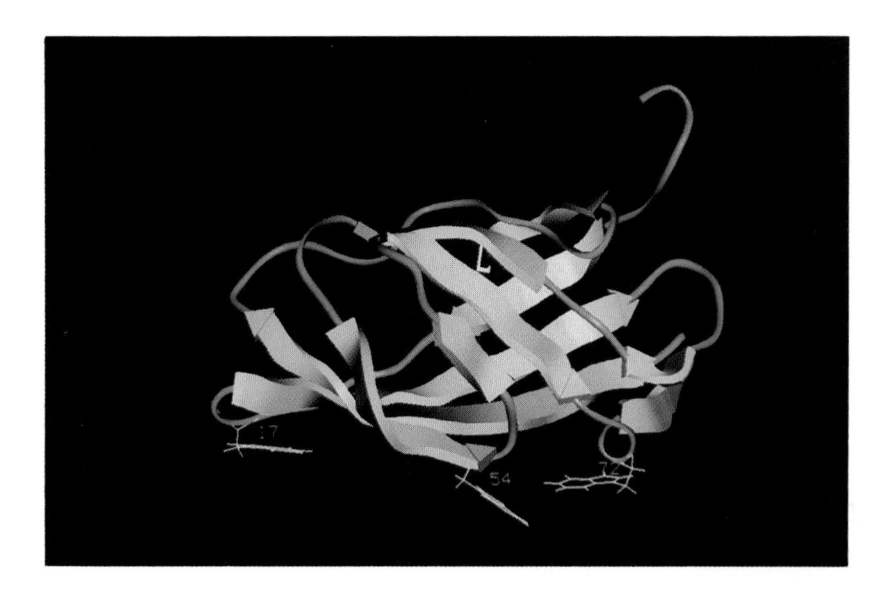

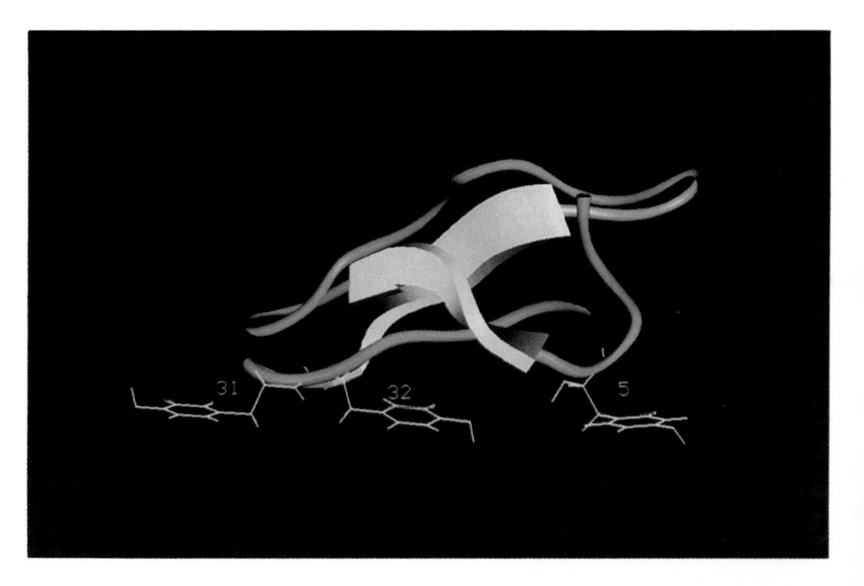

Figure 2.5 NMR minimized average structures of the (top) Type 1 and (bottom) Type 2 CBDs from *Cellulomonas fimi* exoglucanase/xylanase (Cex) and *Trichoderma reesei* CBH I. The three putative cellulose-binding aromatic residues, Trp17, Trp54, and Trp72 (*C. fimi)* and Tyr5, Tyr31 and Tyr32 (*T. reesei*) are located at the flat surfaces of the respective CBDs. The figure was obtained using the data deposited in the Brookhaven Protein Bank (1exg and 1exh, respectively) and visualized using Insight ver. 2.3.0 software (Biosym Technologies, San Diego).

```
Cfi CenA      1    APGC--RVDYAVTNQWPGGFGANVTITNLG-DPVSSWKLDWTYTA-GQRIQQLWNG
Cfi Cex       338  PAGC--QVLWGV-NQWNTGFTANVTVKNTSSAPVDGWTLTFSFPS-GQQVTQAWSS
Cfu CenB      910  TPSC--TVVYS-TNSWNVGFTFSVKITNTGTTPL-TWTLGFAPPS-GQQVTQGWSA
Cfi CenD      599  TGSC--VVTYT-ANGWSGGFTAAVTLTNTGTTALGGWTLGFAGGS-GQTLTQGWSA
Cce EngD      377  QSAV--EVTYAITNSWGSGASVNVTIKNNGTTPINGWTLKWTMPI-NQTITNMWSA
Pfl EGA       862  GGNC----QYVVTNQWNNGFTAVIRVRNNGSSAINRWSVNWSYSD-GSRITNSWNA
Pfl EGB       29   AAVC----EYRVTNEWGSGFTASIRITNNGSSTINGWSVSWNYTD-GSRVTSSWNA
Pfl CelC      1      GC----EYVVTNSWGSGFTAAIRITNSTSSVINGWNVSWQYNS--NRVTNLWPN
Pfl XynA      1    TATC----SYNITNEWNTGYTGDITITNRGSSAINGWSVNWQYAT--NRLSSSWNA
Pfl XynB/c    1      AC----TYTIDSEWSTGFTANITLKNDTGAAINNWNVNWQYSS--NRMTSGWNA
Bfi End1      457  GALK---AEYTI-NNWGSGYQVLIKVKNDSASRVDGWTLKISKSF--VKIDSSWCV
Mbi CelA      361  GRAC--EATYALVNQWPGGFQAEVTVKNTGSSPINGWTVQWTLPS-GQSITQLWNG
Tfu EG2       309  RLLGGVHGDVHDANEWNDGFQATVTVTANQN--ITGWTVTWTFTP-GQTITNAWNA
Tfu EG5       1    AGLT--ATVTKESS-WDNGYSASVTVRNDTSSTVSQWEVVLTLPS-GTTVAQVWNA
Sli CelA      1    ATGC--KAEYTITSQWEGGFQAGVKITNLG-DPVSGWTLGFTMPPAGQRLVQGWNA

Cfi CenA      52   TASTNGGQVSVTSLPWNGSIPT-GGTASFGFNGSWA-GSNPTPASFSLNGTTCTGT
Cfi Cex       389  TVTQSGSAVTVRNAPWNGSIPA-GGTAQFGFNGSHT-GTNAAPTAFSLNGTPCTVG
Cfi CenB      960  TWSQTGTTVTATGLSWNATLQP-GQSTDIGFNGSHP-GTNTNPASFTVNGEVCG
Cfi CenD      650  RWAQSGSSVTATNEAWNAVLAP-GASVEIGFSGTHT-GTNTAPATFTVGGATCTTR
Cce EngD      439  SFVASGTTLSVTNAGYNGTIAANGGTQSFGFNINYS-GVLSKPTGFTVNGTECTVK
Pfl EGA       912  NVTGNNPY-AASALGWNANIQP-GQTAEFGFQGTKGAGSRQVPA---VTGSVCQ
Pfl EGB       79   GLSGANPY-SATPVGWNTSIPI-GSSVEFGVQGNNGSSRAQVPA---VTGAICGGQ
Pfl CelC      48   NLSGSNPY-SASNLSWNGTIQP-GQTVEFGFQGVTNSGTVESPT---VNGAACTG
Pfl XynA      50   NVSGSNPY-SASNLSWNGNIQP-GQSVSFGFQVNKNGGSAERPS---VGGSICSGS
Pfl XynB/C    48   NFSGTNPY-NATNMSWNGSIAP-GQSISFGLQGEKNGSTAERPT---VTGAACNSA
Bfi End1      506  NIAEEGGYYVITPMSWNSSLEP-SASVDFGIQGS---GS-IGTS---VNISVQ
Mbi CelA      413  DLSTSGSNVTVRNVSWNGNVPA-GGSTSFGFLGS---GTGQLSSS-----ITCSAS
Tfu EgE2      361  DVSTSGSSVTARNVGHNGTLSQ-GASTEFGFVGL
Tfu EgE5      52   QHTSSGNSHTFTGVSWNSTIPP-GGTASSGFIASGS-GEPTHCT---INGAPCDEG
Sli CelA      53   TWSQSGSAVTAGGVDWNRTLAT-GASADLGFVGSFT-GANPAPTSFTLNGATCSGSV
```

Figure 2.6 Comparison of deduced amino acid sequences of Type I CBDs from bacterial β-1,4-glycanases. The CBDs (or putative CBDs) are from the following: *Cellulomonas fimi* CenA (Cfi CenA) [224,416], Cex (Cfi Cex) [195], CenB (Cfi CenB) [417], and CenD (Cfi CenD) [440]; *Clostridium cellulovorans cellulase* D (Cce EngD) [441]; *Pseudomonas fluorescens* subsp. *cellulosa* cellulase A (Pfl EGA) [420], cellulase B (Pfl EGB) [190], cellodextrinase (Pfl CelC) [442], xylanase A (Pfl XynA) [443], and xylanases B and C (Pfl XynB/C) [421]; *Butyrivibrio fibrisolvens* cellulase 1 (Bfi End1) [415]; *Microbispora bispora* cellulase A (Mbi CelA) [444]; *Thermomonospora fusca* cellulase 2 (Tfu EG2) and cellulase 5 (Tfu EG5) [445]; and *Streptomyces lividans* cellulase A (Sli CelA) [446]. All sequences are numbered from the first amino acid residue in the mature enzymes. Amino acids in bold denote identical residues in at least eight of the fifteen sequences, whereas shaded residues denote complete identity.

former linker sequence is rich in asparagine (45%) and glutamine (26%), whereas the *N. patriciarum* linker region is comprised of twelve tandem repeats of sequence that contains Thr-Leu-Pro-Gly as the core followed by an octapeptide that is repeated in tandem forty-five times [230]. These linker sequences are thought to provide the flexibility required for the CBD to bind and presumably destabilize the local hydrogen-bonding network within crystalline cellulose [231] while permitting hydrolysis by the catalytic domain [218,219]. The glycosylation associated with the linkers not only provides protection from proteolysis [111], as discussed above, but it also appears to play a more active role in substrate binding. Studies with a β-1,4-glycanase (Cex) from *C. fimi* suggest that the glycan chains on the linker region aid in the association of this bacteria enzyme with crystalline cellulose [110]. However, it has not been established whether the glycosylated linker directly participates in the association with cellulose or induces the appropriate conformation of the complete enzyme for productive binding. Indeed, the interdomain peptide present in an engineered mutant form of *T. reesei* cellobiohydrolase that lacked its CBD did not appear to participate in adsorption to crystalline cellulose [232]. This sequence was, however, found to enhance stability of this fungal enzyme.

2.4.3 REPEATED SEQUENCES

A number of bacterial cellulolytic and heteroxylanolytic enzymes contain domains of reiterated amino acid sequences that vary in length from approximately 20 to 150 residues. The specific role, if any, of these repeated sequences remains unknown. However, they are observed in enzymes that aggregate into cell-surface–associated multienzyme complexes, and emerging evidence suggests that these sequences maintain the quaternary structure of the complexes. For example, a twenty-four–amino acid repeat sequence is present in many of the cellulolytic and xylanolytic enzymes and associated proteins that comprise the multienzyme complex produced by *C. thermocellum* [233–236]. Grafting the duplicated amino acid repeat sequence from cellulase D of *C. thermocellum* to the C-terminus of the cellulase C resulted in the production of cytoplasmic inclusion bodies containing the active enzyme [233], and antibodies directed against this same duplicate repeat sequence were observed to cross-react with a set of proteins associated with the multicomponent cellulolytic complex of *C. thermocellum* [234]. The duplicated twenty-four–amino acid sequences on the enzymes appear to interact directly with the seven-repeat sequence of the noncatalytic scaffolding protein (scaffoldin) [235,236]. These data thus suggest that the repeat sequences confer specificity for the formation of a complete and functional multienzyme complex.

2.4.4 MULTICATALYTIC DOMAINS

In addition to the enzymes that possess both catalytic and noncatalytic CBDs, some cellulolytic and xylanolytic enzymes are multifunctional (Table 2.9). These latter enzymes are comprised of at least two catalytic domains that are linked together by short sequences. Of the four genera that encode multifunctional β-glycanases, three (*Fibrobacter, Neocallimastix*, and *Ruminococcus*) are commensal to the rumen. This would suggest the environmental conditions and/or the nature of the available substrates provide a selection pressure. Interestingly, however, there is little similarity between the various enzymes, as exemplified by the three ruminal xylanase and

TABLE 2.9. Multifunctional Cellulolytic and Xylanolytic Enzymes.

Organism	Enzyme	Function	GenBank Accession Number/Reference
Caldocellum saccharolyticum	CelB	CBH/Cellulase	X13602
Caldocellum saccharolyticum	ManA	β-Mannanase/cellulase	L01257
Caldocellum saccharolyticum	CelA	Cellulase/CBH	[428]
Fibrobacter succinogenes	XynC	Xylanase/xylanase	[239]
Neocallimastix patriciarum	XylA	Xylanase/xylanase	X65526
Neocallimastix patriciarum	CelD	Cellulase/CBH/ xylanase	[429]
Ruminococcus flavefaciens	XynD	Xylanase/β-1,3- glucanase	[430]
Ruminococcus flavefaciens	XynA	Xylanase/xylanase	Z11127 [229,238]

xylanase enzymes. The two catalytic domains of the *N. patriciarum* XynA have similar substrate specificities and release similar hydrolysis products. In fact, it was speculated that this enzyme originated as a result of tandem duplication of an ancestral gene [237]. In contrast, the two catalytic domains (XylA-A and XylA-C; XynC-A and XynC-B) of the *R. flavefaciens* [229,238] and *F. succinogenes* [239,240] xylanases, respectively, have different specificities and modes of action. Moreover, although the two domains of *F. succinogenes* XynC belong to the same β-glycanase family (Family 11) [239], the catalytic domains XylA-A and XylA-C of the *R. flavefaciens* enzyme comprise different families (Families 11 and 10, respectively) [238].

The two catalytic domains of *R. flavefaciens* and *F. succinogenes* enzymes were expressed separately in *Escherichia coli* as truncated gene products that permitted their detailed characterization. Neither catalytic domain of the *R. flavefaciens* bifunctional xylanase was active against xylopentaose or shorter xylo-oligosaccharides, but their differences in mode of action pertained to the presence of substituents on the xylan backbone. With arabinoxylan as substrate, the Xyl-A domain is incapable of hydrolysing β-1,4-xylosidic linkages in the vicinity of arabinofuranosyl substituents, whereas the activity of the Xyl-C domain is unrestricted by this side-chain sugar [238]. The modes of action of the two catalytic domains of the *F. succinogenes* bifunctional enzyme are likewise differentiated by the presence of arabinofuranosyl substituents [240].

The discovery of naturally occurring multifunctional domain enzymes involved in the biodegradation of cellulose and heteroxylans may provide the impetus for the development of artificial multifunctional enzymes with optimized properties through genetic engineering techniques. Initial studies involving the fusion of a cellobiohydrolase with a cellulase have proved the viability and potential for this course [241], but more research is required to establish if there is any true benefit to multifunctional enzymes.

CHAPTER 3

Interactions and Associations

3.1 MULTIPLICITY

INDIVIDUAL bacteria and fungi have been observed to produce more than one form of each of the cellulolytic and xylanolytic [172] enzymes. Suggested explanations for this multiplicity include the expression of more than one gene, post-translational modifications involving differential glycosylation, and partial proteolysis at N-termini.

The genomes of all cellulolytic microorganisms studied to date encode for a set of cellulolytic and xylanolytic enzymes [30,200]. The number of distinct enzymes varies from one organism to another and may even change under different culture conditions, but generally, the cellulolytic system of a given fungus will comprise four to eight cellulases, one to three cellobiohydrolases, one to two β-glucosidases, and three to four xylanases. One of the greatest genetic-borne multiplicities has been observed in the bacterium *Clostridium thermocellum*, which codes for at least fifteen cellulase and cellobiohydrolase genes, two β-glucosidase genes, and two xylanase genes [447,448]. The reason for this diversity of genes encoding enzymes with apparently similar activities is not known. However, Wood has proposed the need for at least two forms of both cellulase and cellobiohydrolase based on stereochemical considerations [449]. He has argued that cellulose is a polymer of β-(1→4)-linked cellobiose that presents an alternate orientation of two adjacent glucosidic linkages along the horizontal axis of a single exposed cellulose chain (Figure 3.1). To ensure random attack at the two stereospecific linkages along the accessible surface of crystalline cellulose, two different cellulases and cellobiohydrolases would be required. Indeed, the synergy reported between two distinct cellobiohydrolases (CBHI and CBHII) of *Trichoderma reesei* has been attributed to the different orientation of the nonreducing end groups generated by two stereospecific cellulases (see below). The concerted action of both types of cellobiohydrolases would

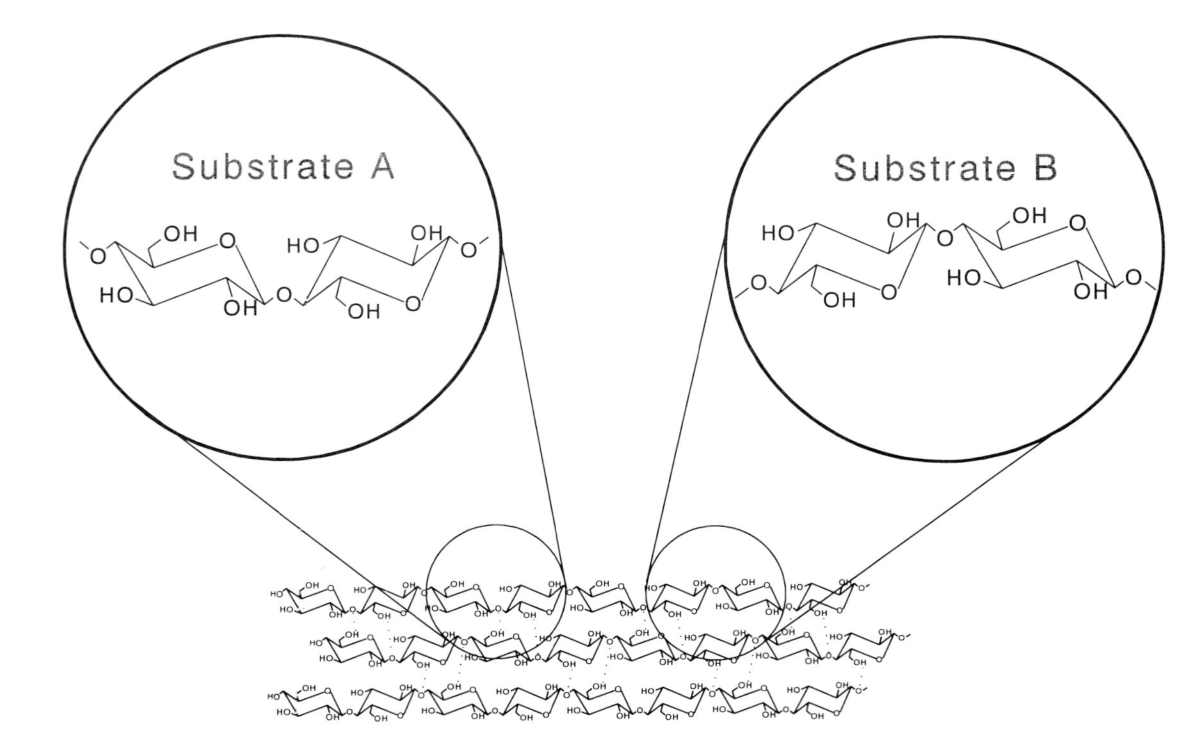

Figure 3.1 Alternate orientation of cellobiose residues within a cellulose chain arising from the stereospecificities of β-(1→4) linkages.

explain the observed increase in the overall efficiency of hydrolysis of crystalline cellulose [450].

Multiple genes certainly account for some of the observed multiplicity of the cellulolytic enzymes, but many microorganisms appear to produce more forms of the individual enzymes than apparently encoded by their respective genomes. The white-rot fungus *Schizophyllum commune* produces and secretes a multiplicity of cellulolytic enzymes but apparently encodes for only one of each type of enzyme. In this case, multiplicity was shown to arise, at least in part, at the level of transcription. Two distinct mRNA-directed products for each of the cellulases, cellobiohydrolases, and β-glucosidases have been detected [196]. This could arise from multiple initiation of transcription or by differential RNA splicing. The mechanism for the apparent transcript heterogeneity could be elucidated with the cloning and sequencing of the genes encoding the cellulolytic enzymes, but, unfortunately, these investigations have not been initiated. However, preliminary Southern blot analysis of the *S. commune* genomic DNA using a cloned cDNA fragment corresponding to the cellulase as a probe, suggested the presence of only one gene for the enzyme [196].

The extensively studied cellulolytic system of the fungus *T. reesei* QM9414 also exemplifies the phenomenon of nongenetic-borne multiplicity of cellulolytic enzymes. Genes encoding the two *T. reesei* cellobiohydrolases, CBHI and CBHII, along with those encoding two cellulases (EGI, EGII) have been detected and isolated [451]. Nevertheless, reports originating from different laboratories describe a greater number of these enzymes present in culture fluids [269,375,452–454]. For example, seven cellulases with similar activity profiles were detected in a culture fluid reacted with a polyclonal antibody raised against EGI [455]. Proteolysis was not considered to be responsible for the apparent multiple forms of EGI, and, moreover, only the three largest observed forms of the enzyme reacted with a monoclonal antibody raised against homogeneous EGI [456]. It is possible that the smaller forms of the cellulase are indeed proteolytic fragments lacking the epitope recognized by the monoclonal antibody. Alternatively, the lower M_r enzymes may represent a unique but homologous gene product(s) from which the encoding gene(s) has yet to be detected.

A further explanation for the lack of cross-reactivity of the monoclonal antibody with the multiple forms of EGI and one that may also account for some of the observed multiplicity of enzyme forms [456] could involve the nonspecific adsorption of components from the culture medium to the surface of the enzymes. Variations in the molecular masses of EGI have been noted upon cultivation of *T. reesei* under different conditions. In one study, an EGI with the highest molecular mass was observed when the fungus was grown on cellulose [456], whereas in separate studies, chromatofocusing [457] and isoelectric focusing [458] of culture supernatants showed that multiple enzyme forms appear when the culture grows older. These observations, together with the occurrence of an unusually high-glucose content associated with an isolated and purified EGI preparation [456], has led to the proposal that the EGI strongly adsorbs components from the culture fluids, such as cello-oligosaccharides derived by the cellulolytic enzymes [131,456]. Hence, a few of the observed EGI forms are thought to represent enzyme glycoproteins containing different amounts of attached cello-oligosaccharides. Nonspecific adsorption to different enzymes may also account for some of the multiplicity that has been observed by isoelectric focusing of both *T. reesei* xylanases and β-glucosidases. A p*I* 5.7 complex of *T. reesei* enzymes was found to contain a xylanase with a p*I* of 6.0 when the cellulase and β-glucosidase enzyme components were subsequently separated from the complex [459]. In the case of the β-glucosidase, it was thought that the low ionic strength of the chromatofocusing buffer combined with a high proportion of hydrophobic amino acids in the enzyme may have accounted for its aggregation with other proteins [460].

Although the observed multiplicity for at least some of the cellulolytic enzymes is apparently no more than an artifact of the growth conditions, extracellular postsecretory proteolysis of fungal enzymes has been shown to occur. Most fungi secrete proteases in addition to the membrane-associated enzymes used to remove the signal peptides of secreted proteins, and at least one, that from the fungus *Sporotrichum pulverulentum*, has been isolated [461]. Prolonged exposure of secreted enzymes to these proteases would increase the potential for nonspecific cleavages, and, indeed, increasing multiplicity as a function of culture conditions or age has been clearly demonstrated with *T. reesei.* When cultivated at a pH above five and in media limited in organic nitrogen content, no detectable proteolytic degradation of a *T. reesei* cellobiohydrolase was observed [462], which is consistent with previous observations concerning the production of this and other cellulolytic enzymes by the fungus [455,463]. However, a truncated form of a *T. reesei* cellulase has been isolated from culture medium with low pH and rich in organic nitrogen [464], and subsequent studies revealed that these conditions favour the production of an extracellular protease [462]. Further investigation confirmed the increase in multiplicity of extracellular cellobiohydrolases concomitant with the production of this protease [462].

Amino acid sequencing studies proved unequivocally that an extracellular protease is responsible for the production of, at least, the two forms of cellulase and β-glucosidase from *S. commune.* Both enzymes are secreted into their environment, initially as larger forms and as the culture matures, lower forms arise [465]. Two cellulases were isolated from the culture medium, and they were shown to have similar chemical, physical, and catalytic properties. Amino acid sequencing revealed that their amino terminal sequences (fifty-five residues) are identical except for an initial sixteen-residue alanine-rich sequence in the larger form, EGI [265]. The alanine-rich sequence is also present at the N-terminus of the larger form of the *S. commune* β-glucosidase, suggesting that it plays a role in secretion [265].

The relevance of proteolytic processing of cellulolytic enzymes has recently been reevaluated with the discovery of the multidomain structure of the enzymes. Separation of the cellulose-binding domains from the catalytic domains leads to enzymes with altered substrate activity profiles. Such proteolysis in effect generates novel enzymes, and although strictly conjecture at this point in time, it has been suggested that optimal saccharification of cellulose may be gained by the cellulolytic enzymes initially acting on the insoluble cellulose followed by the hydrolysis of soluble fragments catalyzed by the proteolytically derived core enzymes [444,451,466].

Differential glycosylation has also been reported to account for some of the observed multiplicity among the cellulolytic enzymes. For example, four different forms of the *T. reesei* cellobiohydrolase I are immunologically cross-reactive and were found to differ from one another only in the content and/or composition of the neutral carbohydrate [295]. Detailed analytical studies have since characterized the carbohydrate composition of this enzyme and confirmed the heterogeneity in glycosylation [467]. Similarly, four cellulases isolated from *Talaromyces emersonii* differ solely in their carbohydrate content [468], and the *S. commune* cellulase and β-glucosidase have been fractionated on Concanavalin A-Sepharose to yield separate pools of the respective enzymes, indicating heterogeneity in glycosylation [196].

3.2 SYNERGY

3.2.1 CELLULOLYTIC ENZYMES

According to the first hypothesis explaining the biodegradation of cellulose by fungi, which was advanced in 1950, crystalline cellulose was proposed to be hydrolysed by the concerted action of at least two enzymes. A nonhydrolytic component, termed C_1, was thought to disrupt the hydrogen-bonding pattern in crystalline cellulose to generate amorphous regions that could be randomly attacked by one of several cellulases (C_x) to release soluble cellodextrins. In the following decades, the hydrolysis of cellulose was repeatedly observed to be affected by mixtures of different proteins, and eventually it was discovered that the C_1 component was in fact cellobiohydrolase [293]. However, a minority view persisted in which it was thought that the cellobiohydrolase and C_1 component, the latter being defined as an enzyme responsible for the destabilization of crystalline cellulose, are separate enzymes that cochromatograph and purify and that C_1 acts by cleaving a few covalent bonds of the substrate surface as a preliminary to the action by the cellulases and cellobiohydrolases [469]. With our current understanding of the structure of many cellulases and cellobiohydrolases involving both catalytic and cellulose-binding domains, it is perhaps not too surprising that the early workers were embroiled in controversy. Nevertheless, these studies did uncover the concept of synergism among the fungal enzymes for the biodegradation of cellulose.

The most clear demonstration of the synergistic action of cellulolytic enzymes was conducted with *Trichoderma koningii* enzymes [105]. Purified cellobiohydrolase and cellulase were both incapable of solubilizing recalcitrant cellulose, but in combination they yielded the same degree of hydrolysis as crude, unfractionated culture fluids. Synergism was subsequently observed with artificial admixtures of enzymes of different origin and prompted speculation that cellulolytic enzymes may exist and act as complexes [293,470]. Synergistic actions of the cellulolytic enzymes were found to depend on the ratio of the individual enzymes [213,471] and on the degree of substrate saturation, with the greatest synergism observed when each enzyme is present at nonsaturating concentrations [472]. It has also been shown that both the degree of synergism and the optimal ratio of the individual enzymes are influenced by the nature of the substrate [471]. A plethora of studies involving the synergistic activities of cellulolytic enzymes from the same or different microbial sources, both fungal and bacterial, have been conducted and they were recently reviewed [473]. The following briefly summarizes the major observations and describes more recent findings.

From the continued studies over the past forty-plus years on the fungal cellulolytic enzymes, two types of synergism have emerged. Endo-exo synergism entails a sequential mechanism of enzymatic action with an initial attack of amorphous regions of cellulose by cellulases that provide new nonreducing chain ends for the subsequent action of cellobiohydrolases [376,471,474]. The degree of this synergism was found to be high when the substrate is highly ordered crystalline cellulose but low with amorphous cellulose and absent with soluble substrates, such as carboxymethyl cellulose [474]. These studies were performed over long periods of time, but when the early stages of reactions are monitored, endo-exo synergism may also be observed with soluble substrates until their sizes fall below the critical value of 4,000. At this point, the relative action of cellulases is negligible [475]. A second version of the endo-exo model invokes competition among the individual cellulolytic enzymes for available adsorption sites on the insoluble cellulose [476,477], but it is not entirely clear how this accounts for the observed synergistic activity.

Two mechanistic models have also been proposed for the second form of synergism, exo-exo. The first is based on experimental evidence obtained with the cellobiohydrolases (CBHI and CBHII) of *Trichoderma reesei*

[213,450,471,476,478]. CBHI was observed to cooperate with CBHII to catalyze the solubilization and hydrolysis of cotton. The two enzymes are thought to form a loose complex in solution, but maximal adsorption to cellulose is achieved with optimal admixtures of the individual CBHI and CBHII enzymes. The CBHI and CBHII from *Penicillium pinophilum*, alone or together, are unable to hydrolyse cotton, but they do interact synergistically against Avicel [479]. In this case, the exo-exo synergism was ascribed to the different orientations of the nonreducing end groups in crystalline cellulose, thus requiring two cellobiohydrolases with different stereochemical specificities to effect hydrolysis. However, the exo-exo form of synergism is suspect in the light of recent findings involving both the homogeneity of the enzymes used in the studies and their mode of action. When apparently homogeneous preparations of the *P. pinophilum* cellobiohydrolases were subjected to further purification by affinity chromatography, they were incapable of solubilizing crystalline cellulose either alone or together. Extensive hydrolysis with these highly purified cellobiohydrolases was only observed upon the addition of cellulases [480]. Likewise, cellulase-like activity appears to be associated with the CBHI from *T. reesei*. Treatment of native cotton with CBHI in the absence of any other enzyme releases short insoluble fibers in addition to soluble sugars [294]. Moreover, although it is clearly different from CBHI, CBHII appears to possess additional enzymatic activities to those of typical cellobiohydrolases (reviewed in Reference [131]). Whether the cellulase properties observed with these cellobiohydrolases are inherent or the consequence of minor contamination remains to be established.

In a recently conducted study with *T. reesei* cellulolytic enzymes, the simultaneous addition of enzymes in both exo-exo and endo-exo reactions was not required to effect synergism. Thus, the activity of CBHI appears to prepare the cellulose substrate for CBHII, and vice versa [481]. This would argue against the need to form prerequisite CBHI-CBHII complexes for synergistic activity as previously thought [478]. In contrast, the observations made with EGIII-CBHI and EGIII-CBHII synergistic systems support the previously held theory of a sequential mechanism of enzymatic action [474]. Synergism was observed following the addition of either CBHI or CBHIII to crystalline cellulose after its prior treatment with EGIII. However, EGIII activity is not influenced by pretreatment of substrate with either cellobiohydrolase [481]. Insights into the physical effects of enzymatic pretreatments on crystalline cellulose have been provided by electron microscopy studies. Morphological observations indicated that both exo- and endo-type cellulolytic enzymes from *Irpex lacteus* are strongly adsorbed on the internal microfibril of cotton fiber before enzymatic hydrolysis. Activity follows with hydrolysis toward the internal cellulose microfibrils with the retention of their original shape, but the two enzymes have distinct modes of action.

The cellulases appear to cause swelling followed by severe internal erosion and cavitation along microfibril layers inside fibers, whereas cellobiohydrolases create deep, transverse cracks from the surface to the lumen structure inside the cotton fibers [482]. Hence, by analogy with this *I. lacteus* system, it is conceivable that the endo-type activities of the *T. reesei* cellulases and cellobiohydrolases would cause swelling and exposure of microfibrils for subsequent action by added cellobiohydrolases in the sequential synergistic studies. Indeed, discrete interphases of cellulose microfibril degradation have been observed by transmission electron microscopy when *T. reesei* cellulase is added alone or sequentially with cellobiohydrolase [483]. On the other hand, it is unlikely that the surface action and crack formation resulting from pretreatment with a *T. reesei* cellobiohydrolase would greatly aid the activity of cellulases.

3.2.2 HETEROXYLANOLYTIC ENZYMES

Synergy is also observed among the xylanolytic enzymes, but in view of the diverse variety of enzymes involved, it is more complicated than that observed with the cellulolytic enzymes. A classification scheme has been established in an attempt to distinguish between the synergistic reactions involving the main-chain and side-chain hydrolytic enzymes [166]. *Homeosynergy* is defined as the synergistic interactions between two or more main-chain–cleaving enzymes or two or more different types of side-chain–cleaving enzymes, whereas *heterosynergy* concerns the reactions between a side-chain– and a main-chain–cleaving enzyme. Homeosynergy is further divided into uniproduct homeosynergy and biproduct homeosynergy. An example of uniproduct homeosynergy would be the release of unobstructed acetyl substituents by acetylxylan esterase, thereby allowing access of α-glucuronidase to its site of action. Alternatively, the release of a ferulic acid moiety from an arabinosyl side chain by a ferulic acid esterase would expose the arabinosyl residue to α-L-arabinofuranosidase (Figure 3.2). This latter form of uniproduct homeosynergy has been observed with the combined action of a ferulic acid esterase from *Aspergillus oryzae* with an arabinofuranosidase from *Penicillium capsulatum* on feruloylated arabinoxylan [346]. With these types of reactions, the rate of activity of only one of the two enzymes (i.e., the second enzyme) is affected by the activity of the other. Biproduct homeosynergy defines the situation whereby both reaction rates of the two enzymes involved are enhanced by the synergistic interaction. Examples of homeosynergistic reactions among xylanolytic enzymes would include the interactions of xylanase from *Neurospora crassa* with its β-xylosidase [182] and the activities of the xylanases from *Streptomyces exfoliatus* [398]. These homeosynergistic reactions are usually of the biproduct type.

As to be expected, there are many examples of heterosynergy in the

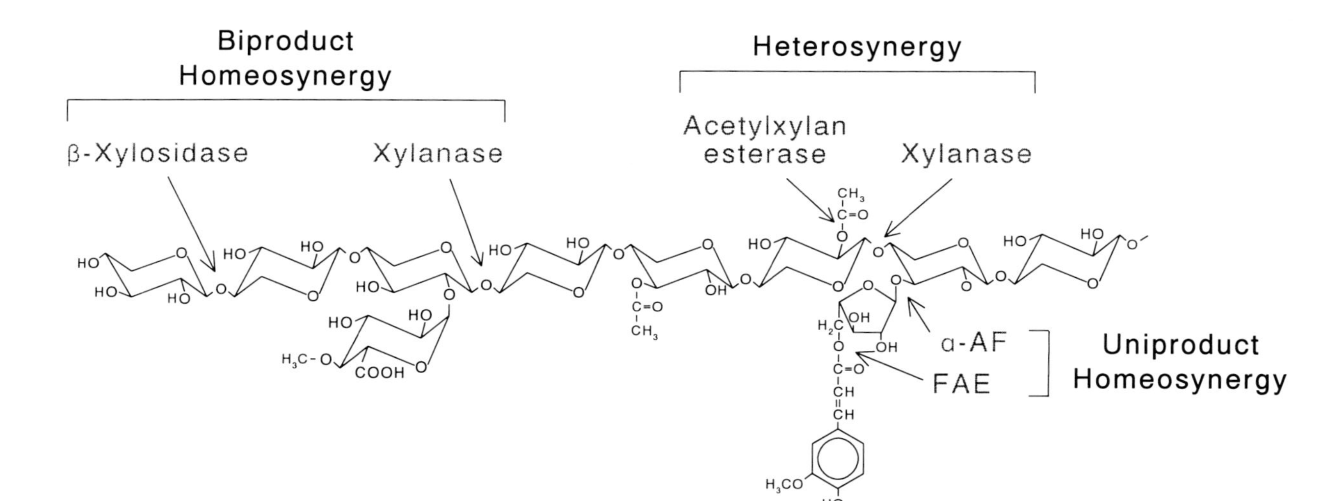

Figure 3.2 Synergistic reactions of heteroxylanolytic enzymes. A hypothetical heteroxylan substrate is depicted to provide examples of the following: uniproduct homeosynergy between α-arabinofuranosidase (α-AF) and ferulic acid esterase (FAE); biproduct homeosynergy between β-xylosidase and xylanase; and heterosynergy between acetylxylan esterase and xylanase.

degradation of heteroxylans. The release of ferulic acids by the ferulic acid esterases from the fungi *Neocallimastix frontalis* [484], *Penicillium pinophilum* [485], and *Schizophyllum commune* [246] and the bacterium *Streptomyces olivochromogenes* [486] apparently occurs only in the presence of xylanase. Similarly, xylanases greatly enhance the release of arabinose from heteroxylans by the α-arabinofuranosidases from *Clostridium acetobutylicum* [487], *P. capsulatum* [346], *Ruminococcus albus* [488], *Thermonospora fusca* [489], and *Trichoderma reesei* [490]. Such synergy, in fact, represents bidirectional heterosynergy [166] because the initial action of xylanases, followed by side-chain–releasing enzymes would generate xylo-oligosaccharides, which could then be re-attacked by the xylanases. Bidirectional heterosynergy has also been observed with the interactions between the xylanases and acetylxylan esterases from both *T. reesei* and *S. commune* [491]. The production of deacetylated xylo-oligomers, xylose, and acetic acid by treatment of acetylxylan with a combination of the two enzymes was ascribed to the initial action of the esterases to create unblocked sites on the xylan backbone for attack by the xylanases. The further hydrolysis catalyzed by the xylanases produced shorter substituted xylo-oligomers, which are the preferred substrates for the esterases [491]. These data and conclusions were confirmed by subsequent studies that revealed that a mixture of xylanase and β-xylosidase was ineffective in hydrolysing acetylxylan in the absence of an acetylxylan esterase [412,492].

Antisynergy is encountered in which the activity of one type of enzyme inhibits or prevents the action of a second. For example, a xylanase from *Bacillus subtilis* [164] and *Gloeophyllum trabeum* [165] hydrolyse heteroxylans preferentially at xylosyl residues vicinal to residues substituted with 4-*O*-methyl-glucuronic acids, and, hence, release of these substituents by α-glucuronidase would retard, if not preclude, the activity of the xylanases. However, it is unlikely that this situation arises in vivo because these xylanases with high substrate specificity are probably produced to accommodate thc lack of a debranching type enzyme. Nevertheless, if care is not taken, antisynergy may arise in vitro and lead to inefficient hydrolytic processes.

3.3 CELLULOSOMES

The cellulolytic systems of anaerobic bacteria have received considerable attention in recent years in view of their intriguing molecular architecture, in addition to their agricultural and biotechnological relevance. In many ways, these systems closely resemble fungal cellulolytic systems, especially in terms of the enzymes involved, but they greatly differ in their organization. It is now apparent that the anaerobic bacteria, and at least one anaerobic fungus, organize their cellulolytic enzymes into macromolecular complexes, termed *cellulosomes* [493,494]. The cellulosomes thus maintain the cellulolytic enzymes within close proximity to further their synergistic action. The microorganisms currently known to produce these discrete multifunctional, multienzyme cellulosomes are listed in Table 3.1.

The cellulosome was first discovered on *Clostridium thermocellum* [495], and it represents the paradigm of bacterial cellulolysis (for recent reviews, see References [496,497]). It comprises a number of different enzymes that are anchored to a noncatalytic, core subunit termed *scaffoldin* (originally, S1 or S_L or CipA) (Figure 3.3). This polypeptide is relatively large (approximately 1,800 amino acids and a molecular weight of approximately 210 kDa) and contains a Type 3 cellulose-binding domain (CBD). In addition, nine distinct but closely related domains, termed *cohesins*, exist along the protein, which are responsible for integrating the cellulolytic enzymes into the cellulosome complex [498]. The cohesins, which are characterized by a common twenty-three–amino acid sequence, are separated by linker sequences that, like the linker sequences observed in multidomain cellulases and cellobiohydrolases, are rich in proline and glycosylated threonine residues. However, the oligosaccharide structures are very unusual and likely unique to cellulosomes [108,188,499]. The *C. thermocellum* scaffoldin also contains a docking domain (*dockerin*) that is capable of interacting with a cohesin domain. The dockerin of a scaffoldin is thought to mediate either attachment to the host cell surface or oligomerization of other scaffoldin

TABLE 3.1. Cellulolytic Microorganisms Thought to Produce Cellulosome-like Complexes.

Microorganism	Reference
Bacteria	
Acetivibrio cellulolyticus	[499]
Bacillus circulans	[529]
Bacteroides cellulosolvens	[499]
Butyrivibrio fibrosolvens	[530]
Cellulomonas sp.	[499]
Clostridium cellobioparum	[499]
Clostridium cellulolyticum	[109]
Clostridium cellulovorans	[499,525]
Clostridium josui	[531]
Clostridium papyrosolvens	[532]
Clostridium thermocellum	[493]
Fibrobacter succinogenes	[533]
Ruminococcus albus	[499]
Ruminococcus flavefaciens	[534]
Thermomonospora curvata	[535]
Fungi	
Neocallimastix frontalis	[536]

Source: Adapted from Bayer et al. [496].

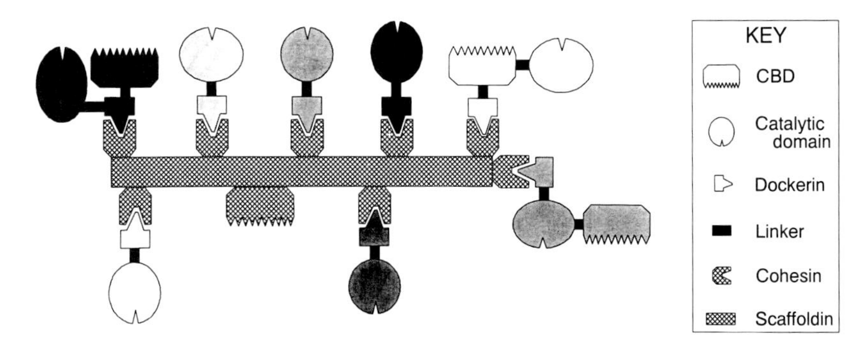

Figure 3.3 Model for the organization of the cellulosome produced by *Clostridium thermocellum* (adapted from Reference [496]).

molecules [500]. The only other characterized scaffoldin is from *C. cellulovorans,* and it shares most of the same features as that from *C. thermocellum.* The notable differences include the general arrangement of the cohesins and the apparent lack of a dockerin domain [501].

The scaffoldins from *C. thermocellum* [235,495,502] and *C. cellulovorans* [501,503] have been cloned and sequenced, and their CBDs are very similar, with approximately 50% identity [502]. In spite of this extensive homology and presumably a similar three-dimensional fold, antibodies raised against the CBD from the *C. thermocellum* scaffoldin did not react with that from *C. cellulovorans*, indicating structural differences between the two [502]. Nevertheless, the binding affinities and capacities of the two CBDs for substrate are similar. In addition, both are capable of forming tight complexes with chitin in addition to cellulose ligands but fail to adsorb to xylan. This is consistent with the fact that *C. thermocellum* does not grow on or utilize xylan as an energy or carbon source. A direct comparison between the CBDs of scaffoldins and all those of the cellulolytic enzymes associated with the cellulosome [474,504,505] has not been made, but it is highly likely that they too will share many common features.

Reports on the number of enzymes associated with the cellulosome differ significantly, in spite of the ease at which the complex can be isolated in a highly pure state [506]. The estimated number of individual polypeptides range from fifteen to thirty-five [495,507,508], but only ten to fourteen have been resolved by sodium dodecyl sulfate polyacrylamide gel electrophoresis [508,509]. These differences reflect the problems associated with dissociating and isolating the individual cellulosome subunits [493,494]. In addition to an insensitivity to most chaotropic reagents, the cellulosome is quite resistant to proteolysis [493], but it was found that proteinase K is capable of degrading cellulosome components [510]. Using the technique of limited proteolysis with proteinase K, a major cellobiohydrolase was released and isolated from the *C. thermocellum* cellulosome [296]. This truncated form of the enzyme was purified and characterized, and it was found to have very similar properties to those of fungal cellobiohydrolases.

The failure to effectively separate the majority of the other cellulosome components has forced researchers to adopt the strategy of cloning the genes encoding the different enzymes and expressing them in *Escherichia coli*. The genes for fifteen cellulases and cellobiohydrolase, two β-glucosidases, and two xylanases are thought to exist in the genome of *C. thermocellum* [447,448] and, to date, the enzymes that have been cloned and characterized include nine cellulases, CelA [511], CelB [512,513], CelC [273,514], CelD [515], CelE [107], CelF [504], CelG [516], CelH [517], and CelM [518], the cellobiohydrolase CelS [519], two β-glucosidases [447], BglA [520], and BglB [521], and a xylanase, XynZ [159,522]. These enzymes belong to six different families (Appendix I), but all, except cellulase C, contain a conserved noncatalytic domain of approximately sixty-five residues that is comprised of two similar segments of twenty-two residues each [234,523]. Several lines of evidence, involving radiolabelling, immunological, and mutation studies, revealed that these noncatalytic domains constitute the dockerins that form associations with the cohesins of the scaffoldin protein [233,234]. Antibodies to the duplicated segments of cellulase D were observed to cross-react with a variety of the cellulosome components, and ^{125}I-labelled cellulase D was found to bind predominantly to scaffoldin. No association was observed to occur between the scaffoldin and a truncated form of the enzyme that lacked the dockerin sequence [234]. Fusion of the genetic element encoding for the dockerin sequence of cellulase D to the cellulase C gene and its subsequent expression in *E. coli* resulted in the production of a hybrid enzyme that could be associated with scaffoldin in a manner similar to cellulase D [233]. Divalent cations also appear to play a role in stabilizing the cellulosome structure [524].

Cellulosomes produced by *C. thermocellum* are packed at the cell surface into protuberance-like organelles, known as protubozymes [507,525]. A fibrous connective matrix serves to retain the cellulosomes within protubozymes. Protubozymes adorn the surface of the bacterial cell being held in place through interactions involving a cell surface protein, termed outer layer protein (formerly, ORF3p) [526]. At this point in time, it is not clear whether the outer layer proteins associate with scaffoldin or the individual enzymes associated with the cellulosome, but there is some evidence to suggest that the latter occurs [236]. The cellulosomes incorporated into the cell wall mediate adhesion of the cell to cellulose. Upon binding, the protubozymes appear to undergo a conformational change to form protracted columns between the cell and substrate. The CBDs of both the scaffoldin and its associated enzymes are thought to promote the destabilization of the substrate to facilitate hydrolysis [527]. These interactions combined with the physical organization of the cellulosome are postulated to promote the synergistic action of the complement of cellulolytic and xylanolytic enzymes [493,495]. To further the cellulolytic potential of *C. thermocellum*, cellulosomes may also be released into the external environment as the cell matures, where they continue cellulolytic activity.

C. thermocellum does not grow on or utilize xylan or derivatives, but it still produces a highly developed xylanolytic system of enzymes [528]. Four major xylanase activities with molecular weights of 170, 84, 67, and 54 kDa have been detected by zymogram analysis, and three minor enzymes (150, 98, and 60 kDa) are also present. Presumably, these enzymes are required to degrade obstructing heteroxylans and thereby expose cellulose to the cellulosome.

Two other species of clostridia, the mesophilic *C. cellulovorans* and *C. cellulolyticum*, seem to produce cellulosomes that are structured very similarly to the cellulosome of *C. thermocellum*, but information concerning them is only now emerging. Few details are known about the multienzyme complexes produced by the other microorganisms listed in Table 3.1, but with further investigation, they too may be characterized as true cellulosomes.

CHAPTER 4

Production and Purification

4.1 INDUCTION

4.1.1 CELLULOLYTIC ENZYMES

IT has long been known that the fungal cellulolytic enzymes are inducible, but the exact nature of the inducer confounded early researchers for many years. Indeed, two pioneers of research into cellulose biodegradation, Mary Mandels and Elwyn T. Reese [537] exclaimed more than thirty-five years ago, "Cellulase is an adaptive enzyme, but its substrate cellulose is insoluble: how could induction then occur?" Despite considerable effort during the subsequent years, how the biosynthesis of cellulolytic enzymes is triggered by the extracellular, insoluble cellulose is still poorly understood. What is known about the regulation of the enzymes involved in cellulose degradation by bacteria [538] and by both bacteria and fungi [539,540] has been the subject of comprehensive reviews.

Many of the early investigations pertained to the study of the *Trichoderma reesei* system of cellulolytic enzymes. Such studies led to the assumption that the fungi produce low levels of constitutive enzymes that are secreted and attack the cellulose of their immediate environment, resulting in the production of a soluble inducer of cellulolytic enzymes. The availability of both poly and monoclonal antibodies aided the identification and quantification of the low levels of the individual enzymes associated with *T. reesei* cells. These immunochemical experiments indicated that the low basal levels of secreted enzymes appear to remain associated only with the conidial walls and not on the fungal mycelia [541–543]. This is consistent with the observations that long lag periods are exhibited by *T. reesei* mycelia before they are capable of utilizing cellulose and that such catabolic activities are always accompanied by sporulation [544–546]. At least one study, however, has suggested the presence of constitutive enzymes on the surface of *T. reesei* mycelia [547], but it is likely that the mycelial preparations used in this latter study contained conidia, thereby leading to the apparent contradictory results.

The enzymes associated with the conidial walls are capable of hydrolysing crystalline as well as amorphous cellulose [541], and the resulting soluble product(s) was observed to induce synthesis of enzymes in germinating conidia [544]. More recent evidence has indicated that maximal induction by cellulose is correlated with the level of conidial-bound cellobiohydrolase II [542]. This is somewhat surprising given that this enzyme comprises only 15–20% of the total secreted proteins present in typical *T. reesei* culture fluids, whereas cellobiohydrolase I accounts for up to 60% [131]. Even though cellobiohydrolase I was subsequently shown not to be essential for induction of conidial cellulolytic enzymes, it does significantly contribute to the formation of lower molecular mass inducers [548] and therefore strongly implicates the identity of the inducer molecule as cellobiose. A cellobiose permease is present in the cell membrane of *T. reesei* [549], and induction by cellobiose has been found to be concentration dependent, with higher concentrations actually having an inhibitory effect [550]. This is presumed to arise from the intracellular hydrolysis of cellobiose to glucose and thereby causing catabolite repression (see below). Cellobiose is also known to be an effective inducer of the cellulolytic enzymes of other fungi, such as *Schizophyllum commune* [465,551]. Other compounds found to induce cellulolytic activities of *T. reesei* include water-soluble lactone derivatives of cellobiose and glucose, such as cellobiono-1,5-lactone and 2,3,4,6-tetramethyl glucono-1,5-lactone [552–554]. Cellobiono-1,5-lactone is particularly effective as an inducer of cellobiohydrolase I, and its production is thought to occur by the action of cellobiohydrolase-catalyzed hydrolysis of naturally oxidized cellulose [554].

Of the various known inducers, sophorose, a disaccharide of β-1,2-linked glucose (Figure 4.1) previously known to induce the cellulolytic enzymes of *T. viride* [555], is the most potent [545]. Sophorose has been isolated from a *Trichoderma* culture medium [545], and its formation in vivo has been demonstrated [556], leading to speculation that it is the product of transglycosylation reactions involving either β-glucosidase [557] or cellulase [558]. Other studies have provided support for this hypothesis. A constitutive β-glucosidase concentrated in the cell membrane has been shown to cause the synthesis of sophorose from cellulose [149], whereas a comparison of the induction of cellulase and β-glucosidase in conidia, mycelia, and protoplast revealed that cellulose induces cellulase in conidia but sophorose can induce this enzyme in all three cell forms. On the other hand, β-glucosidase is produced in conidia, mycelia, and protoplasts after induction with both inducers [544]. More direct evidence was provided by inhibition studies. Inhibition of β-glucosidase activity by the specific inhibitors of the enzyme, nojirimycin and glucono-β-lactone, resulted in the concomitant inhibition

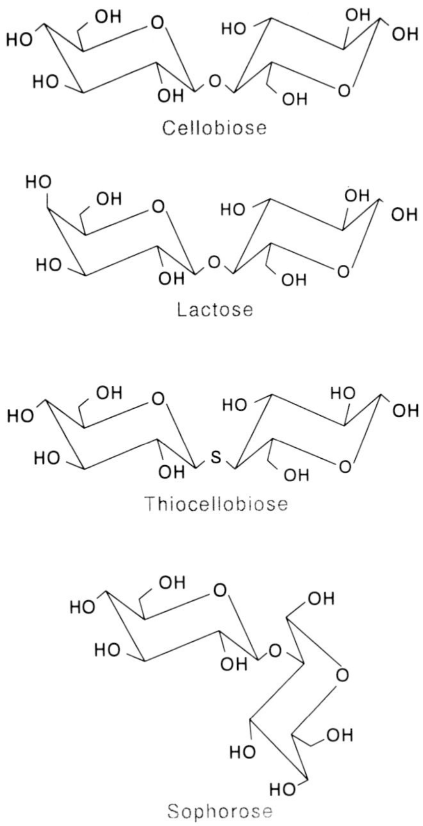

Cellobiose

Lactose

Thiocellobiose

Sophorose

Figure 4.1 Structures of some of the known inducers of cellulolytic enzymes.

of de novo cellulase synthesis when the fungus was cultured in the presence of cellodextrins but not on sophorose [544]. Hence, with *T. reesei*, the cellulase bound to the conidia acts upon cellulose to release soluble fragments that are subsequently converted to a soluble inducer by the action of β-glucosidase.

T. reesei produces two forms of β-glucosidase that apparently have two different roles in the induction process. One form of the enzyme is extracellular and participates with cellulase and cellobiohydrolase in the concerted hydrolysis of cellulose, whereas the second form remains associated with the cell membranes (reviewed in Reference [539]). It is this latter form that is thought to catalyze the transglycosylation reactions converting cellobiose to sophorose. Conidia treated with β-octylglucoside exhibited very poor growth on cellulose because of the removal of the wall-bound enzyme by the action of the mild detergent, but the fungus grew normally on glycerol [541]. That both forms of the enzyme are required for the cellulose-stimulated induction of cellulolytic enzymes was apparent from mutant studies.

T. reesei mutant D6-13 produces a wall-bound β-glucosidase but lacks extracellular activity [539]. The mutant was found to contain basal levels of cellulase activities when grown on crystalline cellulose, and induction did not occur unless the cultures were supplemented with cellobiose or sophorose. A model to account for these observations is presented in Figure 4.2. The basal levels of constitutive extracellular cellulolytic enzymes act to release cellobiose and glucose from crystalline cellulose. The resulting concentrations of both metabolites at the cell wall are sufficient to promote the vectorial transglycosylation reactions of the wall-bound β-glucosidase, which releases its reaction product sophorose intracellularly.

Although sophorose has been shown to be a powerful inducer of the cellulolytic enzymes of *T. reesei*, electrophoretic analysis of the spent media recovered from cultures of the fungus supplemented with the disaccharide revealed a reduced expression of some of the enzyme components relative to others [559]. Specifically, the production of some cellulases did not appear to be induced by sophorose. Moreover, this sugar does not have a universal effect on all cellulolytic microorganisms because there is at least one example of an enzyme system, that of *Sporotrichum pulverulentum* [560], that is not induced by sophorose.

Northern Blot hybridization experiments have revealed, at least for *T. reesei* cellobiohydrolases and cellulases, that regulation of gene expression occurs at the pretranslational level. Transcription of genes coding for cellulolytic enzymes was observed to follow within twenty minutes of addition of inducer [547,561,562], and the fact that biosynthesis of a constant ratio of the three major enzymes, cellobiohydrolases I and II and cellulase I, was strongly promoted suggested their expression is coordinately regulated. Such regulation may involve cAMP, although in this regard there are conflicting data. In one study, no correlation between the intracellular level of cAMP and the rate of cellulolytic enzyme synthesis was observed during growth of *T. reesei* on cellobiose or sophorose [563]. On the other hand, a rise in intracellular cAMP levels preceding the onset of cellulase synthesis has been observed when the same fungus was cultivated on lactose [564] or sophorose [565], suggesting that cAMP serves as the positive effector of the induction.

As stated above, cellobiose has been demonstrated to be an inducer of cellulolytic enzyme expression in *S. commune* [551], and it is presumed to be produced by the direct action of basal levels of constitutive cellobiohydrolase activity on cellulose. Surprisingly, neither sophorose nor lactose enhance the enzyme production in this fungus, but carboxymethyl cellulose, xylan, and synthetically produced thiocellobiose (4-*S*-(β-D-glucopyranosyl)-4-thio-D-glucopyranose) (Figure 4.1) serve as inducers [551]. The latter compound has proved to be the most effective inducer of the *S. commune* enzymes, presumably because it is not catabolized within the cell to produce

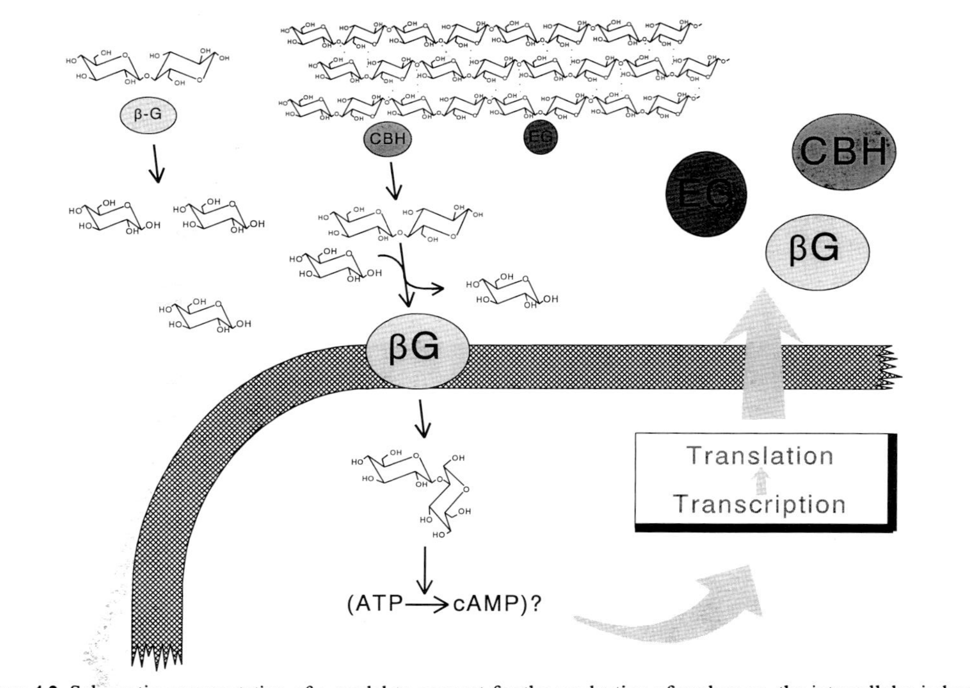

Figure 4.2 Schematic representation of a model to account for the production of sophorose, the intracellular inducer of cellulolytic enzymes. Low levels of constitutive, extracellular cellulase (EG), cellobiohydrolase (CBH), and β-glucosidase (β-G) act in a concerted fashion to release cellobiose and glucose from insoluble cellulose. These released sugars serve as substrates for the production of sophorose by a transglycosylation reaction involving a membrane-bound form of a β-glucosidase. Sophorose acts as the intracellular inducer, possibly by stimulating the production of cAMP, for the de novo biosynthesis of extracellular cellulolytic enzymes.

higher levels of glucose, a known repressor of cellulolytic enzymes (see below). Induction within *S. commune* cells appears to lead to the coordinate expression of cellulases and cellobiohydrolases, but β-glucosidase is under separate regulation [465]. Other cellulolytic systems that appear to be directly induced by cellobiose include those produced by *Aspergillus nidulans* [566], *Penicillium funiculosum* [567], and the bacteria *Cellvibrio gilvus* [568], and *Thermomonospora fusca* [569].

4.1.2 XYLANOLYTIC ENZYMES

It is apparent that the production of xylanases may also be induced by either large polymeric substrates or small molecular weight metabolites, but as with the situation concerning the cellulolytic enzymes, the regulatory mechanisms remain relatively obscure. The xylanolytic enzymes of the yeast, *Cryptococcus albidus*, represent one of best-studied systems. Production of extracellular enzymes increases approximately two orders of magnitude when cells are cultured in the presence of wood xylans or xylo-oligosaccharides compared with growth on glucose [570]. Because xylobiose was the only component not degraded by the xylanases, it is assumed to be either the natural inducer or its immediate precursor. Indeed, xylobiose, in addition to xylose and arabinose, has been demonstrated to induce xylanase production in another yeast, *Aureobasidium pullulans* [571]. Other xylanopyranosides have been tested for their xylanase-inducing potential in *C. albidus*, and analogous to the use of thiocellobiose for cellulolytic systems, the nonmetabolizable methyl-β-D-xylopyranoside proved to be the most effective [409]. Further insights into the regulatory mechanism of the induction were gained by the use of this compound, and it appears that the induction is manifested at the level of transcription [572], possibly involving a putative 15-nucleotide cAMP regulatory sequence located upstream from the xylanase gene [573]. Moreover, it has been established that uptake of these low molecular mass inducers increases during incubation in their presence, and that they, in fact, initially induce the production of an active transport β-xyloside permease system [574]. Although methyl-β-D-xylopyranoside was equally effective as an inducer of the *C. albidus* enzymes compared with xylan, this compound was observed to be fifteen to twenty times better than xylan for the induction of xylanase production by *C. flavus* [575].

Production of the xylanolytic enzymes in the fungi may also be induced, and almost without exception, the inducers include xylans or their hydrolysis products. There are many examples in the literature of xylanase induction in fungi cultivated on media containing xylan or its degradation products (see Reference [576] for a recent review), and, in general, there is a direct relationship between the concentration of the inducing material and level of induction. That xylan products and not contaminating cellulose serve as the inducers of *T. reesei* xylanases was unequivocally demonstrated in experiments employing *Acetobacter xylinum* cellulose. Use of this xylan-free bacterial cellulose as a growth medium led to a low level constitutive production of xylanases, but as the concentration of added xylan increased, a higher production of xylanase was observed [577]. Likewise, the *T. reesei* xylanase I is specifically formed upon growth on xylan, whereas only low levels are produced on cellulose. Using a monoclonal antibody to monitor its production and release, both conidia and mycelia were found to produce low constitutive levels of this enzyme, and it was induced in resting mycelia by xylobiose and, to a lesser extent, by sophorose. A mutant with an impaired β-linked disaccharide permease exhibits very little of the xylanase when grown on xylan, and production of the enzyme is not induced by xylobiose [578]. These data suggest that, similar to the induction of cellulolytic enzymes, small metabolites released by the low levels of constitutively expressed xylanases are taken up by the cell to induce transcription of the genes coding for the xylanolytic enzymes.

The situation with *S. commune* represents at least one exception to the general trend discussed above. The levels of xylanase production have been observed to be notably higher when this fungus is cultivated on cellulose products compared with growth on xylans [246,579], suggesting that cellobiose rather than xylobiose serves as the inducer of the xylanases for this microorganism.

An intriguing induction mechanism has recently been delineated for the control of the *Streptomyces lividans* xylanases, which may account in part for the observed multiplicity of xylanase production [580]. Three xylanases, XlnA, XlnB and XlnC, are produced by this bacterium; the former cleaves xylan to yield predominantly small xylo-oligosaccharides [384], whereas the latter two hydrolyse xylans to larger xylo-oligosaccharides [581,582]. Characterization of *S. lividans* mutants with disrupted xylanase-encoding genes has indicated that all three enzymes are constitutively expressed at very low levels in the absence of substrate. In the presence of xylans, XlnB and XlnC act in a concerted fashion, initially to hydrolyse a fraction of the substrate and then to produce large xylo-oligosaccharides (viz., xylodecaose and larger) by transglycosylation reactions. These large xylo-oligosaccharide products serve to signal the availability of the substrate, presumably through binding to an external receptor, and lead to the induction of *xlnB* and *xlnC*. The large xylo-oligosaccharide signal molecules are finally cleaved by XlnA into assimilable products [580]. Further investigation is required to test the validity of the model, to identify the external recognition site, and to delineate the sequence of events within the cell that leads to induction of the xylanase genes. However, this model does provide another purpose for the production of multiple forms of xylanolytic and possibly cellulolytic, enzymes.

4.1.3 HETEROXYLAN SIDE-CHAIN HYDROLASES

Detailed studies of the production and regulation of the heteroxylan side-chain hydrolysing enzymes have not been conducted, but preliminary data do suggest that xylans and/or cellulose induce their production. The acetyl xylan esterases of *T. reesei* and *S. commune* appear to be coproduced with their cellulases and xylanases. Production of the acetyl xylan esterases is increased when deacetylated xylan in the growth medium is replaced by an acetylated xylo-oligomer, but addition of cellulose further enhances their production [579]. Likewise, the production of feruloyl esterase by *S. commune* is induced by growth on cellulose substrates [246]. In direct contrast, the production of acetyl xylan esterases, ferulic acid esterases, α-L-arabinofuranosidases and α-L-*O*-methylglucuronidases of *Aspergillus* and *Streptomyces* species was enhanced by the presence of xylan, and to a lesser extent, by cellulose [247,583]. *A. niger* α-L-arabinofuranosidase production is apparently induced by metabolites of arabinose, particularly L-arabitol [584].

4.2 REPRESSION

As with most metabolic enzymes, the cellulolytic and xylanolytic enzymes are subject to catabolite repression. In an almost universal response, addition of easily metabolizable substrates prevents the further biosynthesis of these enzymes [539]. Thus, glucose represses the biosynthesis of cellulases, cellobiohydrolases, β-glucosidases, and xylanases of both bacterial and fungal origin (for recent reviews, see References [539] and [585]). In *Trichoderma*, glucose and its catabolites, such as glycerol, strongly repress the further production cellulolytic enzymes [560,586] even when added simultaneously with crystalline cellulose [587]. Such repression will continue until the catabolites of cellulose biodegradation are completely consumed by the cells, at which time the inductive effect of the cellulose substrate will be exerted.

The second messenger cAMP is a general regulator of metabolic processes in living cells. In bacteria, one of its functions is the regulation of catabolite repression [588], whereas in fungi, it is implicated in the control of a broad variety of metabolic functions, with the majority involving the phosphorylation of proteins by cAMP-dependent protein kinases. Despite continued effort, the participation and role of cAMP in the regulation of extracellular microbial glycanases remains to be resolved. For example, addition of glucose to wild-type cultures of the thermophilic actinomycete *Thermonospora fusca* led to a significant reduction in the level of cAMP from 24 to 7 pmol/mg dry weight, but only reduced cAMP concentrations in a mutant completely resistant to glucose repression of both growth and cellulase synthesis from 31 to 22 pmol/mg dry weight [589]. These data are thus wholly consistent with cAMP-regulated catabolite repression. That cAMP levels control the biosynthesis of the cellulolytic enzymes of this fungus, at least in part, was demonstrated by the exogenous addition of cAMP to permeabilized cells of *T. fusca,* which led to the stimulation of cellulase production [590]. Moreover, a correlation between cAMP concentrations and levels of cellulase biosynthesis in vivo was established [590]. On the other hand, the biosynthesis of a xylanase in *Cryptococcus albidus* was stimulated by the addition of exogenous cAMP, but it did not relieve the repression caused by growth on xylose [573]. Even more curious, in addition to not relieving catabolic repression, the presence of 10 mM cAMP, in fact, inhibited the biosynthesis of *Bacillus circulans* cellulases and xylanases [591]. An explanation for these latter results may concern the high concentration of added cAMP, because a concentration-dependent stimulation *and* repression of *Trichoderma reesei* cellulase has been described [565]. Thus, addition of low concentrations (μM) of a nonhydrolysable cAMP derivative (dibutyryl cAMP) stimulated the sophorose-induced production of cellulase, but at higher concentrations (mM), its biosynthesis was repressed. cAMP alone was neither capable of inducing cellulase production nor able to relieve catabolite repression caused by glucose. Further investigation of this cellulase system for other effector molecules revealed that addition of hexoses that may undergo phosphorylation increased intracellular levels of glucose-6-phosphate, which in turn resulted in repression of cellulase biosynthesis, whereas addition of exogenous nonphosphorylable hexoses and pentoses, such as 2-deoxyglucose and xylose, caused induction of the cellulase [565].

In an effort to explain the need for an inducer and that cAMP alone is unable to both effect initiation of cellulase synthesis and relieve repression by glucose, a model has been invoked, which is based on an understanding of the regulatory mechanism of the prokaryotic *lac* and *ara* operons in *Escherichia coli*. For both the streptomycete *Acidothermus cellulolyticus* [538] and the fungus *T. reesei* [565], the existence of three regulatory elements operating at the level of transcription are proposed: a cellulase-activating molecule (CAM) that complexes with the inducer; a cAMP receptor protein (CRP); and a repressor protein (RP). The biosynthesis of cellulase is postulated to occur at a very low constitutive level when neither CAM nor CRP are in their complexed state with their appropriate effector molecule, inducer, and cAMP, respectively. Maximal transcription of the cellulase gene(s) would occur with the cooperative, simultaneous binding of active inducer-CAM and cAMP-CRP complexes to the putative promoter site on the DNA. In the presence of a repressor, such as glucose, the intracellular concentrations of cAMP would decrease, diminishing the positive regulation at the promoter. In addition, an active glucose-RP complex would form, which would bind to an operator site downstream from the promoter and thereby block the transcription process. Although this model does account for most of the observations noted above, it is probably too simplistic, and much further investigation is required for its corroboration.

Recently, the regulation of the xylanase-encoding *xlnA* gene of *Aspergillus tubigensis* was investigated at the molecular level. By deletion analysis of the 5′ upstream region of the gene, a 158-base pair sequence involved in the xylan-specific induction has been identified [592]. Deletion of this region completely abolished transcription of *xlnA*. Analysis of the sequence revealed an upstream activating sequence (UAS) that is proposed to control the induction of the adjacent *xlnA*. Transcription of *xlnA* is also known to be controlled by catabolite repression because no expression was observed when the system was induced in the presence of glucose. The specific mechanism of this repression remains to be established, but four potential catabolite repression (CRE A) binding sites are located upstream of the 158-base pair fragment. These observations suggest that carbon catabolite repression of *xlnA* is controlled at two levels. The first involves the direct repression of *xlnA* expression, whereas the second concerns the repression of the expression of a transcriptional activator [592]. A study of the regulation of expression of the xylanases in *A. niger* and *A. nidulans* using this 158-base pair fragment suggested a similar regulatory pattern and prompted speculation that all fungal xylanolytic organisms may be under a similar mechanism of control [592].

It is also highly unlikely that one mechanism of regulation is operative in all cellulolytic microorganisms. Indeed, as with the induction systems discussed above, exceptions to the general observed trends of enzyme repression also exist. For example, carbon catabolites of cellulose biodegradation, such as glucose and lactose, do not repress the production of the cellulolytic enzymes from *Penicillium funiculosum* [567]. However, as more cellulolytic systems are subjected to investigation, it will likely be found that this insensitivity to glucose repression is highly atypical.

4.3 PRODUCT INHIBITION

Almost without exception, the cellulolytic enzymes have been found to be subject to product inhibition. This was soon realized during the early studies with *Trichoderma reesei*. Thus, glucose and cellobiose were found to inhibit the crude preparations of the cellulolytic activities of *T. reesei* strains [593–598]. Further characterization of the individual enzymes revealed that cellobiose, and to a lesser extent, glucose, inhibits cellobiohydrolase [596–599], whereas glucose inhibits β-glucosidase [594,598,599]. Of the two, cellobiose is by far the more potent inhibitor [599], and two mechanisms of reversible inhibition have been proposed: competitive [600] and noncompetitive inhibition [599]. This trend is found in most cellulolytic microorganisms, but there are a few exceptions. For example, glucose is not a very effective inhibitor of the β-glucosidase from *Microbispora bispora* [601].

Substrate inhibition has also been observed to occur, at least with *T. reesei* enzymes. Using crystalline cellulose as a substrate, increasing substrate concentrations resulted in increasing rates of solubilization up to an optimum above which the rates apparently decreased asymptotically [602]. Levels of up to 70% substrate inhibition relative to optimum substrate concentrations were reported. Unfortunately, these studies were conducted with a complete enzyme system, and which enzyme(s) responsible for the apparent substrate inhibition remain unknown.

4.4 DETECTION

The search for cellulolytic and xylanolytic microorganisms provided the impetus for the development and improvement of rapid screening procedures. This need for such methods became even more acute with the application of molecular biological techniques to the study the cellulolytic and xylanolytic enzymes, where hundreds to thousands of clones need to be rapidly screened for expressed activities. Detailed procedures for the screening and detection of cellulolytic and xylanolytic enzymes may be found in a recent review by Wilson [603].

Screening for cellulase-producing colonies is readily accomplished by plating cells on agar containing a water-soluble substrate, usually 0.5%–1% carboxymethyl cellulose. After appropriate incubation, the agar is stained with Congo Red, and the size of the clear (yellow) halo against the dark background reflects the extent of substrate hydrolysis [604]. Unbound dye molecules may be removed by washing the plates with 1 M sodium chloride. In some cases, and especially for the detection-cloned enzymes that usually remain confined intracellularly, a drop of chloroform may be placed under the lid of the plate after colony development to facilitate enzyme release (for example, [162]). In this case, the plates require further incubation to allow time for the released enzymes to act upon the substrate. To further facilitate detection, colonies can be washed off the plates prior to addition of stain. Xylanase production can be screened for in the same manner, with the replacement of carboxymethyl cellulose with an appropriate xylan. Alternatively, cells can be cultured on plates containing a dye-bound xylan, such as the commercially available Remazol-Brilliant Blue xylan [605]. Colonies expressing and secreting xylanase are readily detected as those producing halos [606,607]. Detection of enzymes may also be achieved by overlaying agar plates with 1% agarose containing a dyed substrate. Further incubation to allow diffusion of secreted enzymes into the agar and the subsequent hydrolysis of the substrate will again result in the production of identifying halos. If necessary, the plates may be destained with a solution of buffered 95% ethanol.

The procedures described above work well for the detection of cellulases, xylanases, and to a lesser extent, β-glucosidases. Cellobiohydrolases are typically active on only crystalline cellulose substrates, and, unfortunately, there are no simple enzymatic assays to detect this activity. For this reason, screening for cellobiohydrolase-producing colonies has proved to be more difficult, and researchers have resorted to using either antibody or genetic probes [608].

Zymogram analyses have also been applied to the detection of both cellulolytic and xylanolytic enzymes. This technique has proven to be particularly useful for determining the extent of multiplicity of enzymes produced by a given organism, in addition to monitoring and analyzing stages of purification. As with the screening of agar plates, detection of enzymes in polyacrylamide gels can either involve the inclusion of substrate within the separation gel [609] or agarose overlays [610–612]. In the former case, an appropriate substrate may be added to the acrylamide solution prior to casting the separation gel (for example, 1% carboxymethyl cellulose for cellulases). Following isoelectric focusing or electrophoresis and in the case of denaturing electrophoresis, renaturation, the gel slabs may be stained with Congo Red to permit visualization of the clear zones of hydrolysed substrate [609]. In a recent development of the technique, a chromogenic substrate has been incorporated into separation gels, thereby negating the need for staining and moreover, facilitating the zymogram analysis of β-glucosidases. Thus, the location within a gel of the β-glucosidase-catalyzed release of methylumbelliferone from the 4-methylumbelliferyl-β-D-glucopyranoside substrate was visualized under ultraviolet light [613].

The substrates incorporated into agarose for use in the overlay method may either be native [610,612] or prestained [611]. Congo Red is again used to stain agarose overlays that contain nonchromogenic, soluble substrates such as carboxymethyl cellulose or xylans. However, the detection of xylanases and cellulases within electrophoresis gels is more readily achieved with agarose overlays containing the dyed substrates Remazol-Brilliant Blue xylan and Ostazin Brilliant Red-hydroxyethyl cellulose, respectively [605]. An advantage of this method, in addition to its relative ease, is that the dyed fragments liberated from the stained detection agarose by the action of the separated enzyme(s) diffuse into the acrylamide separation gel where they mark the position of the active enzyme(s). Such immediate and direct identification facilitates the recovery of the enzyme by, for example, electroelution techniques [614]. The use of prestained substrates also enables direct observation of the time course of reaction that has in at least one case permitted the preliminary differentiation of the multiple forms of the enzymes present in an isolate [582].

The usefulness and value of the zymogram techniques developed over the past ten years is reflected in their widespread adoption. Having said this, it is important to recognize that this technique will only detect enzymes capable of renaturing following the harsh conditions of electrophoresis, particularly under denaturing conditions. Perhaps the reason that this technique has gained so much popularity and use is the fact that most cellulolytic and xylanolytic enzymes are resilient to irreversible denaturation in the presence of detergents such as sodium dodecyl sulfate.

4.5 ASSAY OF ENZYME ACTIVITY

4.5.1 CELLULOLYTIC ENZYMES

The assay of cellulolytic enzymes has been complicated by the fact that the homopolymer of β-1,4-linked glucose is naturally ordered and structured into a number of forms of cellulose that in turn are manipulated into a variety of commercial products. Compounding this issue is the fact that a number of diverse analytical chemical and physical techniques can be exploited to monitor the hydrolysis of cellulose. Finally, the multiple forms of each of the three major cellulolytic enzymes, in addition to those isolated from different sources, will often possess different substrate specificities. As a result, it is often difficult to compare and interpret the wealth of data available in the literature, and caution should be exercised.

Pure cellulose is commercially available in several forms (cotton, filter paper, Avicel). These highly structured forms of cellulose are usually used only to assess the efficiency of hydrolysis by complete cellulolytic systems because the individual enzymes comprising these systems usually are only weakly active on these substrates. Furthermore, the physical heterogeneity of these substrates (viz., degree of crystallinity, pore size, available surface area) produced by the various manufacturers complicates detailed enzymological studies. A more popular choice of substrates for such studies has involved the more amorphous forms of cellulose, such as acid-swollen cellulose or the covalently modified soluble forms, e.g., carboxymethyl cellulose (CM-cellulose). These soluble substrates are typically hydrolysed by isolated cellulolytic components at faster rates compared with insoluble forms of cellulose, greatly facilitating their characterization. Although the covalent modifications aid in the solubilization of cellulose substrates, the rates of enzymatic hydrolyses are inversely proportional to the degree of substitution that often results in a rapid departure from linear kinetics. Procedures for the preparation of the various substrates have been reviewed by Wood [26], and a list of commercial suppliers for some of these and the other substrates discussed below may be found in Appendix II.

A list of the more commonly used assays for the cellulases, cellobiohydrolases and β-glucosidases is presented in Table 4.1. Of the substrates used, CM-cellulose, Avicel, acid swollen cellulose and *p*-nitrophenyl-β-D-glucose are probably the most popular for kinetic studies. Indeed, some workers have coined *phenomenase* terminology (e.g., CMCase, Avicelase) to identify a particular enzyme with a characteristic type of activity. Other than the use of spectrophotometry for chromogenic substrates (see below), the most commonly used assay technique for monitoring the hydrolysis of the various substrates involves measuring the increase of reducing sugars, expressed as glucose equivalents, over a period of time. Two basic assays for reducing sugars have generally been used: the 2,4-dinitrosalicylic test [615] and the Nelson-Somogyi procedure [616,617]. Although these single-time point assays are simple in practise and allow for the assay of a number of samples in one experiment, they tend to suffer from nonlinearity, and factors that are thought to contribute to this phenomenon include the following: the practise of diluting reaction products before quantification of reducing sugars; insufficiency of substrate; and high-substrate concentrations (leading to product inhibition and/or transglycosylation reactions) [618]. To compound these problems, the colour response of the dinitrosalicylic assay is non-stoichiometric with the concentration of oligosaccharides with increasing degrees of polymerization. This may not pose a problem for the assay of β-glucosidases or cellobiohydrolases, but the endoaction of cellulases produces reaction products with varying degrees of polymerization. Hence, caution should always be exercised with the implementation of this assay. In this regard, the Nelson-Somogyi reducing sugar assay is more reliable and, furthermore, it is usually found to be more sensitive than the dinitrosalicylic assay. The major disadvantage with the Nelson-Somogyi procedure, however, is that the copper reagent causes precipitation of residual substrate, and this is most pronounced in the reaction blanks. This assay thus requires the added step of centrifugation of the final reaction solutions before absorbance measurements may be taken.

Reducing sugar assays are extremely popular and have received widespread use, but two other major concerns need to be addressed. These assays are totally nonspecific and will reflect the activities of any and all glycosidases present in the sample that are capable of hydrolysing the provided substrate. This, of course, is a nonissue if homogeneous preparations of enzyme are being investigated, but it is for this reason that quite often, specific activities of enzymes appear to *decrease* upon purification from a crude preparation. The second major concern of reducing sugar assays, and one that tends to be neglected, is that they do not provide accurate and useful kinetic data. Initial rates of reaction (initial velocities) are required for kinetic analyses, which may be acquired through time-course experiments, but the reducing sugar assays are not easily amenable for this purpose. With insoluble substrates, the required continuous data may be obtained turbid-

TABLE 4.1. Methods for Measuring Activities of Cellulolytic Activities.

Enzyme	Substrate	Product Detected	Reference
Cellulase	Carboxymethyl cellulose	Reducing sugar	[293,664]
		Reduction in viscosity	[293]
	O-Hydroxyethyl cellulose	Reducing sugar	[664]
	Phosphate swollen cellulose	Reducing sugar	[452]
	Filter paper	Disintegration	[665]
		Reducing sugar	[666]
	Dyed cellulose	Release of dye	[667,668]
	Cotton thread	Break point	[669]
	Soluble cello-oligosaccharides	Reducing sugar	[452]
		Oligosaccharides/HPLC	
	4-Methylumbelliferyl-β-D-cellobioside	4-Methylumbelliferol	[624–626]
	4-Methylumbelliferyl-β-D-cellotrioside	4-Methylumbelliferol	[624–626]
Cellobiohydrolase	Cotton	Loss of weight	[670]
	Avicel	Reducing sugar	[622]
		Light scattering	[671,672]
	Bacterial cellulose	Reducing sugar	[673]
	Filter paper	Disintegration	[665]
	Phosphate-swollen cellulose	Loss of weight	[293]
		Reducing sugar	[293]
		Soluble sugar	[128]
	Soluble cello-oligosaccharides	Reducing sugar	[622]
	Insoluble cello-oligosaccharides	Reducing sugar	[674]
	p-Nitrophenyl-β-glucoside/lactoside	*p*-Nitrophenol	[675]
	4-Methylumbelliferyl-β-D-cellobioside	4-Methylumbelliferol	[624–626]

(continued)

TABLE 4.1. (continued).

Enzyme	Substrate	Product Detected	Reference
β-Glucosidase	*p*-Nitrophenyl-β-glucoside	*p*-Nitrophenol	[622]
	o-Nitrophenyl-β-glucoside	*o*-Nitrophenol	[623]
	4-Methylumbelliferyl-β-D-glucoside	4-Methylumbelliferol	[624–626]
	Salicin	Glucose	[676]
		Reducing sugar	[676,677]
	Esculin	Glucose	
		Reducing sugar	
	Cellobiose	Glucose	[670]
	Insoluble cello-oligosaccharides	Reducing sugar	[674]

Source: Adapted from Reference [249].

ometrically [619], gravimetrically [620], or colorimetrically [621]. It is worth noting that in spite of its profound influence on the kinetics of hydrolysis, very few studies have concerned the physicochemical properties of the insoluble substrate. Such methods could include monitoring the index of crystallinity, degree of polymerization, changes in surface area, or formation of smaller particles.

Soluble chromogenic and fluorogenic substrates have proven to be more practical and convenient for routine assays. These substrates are comprised of a coloured or fluorescent compound linked to the anomeric carbon of glucose, cellobiose, or higher cello-oligosaccharides [605,622–626]. Two examples of such compounds that are both widely used and commercially available are *p*-nitro-β-D-cellobioside and methylumbelliferyl-β-D-cellobioside. These cellobiose derivatives serve as substrates for cellobiohydrolases and cellulases, and the analogous glucosides are available for the assay of β-glucosidases. Kinetic data are readily obtained with these compounds by continuously monitoring the release of chromophore (for an example, see Reference [315]), but they have to be viewed with caution because most of the chromogenic aglycon structures bear little resemblance to the structure of glucose. After all, it is highly unlikely that the interactions between the chromophore and the binding site of the enzyme closely mimic a glucose residue.

Ostazin Brilliant Red-hydroxyethyl cellulose [605] represents a different form of chromogenic substrate in that the dye is randomly associated with the cellulose. It is released from the polymer with oligosaccharide fragments and therefore does not directly correlate with hydrolytic events. However, the polymeric substrate is precipitated by ethanol while the released fragments remain soluble, and it is these properties that form the basis of the assay. After incubation with enzyme, residual polymer is ethanol precipitated from reaction mixtures, and the amount of enzyme-released cellulose fragments is determined spectrophotometrically.

Soluble cello-oligosaccharides comprising three to eight glucosyl residues probably represent the most ideal substrate for enzymological studies involving individual enzymes. These can be obtained commercially or prepared by partial acid hydrolysis of cellulose and their subsequent purification by chromatography [627–629]. The quantification of either substrate loss or product formation over time is best achieved by high-pressure liquid chromatography, as utilized for the assessment of the action pattern of cellulase [630] or cellobiohydrolase [290] or the assay of xylo-oligosaccharide hydrolysis by xylanases [270,385,631,632]. This technique, which may involve either normal-phase silica chromatography with refractive-index detection [630], gel permeation chromatography with refractive-index detection [632], or high-performance anion-exchange chromatography coupled with pulsed-amperometric detection [270,311,631], permits the simultaneous quantification of both substrate loss and the production of each reaction product, thereby providing information on the action pattern of hydrolysis in addition to overall enzymatic rates.

A complication that may arise, especially when high concentrations of substrate are used, is the possibility of enzyme-catalyzed transglycosylation reactions. Because the products build up during an incubation, the reversibility of the enzyme-catalyzed reaction leads to both the continued formation of new bonds, in addition to the continued hydrolysis of the original substrate. One assay method that both permits the continuous monitoring of hydrolytic activity and the impeding of transglycosylation reactions involves the oxidation of product as it is formed. Coupled assays using cellobiose oxidase, an enzyme capable of oxidizing cellobiose and higher cellodextrins, have been established and suitably refined for kinetic studies [633].

4.5.2 XYLANOLYTIC ACTIVITY

Most of the techniques, and associated problems, for the assay of cellulolytic activity noted above also pertain to the assay of xylanolytic activity. Heteroxylans from different terrestrial or marine plants have been used in the assay for xylanase activity, with L-arabinoglucuronoxylan (oat spelts, rye flour), 4-*O*-methylglucuronoxylan (beech wood, larch wood), and rhodomenan (a linear homoxylan from the seaweed *Rhodymenia stenogona*) serving as the most commonly used insoluble substrates. Soluble fractions of these xylans can be readily prepared by simple extraction with water (for example, Reference [631]) or alkali (1 M sodium hydroxide) followed by neutralization [634]. Although not commercially available, *O*-(carboxymethyl)-D-xylan can be synthesized from a suitable heteroxylan starting material, such as 4-*O*-methylglucuronoxylan [334,635]. The activity of xylanases toward these soluble heteroxylans can be determined viscometrically (for example, Reference [636]) or by detecting the release of reducing sugar [615–617]. In the latter case, the unit of xylanase activity is usually defined as the amount of enzyme that causes the liberation of reducing-end groups corresponding to 1 μmole of D-xylose per minute under specified conditions of pH, ionic strength, and temperature.

A chromogenic assay analogous to that described above for cellulases has also been developed for xylanases. Treatment of a buffered solution of Remazol-Brilliant Blue-xylan [605] with xylanase results in the release of ethanol-soluble dyed fragments, which may be quantified spectrophotometrically (595 nm) after precipitation of residual substrate with the addition of ethanol [637]. A variation of this assay involves the insoluble, cross-linked derivative of the dyed xylan (AZCl xylan). This substrate forms a colloidal suspension that can be dispensed directly into spectrophotometer cuvettes. Following incubation with agitation, the residual substrate is allowed to settle, thereby permitting the direct spectrophotometric measurement of the soluble released fragments. This latter protocol was found to be particularly useful for the assay of xylanases at elevated temperature [638].

4.6 PRODUCTION

The filamentous fungi have attracted considerable attention over the years since research into the production, function, and use of the cellulolytic and xylanolytic enzymes was initiated for two fundamental reasons: the enzymes are produced in relatively high levels, and they are released extracellularly. Both factors facilitate the isolation and purification of the enzymes in homogeneous form, and for some applications, the enzyme of interest is present in sufficient concentration and state not to warrant such manipulations.

Historically, the production of cellulolytic and xylanolytic enzymes has been conducted by liquid-state fermentations (where the term fermentation refers to the physical process of cell cultivation and does not necessarily imply the anaerobic means of energy generation). In liquid-state fermentation, the insoluble substrates are suspended as fine particles in a large volume of water. In most cases, substrate concentrations range from 0.5 to 6% (wt/vol) with the upper limit being imposed by the rheological problems associated with the different substrates. For example, the maximum concentration of substrate that can be effectively managed by conventional fermenters is approximately 2% (wt/vol) for wood pulp and 6% (wt/vol) for a crystalline cellulose, such as Avicel. The activity of cellulolytic enzymes per unit volume can be increased by increasing the cellulose concentration [639], but this does not necessarily equate to increased efficiency of substrate utilization. For example, studies with *Trichoderma reesei* QMY-1 showed that the cellulase titer and yield obtained by liquid-state fermentation in 0.4% cellulose after seven days were 1.65 IU/ml and 412 IU/g of cellulose, respectively (Table 4.2). Increasing the cellulose concentration to 2% (wt/vol) and incubation for the same time period resulted in a comparable

TABLE 4.2. Production of Cellulase on Wheat Straw by *T. reesei* QMY-1.

Cellulose [% (wt/wt)]	Incubation Time (days)	Cellulase Titer (IU/ml)	Cellulase Yield (IU/g cellulose)
0.4	5	1.44	360
	7	1.65	412
2.0	7	1.3	65
	11	6.0	300
	14	5.5	275

Data obtained from Reference [640].

TABLE 4.3. Cellulase Production of *T. reesei* Strains as a Function of Cellulose Concentration.

Strain	Cellulose [% (wt/wt)]	Cellulase Yield (IU/g cellulose)	Reference
QM9414	0.75	240	[678]
	2	200	[679]
	2.5	172	[680]
	6	166	[681]
Rut-C30	5	290	[680]
	5	160	[271]
	6	233	[681]
	15	200	[271]
MCG77	2	195	[679]
	6	183	[681]
NG14	2	180	[679]
	6	250	[681]

Source: Adapted from Reference [640].

TABLE 4.4. Production of Cellulase by Strains of *Trichoderma* Grown on Different Cellulosic Substrates.

Strain	Cellulose [% (wt/wt)]	Cellulase Titer (IU/ml)	Cellulase Yield (IU/g cellulose)
T. reesei QM-9414	Solka-Floc	1.85	36.1
	SE[a]	0.70	25.4
	SEWA[b]	0.96	21.1
T. reesei Rut-C30	Solka-Floc	5.56	111
	SE[a]	1.57	57.1
	SEWA[b]	2.10	46.1
T. harzianum E.58	Solka-Floc	1.93	38.6
	SE[a]	0.62	22.5
	SEWA[b]	0.51	11.2

[a]SE, steam-exploded cellulose.

[b]SEWA, steam-exploded, alkali-extracted cellulose.

Source: Adapted from Reference [640].

enzyme titer (1.3 IU/ml) but a drastic decrease in yield to 65 IU/g of cellulose. Prolonged incubation to eleven days did increase the enzyme titer to 6.0 IU/ml, but the cellulase yield was less than 75% of that achieved with the lower cellulose concentration [640]. Decreases in cellulase yields by other strains of *T. reesei* cultivated in higher cellulose concentrations are evident from the compiled data presented in Table 4.3. A plausible explanation provided to account for the poorer cellulase yields invoked poor mass transfer in the thicker slurries of substrate [640]. Given that a major obstacle to the development of an economical process for the production of cellulolytic enzymes is the high cost of the substrates, attention to substrate concentrations would thus appear to be important.

Another factor that significantly affects the titer and yield of enzyme production is, not surprisingly, the nature of the substrate. As indicated in Table 4.4, both production parameters for two different strains of *T. reesei* and a strain of *T. harzianum* were highest when the fungi were cultivated on pure cellulose (Solka Floc) compared with steam-exploded wood (55% cellulose) and steam-exploded, water-extracted, and alkali-treated wood (91%) [640].

Solid-state fermentation represents an interesting and promising development for the optimal production of cellulolytic and xylanolytic enzymes. This process involves the growth of an organism on insoluble substrates with sufficient moisture but in the absence of free water. Increases of greater than 70% were recorded for the production of *T. reesei* QMY-1 enzymes when grown on wheat straw in a solid-state fermenter compared with liquid-state fermentation with the same substrate [640].

4.7 ISOLATION

4.7.1 FUNGI

Perhaps the only drawback to the extracellular production of cellulolytic and xylanolytic enzymes by the filamentous fungi is that the enzymes need to be recovered from the large volumes of the spent culture fluids. This may not pose a serious problem if small cultures have been prepared, but technical difficulties may be encountered during large-scale productions. The technique of salting out, usually employing ammonium sulfate, has been widely used by researchers for the laboratory-scale isolation of cellulolytic and xylanolytic enzymes. In addition to rapidly concentrating the enzymes, fractional ammonium sulfate precipitations may also effect an initial purification. Other researchers have replaced salting out with organic solvent precipitations. Thus, fractional precipitations have been conducted with isopropanol [334] and acetone [641,642] to isolate xylanase, whereas ethanol has been successfully used to fractionally precipitate xylanases [334,349, 634,643,644], cellulases [264,643,645], and β-glucosidase [315,645]. With both ammonium sulfate and organic solvent precipitations, the appropriate volume of reagent is added to the culture supernatant and allowed to stand at low temperature for a period of time to permit precipitation of the desired enzyme. Although the method is simple and effective, it does suffer from losses due to problems with resolubilization of the isolated enzyme. Ultrafiltration is an attractive alternative for the concentration of extracellular enzymes, and this technique has been applied to the isolation of xylanases [158,385,394,400,528], cellulase [271,536], and β-glucosidase [148].

4.7.2 BACTERIA

As noted in Chapter 2, many of the bacterial cellulolytic and xylanolytic enzymes remain cell associated, either freely or in multienzyme complexes. In some cases, the enzymes or complexes can be liberated from cell surfaces by simple extraction of a harvested cell culture with water or salt buffers (for example, Reference [240]). For the enzymes that either remain cell bound or are located within periplasmic gels, disruption of the cellular membranes is required. This would also apply to those enzymes cloned into bacterial hosts, where the enzymes are generally located intracellularly. This is especially the case for the production of cellulase-free xylanases from cloned systems [161,391,522,646–648]. Enzymes located either on outer membranes or within periplasms are ideally recovered by osmotic shock techniques that retain the integrity of the cytoplasm and, hence, release relatively few proteins from which the desired enzyme will have to be purified. Intracellular enzymes can only be isolated from totally disrupted cells, which may be achieved by sonication, treatment with EDTA-lysozyme, or passage through a French Press cell.

4.8 PURIFICATION

Once isolated, the majority of the enzymes studied to date have been separated and purified to homogeneity by conventional chromatographic techniques involving hydrophobic interaction, ion exchange and gel filtration techniques. In some cases, both the cellulose-binding capacity of the various cellulolytic and xylanolytic enzymes and the insoluble nature of the substrates have been exploited to effect the purification of the enzymes by affinity chromatography. As indicated in Table 4.5, a variety of affinity matrices have been employed, some of which are various forms of the substrate cellulose, whereas others are more generic (for example, Procion Red HE3B-Agarose or BioGel P-150). The use of Concanavalin A-Sepharose capitalizes on the fact that most of the fungal and many bacterial cellulolytic enzymes are glycosylated. An example of the success that has

TABLE 4.5. Affinity Chromatography of Cellulolytic Enzymes.

Enzyme	Affinity Matrix	Reference
Cellulase	Avicel (crystalline cellulose)	[103,375,682–685]
	Amorphous cellulose	[251,686]
	p-Aminobenzyl-thio-β-D-cellobioside	[687,688]
	Immunoadsorbent	[689]
	Concanavalin A-Sepharose	[264,690,691]
	Bio-Gel P-150	[662]
Cellobiohydrolase	Avicel	[295]
	Cellulose powder	[673]
	Amorphous cellulose	[686]
	p-Aminobenzyl-thio-β-D-cellobioside	[687]
	Concanavalin A-Sepharose	[135,692]
β-Glucosidase	Concanavalin A-Sepharose	[148,196,315]
	Procion Red HE3B-Agarose	[690]
Xylanase	Cross-linked xylan	[663,693]
	Concanavalin A-Ultragel	[343]
	Procion Blue-Agarose	[693]

been achieved with the separation and purification of the various types and multiple forms of the cellulolytic enzymes using these techniques is illustrated in Figure 4.3. The six cellulases, three cellobiohydrolases, and β-glucosidase that comprise the commercial *Trichoderma viride* enzyme preparation, Maxazyme, have been purified by a combination of gel filtration, anion exchange, and affinity chromatographies [375].

Many cellulolytic and xylanolytic enzymes have been purified to homogeneity, and examples can be found in the reviews of Dekker and Richards [99], Senior et al. [576], and Woodward [649] for xylanases and for cellulolytic enzymes, Enari et al. [131], Ghosh and Ghosh [650], and Wood and Kellogg [651]. Some other enzymes that have been more recently purified are listed in Table 4.6. In most cases, conventional low-pressure chromatography was utilized, but fast protein liquid chromatography (FPLC) [285,296,326,330,385,394,402,489,652–656] and high-performance liquid chromatography (HPLC) [145,379,380,404,534,632,657–660] techniques have also been employed. A study involving *Thermotoga maritima* provides a recent example of the use of FPLC for the isolation of the cellulolytic and xylanolytic enzymes [285]. As the flow chart presented in Figure 4.4 shows, the crude culture filtrate was initially fractionated on an anion exchange resin, and subsequent chromatographies involving hydrophobic interaction, anion exchange, chromatofocusing, and gel-filtration employed FPLC technology for the purification of the cellulase and cellobiohydrolase. In addition, separate samples enriched with arabinofuranosidase, laminarinase, β-xylosidase, and xylanase were obtained. Typically, however, the use of

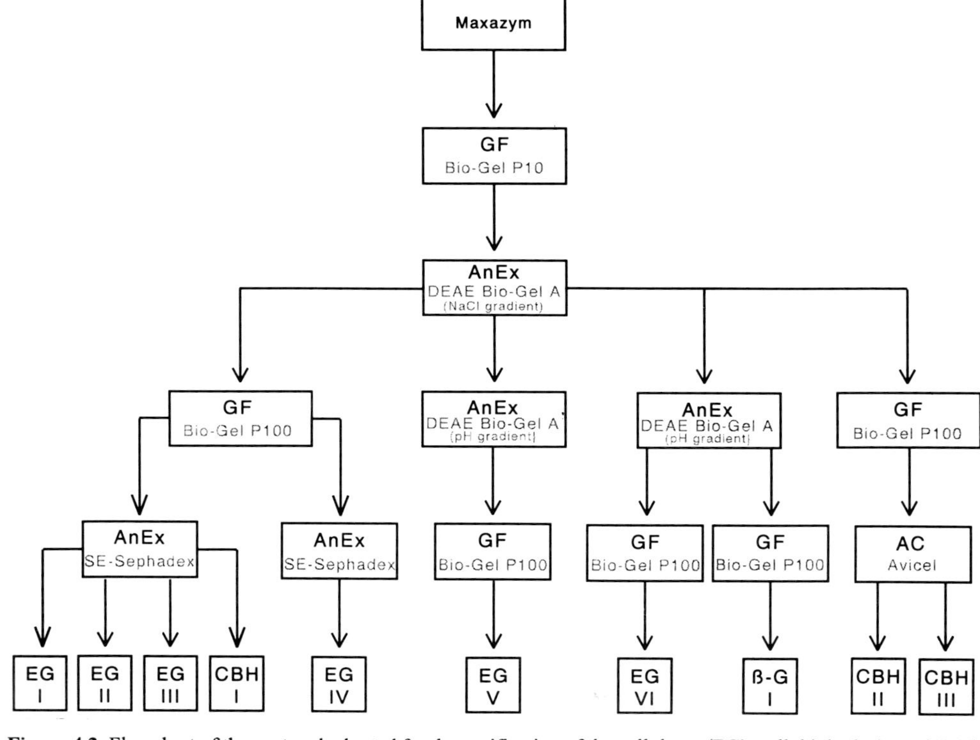

Figure 4.3 Flowchart of the protocol adopted for the purification of the cellulases (EG), cellobiohydrolases (CBH) and a β-glucosidase (β-G) from *T. viride* employing conventional liquid chromatographic procedures. AC, affinity chromatography; AnEx, anion exchange chromatography; GF, gel filtration (adapted from Reference [375]).

TABLE 4.6. Partial List of Recently Purified Cellulolytic and Heteroxylanolytic Enzymes.

Enzyme	Organism	Purification Protocol	Reference
Cellulase	*Trichoderma reesei*	ASP/AnEx(DEAE-Ss)/GF(UG AcA-44)/AnEx(DEAE-Ss)	[630]
	Acidothermus cellulolyticus	HIC(Phenyl Ss)/AnEx(Q Ss)/GF(Superdex 200)	[654]
	Bacillus sp.	UF/ASP/AnEx(DEAE-Sl)/GF(BG A)/GF(Super 6)	[271]
	Clostridium cellulolyticum	SSP/Asp/HIC(Phenyl Ss)/AnEx(DEAE trisacryl)/ AnEx(TSK DEAE 5PW)/HIC(Phenyl Ss)	[109]
	Thermotoga maritima	AnEx(Q Ss)/HIC(Phenyl Ss)/AnEx(Mono Q)/GF(Superdex 200)	[285]
Cellobiohydrolase	*Trichoderma reesei*		[694]
	Penicillium occitanis	ASP/GF(Sx G-25)/AnEx(DEAE Ss CL-6B)/PrepEI	[290]
	Streptomyces flavogriseus	AnEx(DEAE BG A)/GF(BG P-60)/PrepIEF/AC(ConA)	[135]
	Bacillus circulans	ASP/AnEx(DEAE Toya)/AnEx(DEAE Toya)/ GF(TSK HW50s)/HIC(Phenyl 5PW)/GF(TSK HW50s)	[145]
	Thermotoga maritima	AnEx(Q Ss)/HIC(Phenyl Ss)/CF/GF(Superdex 200)	[285]
β-Glucosidase	*Aspergillus niger*	AnEx(DEAE-Toya)/HIC(Butyl-Toya)/GF(Toya HW-55F)	[302]
	Fusarium oxysporum	GF(Sx G-25)/GF(Sx G-150)/AnEx(PBE-94)	[695]
	Hordeum vulgare (barley)	ASP/GF(Sx G-75)/CaEx(Mono S)	[655]
	Neocallimastix frontalis	AC(Avicel)/GF(Ultrogel AcA 34)/AnEx(DEAE Ss)/ HIC(HA Ultrogel)/HIC(Phenyl Ss)	[311]
	Streptomyces sp. QM-B814	ASP/AnEx(MonoQ)/AnEx(MonoQ)	[330]

(continued)

TABLE 4.6. (continued).

Enzyme	Organism	Purification Protocol	Reference
Xylanase	*Aspergillus ochraceus*	ASF/GF(Sx G-75)	[331]
	Bipolaris sorokiniana	AnEx(SP-Ss)/GF(Sx G-50)/An-Ex(SP-Ss)	[389]
	Gloeophyllum trabeum	AnEx(DEAE-Ss)/PrepIEF/GF(Superdex 100)	[165]
	Humicola lanuginosa	ASP /AP/AnEx(DEAE-Sx)/AnEx(DEAE-Sx)/ CaEx(QAE-Sx)/GF(BG P-30)	[342]
	Neocallimastix frontalis	ASP/GF(UG AcA-44)/AnEx(DEAE BG A)/GF(UG AcA-44)	[696]
	Aeromonas caviae	ASP/HIC(Butyl-Toya)/HIC(Butyl-Toya)/HIC(Butyl-Toya)/ AnEx(DEAE-Toya)/GF(TSK G3000-SW)	[380]
	Aeromonas caviae	ASP/GF(Sx G-75)/GF(Toya HW-55)/AnEx(Mono Q)	[379]
	Bacillus polymyxa	UF/AnEx(DEAE BG A)/CF/AnEx(Mono Q)/AnEx(Mono Q)	[385]
	Bacillus sp.	ASP/AnEx(DEAE-Toya)	[382]
	Bacillus sp.	CaEx(Q-Seph)/HIC(Phenyl-Ss)	[381]
	Bacillus sp.	EtP/GF(BG P-10)	
	Clostridium sterocorarium	ASP/AnEx(DEAE-Toya)/GF(TSK G3000SW)	[659]
	Clostridium thermolacticum	AnEx(Q-Ss)/HIC(Phenyl-Ss)/GF(Ss-200)	[393]
		AnEx(DEAE-Tris)/ GF(UG Ac A54)	[52]
	Streptomyces halstedii	UF/AnEx(Mono Q)	[400]
	Streptomyces sp.	UF/AcP/AnEx(MonoQ)/HIC(Phenyl Ss)	[394]
	Streptomyces thermoviolaceus	AnEx(DEAE-Toya)/CaEx(CM-Toya)/GF(Sx G-75)/ HIC(Phenyl-Toya/AnEx(MonoQ)	[402]
	Thermoanaerobacterium sp.	AnEx (DEAE Ss)/AnEx(Q Ss)/HIC(Phenyl Toya)/ GF(Ss 6)/AnEx(MonoQ)	[404]

TABLE 4.6. (continued).

Enzyme	Organism	Purification Protocol	Reference
Acetyl xylan esterase	*Aspergillus niger*	UF/AnEx(DEAE Trisacryl)/AnEx(DEAE 5PW)	[660]
	Schizophyllum commune	EtP/GF(Sx G-25)/AnEx(DEAE-Ss A50)/CaEx(Q-Ss)/ HIC(Phenyl-Ss HR 5/5)	[652]
	Thermoanaerobacterium sp.	AnEx(DEAE Ss)/HIC(Phenyl Ss)/AnEx(DEAE Sc)/ GF(Super 6)	[656]
Ferulic acid esterase	*Neocallimastix* sp.	IEF/GF(Super 12)/AnEx(MonoQ)/GF(Super 12)	[697]
	Penicillium pinophilum	AnEx/HIC(Phenyl-Ss)	[485]
α-L-Arabinofuranosidase	*Monilinia fructigera*	CF/HIC(Phenyl-Ss)/GF(Sx G-100) / AnEx(PBE-94)	[698]
	Bacillus stearothermophilus	AnEx (MonoQ)/HIC (Phenyl Ss)	[699]
	Bacillus subtilis	AnEx(DEAE-Sx)/AnEx(DEAE-Toya)/GF(UG AcA)/PrepEI	[700]
	Clostridium acetobutylicum	UF/CaEx(CM-Ss)/GF(BG P-150)	[487]
	Ruminococcus albus	UF/GF (Sc S-300)/ASP/PrepIEF	[488]
	Streptomyces lividans	ASP/HIC(Phenyl-Ss)/AnEx(DEAE-Ss)/GF(TSK300)	[661]
	Trichoderma reesei	CaEx/AnEx	[490]

AcP, acid precipitation; AC, affinity chromatography; AnEx, anion exchange; ASP, ammonium sulfate precipitation; BG, BioGel; CaEx, cation exchange chromatography; CF, chromatofocusing; ConA, Concanavalin A; EtP, ethanol precipitation; GF, gel filtration; HIC, hydrophobic interaction chromatography; PBE, polybuffer exchanger; PrepEI, preparatory electrophoresis; PrepIEF, preparatory isoelectric focusing; SI, Sephacel; Ss, Sepharose; Sx, Sephadex; SP, sulfopropyl; SSP; streptomycin sulfate precipitation; Super, Superose; Toya, Toyapearl; Tris, Trisacryl; UF, ultrafiltration; UG, Ultragel.

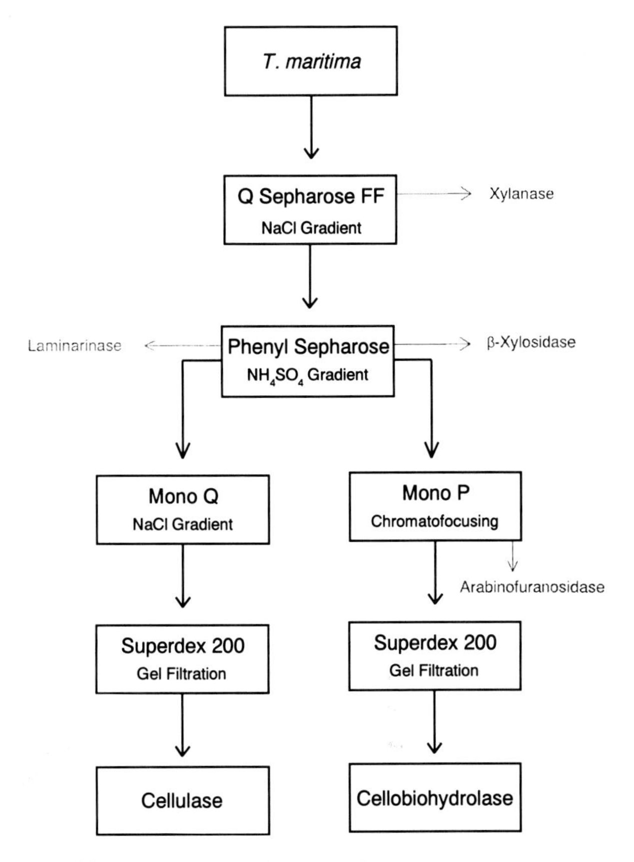

Figure 4.4 Flow of the protocol adopted for the purification of a cellulase and cellobiohydrolase from *Thermotoga maritima* employing fast protein liquid chromatography (FPLC) techniques [285]. Also shown are the steps during the purification protocol where samples enriched with xylanase, β-xylosidase, laminarinase, and arabinofuranosidase were obtained.

FPLC and HPLC is reserved for the final stages of purification protocols because of the limitations imposed by these systems regarding sample volumes (for example, References [145,330,379,380,385,394,402,404,652, 655,656,659–661]). In a few cases, the enzyme of interest was isolated in a single step. Thus, following concentration by ammonium sulfate precipitation, the xylanase from *Aspergillus ochraceus* [331] and the cellulase from *Fusarium lini* [662] were purified to homogeneity by a single passage over a gel filtration matrix. This was made possible by the carbohydrate-binding properties of the enzymes that resulted in a retardation of the enzymes within the matrices. Chromatography on a cross-linked xylan resulted in the single-step purification of a xylanase from the fungal Pectinol A1 enzyme preparation [663], and in a highly unusual occurrence, a xylanase from *Thermoascus aurantiacus* was purified by a single ion exchange chromatography [158].

CHAPTER 5

The Catalytic Mechanism of Action

5.1 STEREOCHEMISTRY

THE most important mechanistic feature of β-glycosidases, and indeed of all carbohydrases, is the stereochemistry of the catalyzed reaction. Hydrolysis of the glycosidic bond between two sugars may occur with either net inversion or net retention of anomeric configuration at the anomeric center of the reducing sugar product (Figure 5.1). Enzymes that hydrolyse the glycosidic bond with net inversion of configuration are termed *inverting enzymes*, and those that catalyze a net retention of anomeric configuration are termed *retaining enzymes*. A variety of rapid procedures have been established to determine the anomeric configuration of reaction products before it becomes obscured by spontaneous mutarotation. One involves the measurement of optical rotation of an enzyme-substrate reaction mixture as the reaction proceeds followed by determining the direction of any sudden change in rotation consequent upon addition of base (for examples, see References [701] and [702]). Another method, first used with the β-glucosidase from sweet almond emulsin [703], involved the rapid derivatization of reaction mixtures by trimethylsilylation and the determination of the anomeric configuration of the adduct by gas-liquid chromatography. More recently, the use of high-field proton NMR to monitor the chemical shift and coupling constant of the anomeric proton on the hemiacetal carbon of reaction products has greatly facilitated the determination of the stereochemical course of hydrolysis catalyzed by glycanases [704]. Reaction conditions are established so that the initially formed anomer of a hydrolytic reaction accumulates sufficiently to permit analysis by proton NMR before mutarotation of the products occurs.

Application of the proton NMR technique has shown that all xylanases studied catalyze hydrolysis with net rentention of anomeric configuration, whereas cellulases and cellobiohydrolases may be either inverting or retain-

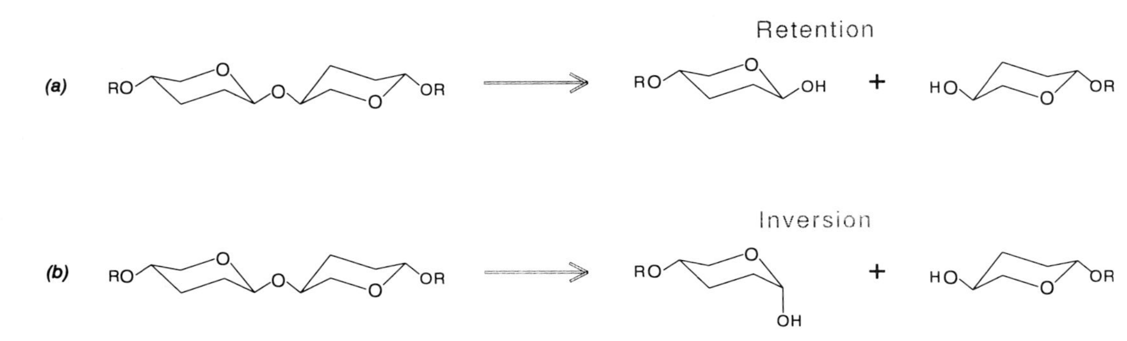

Figure 5.1 Stereochemistry at anomeric center of reaction products obtained by (a) retaining and (b) inverting β-glycosidases.

ing enzymes [704,705]. With respect to β-glucosidases, only two have been characterized, those from sweet almond emulsin [703,704] and *Alcaligenes faecalis* [706], and both are retaining enzymes. Moreover, it became apparent that in all cases tested, representatives of a given family of β-glycosidase catalyze hydrolysis of their substrate with the same stereospecificity [705]. This is not too surprising and perhaps would be expected, given that the two mechanistic pathways for retaining and inverting enzymes are different and would require different active-site structures. This observation does, however, bear out the validity of the classification scheme that is based on homologies of sequence and predicted secondary structure. Not all of the families that contain cellulolytic and xylanolytic enzymes have been tested for enzymatic stereochemistry, but the enzymes of Families 1, 3, 5, 7, 10, and 11 are retaining whereas those of Families 6, 9, and 46 are known to be inverting [206,705].

5.2 CATALYTIC MECHANISM OF RETAINING AND INVERTING GLYCOSIDASES

Mechanisms for these two classes of carbohydrases were proposed by Koshland [707] more than forty years ago, and they have largely withstood the test of time. Inverting glycosidases are thought to function by a single-step mechanism in which a water molecule directly displaces the glycosidic leaving group from the anomeric center as depicted in Figure 5.2. This displacement is facilitated by two active-site amino acid residues serving as general acid-base catalysts. One amino acid residue acts as the general base by deprotonating the nucleophilic water molecule, whereas the second residue participates in a concerted fashion as the general acid, protonating the departing glycosidic oxygen. The reaction is believed to proceed via an oxocarbonium ion-like transition state that may be stabilized, in part, by the deprotonated general acid. Retaining glycosidases are generally believed to utilize a double-displacement mechanism (Figure 5.3). This mechanism, as modified by Sinnott [708], involves the following events: 1) an acid catalyst protonates the substrate; 2) a carboxylate group on the enzyme is positioned on the opposite side of the sugar ring to the aglycone; 3) a covalent glycosyl-enzyme intermediate is formed with the carboxylate in which the anomeric configuration of the sugar intermediate is opposite to that of the substrate; 4) the covalent intermediate may be reached from both directions through transition states involving oxocarbonium ions; and 5) various non-covalent interactions provide most of the rate enhancement.

Although the two-mechanism pathways are distinctly different, involving either one or two chemical transition states, they do share a number of common features. Most noteworthy is the fact that both require the direct participation of two acidic amino acids, and all available evidence obtained

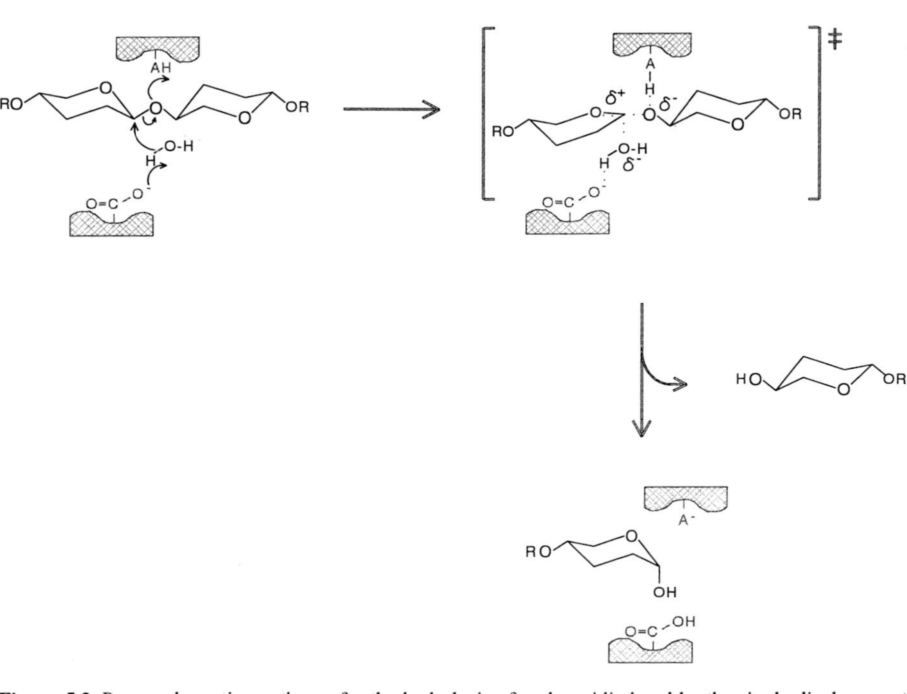

Figure 5.2 Proposed reaction pathway for the hydrolysis of a glycosidic bond by the single-displacement mechanism.

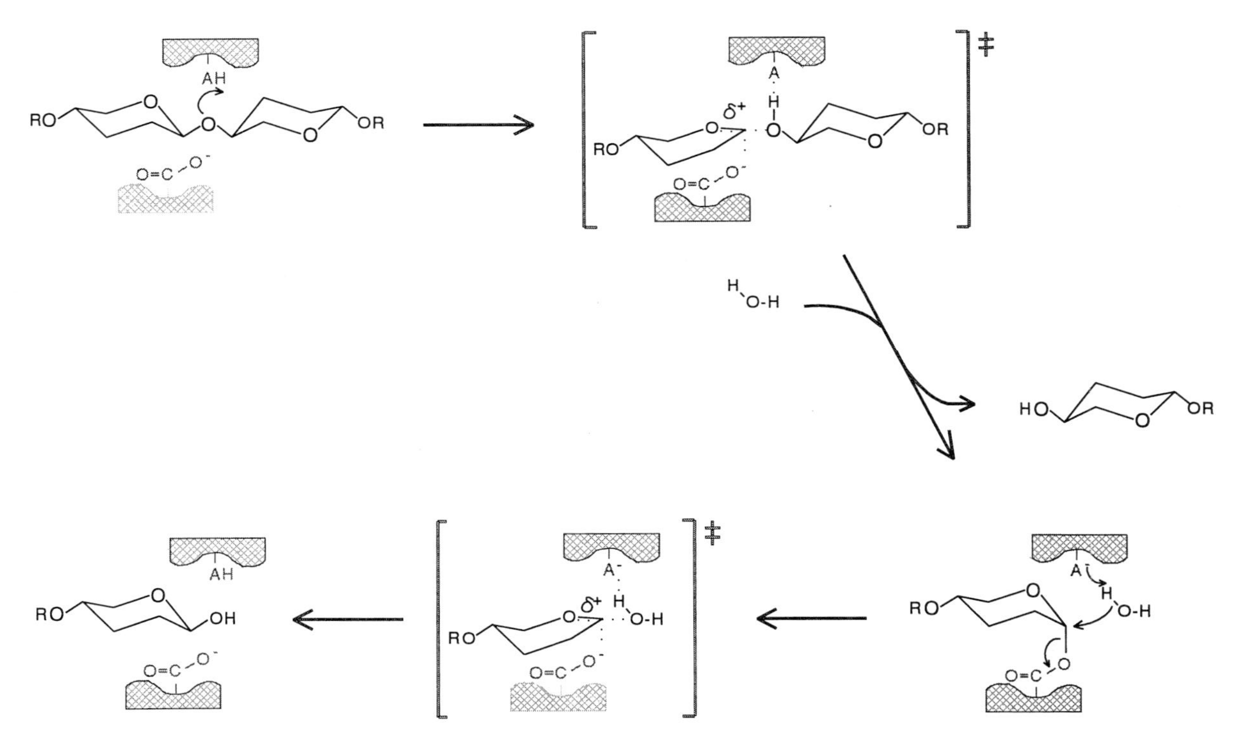

Figure 5.3 Proposed reaction pathway for the hydrolysis of a glycosidic bond by the double-displacement mechanism.

with the cellulolytic and xylanolytic enzymes strongly indicates that this role is assumed by aspartyl and/or glutamyl residues. The early evidence for carboxylic acids serving as the acid catalyst and stabilizing anion-nucleophile for retaining enzymes was founded primarily on the X-ray crystallographic structure of hen egg white lysozyme, the first enzyme to be solved in this manner [709,710]. Glu35 of this lysozyme, with unusually high p*K*s of 6.1 and 6.7 for the free enzyme and enzyme-substrate complex, respectively, was identified as the proton donor for the double-displacement mechanism of this retaining enzyme, whereas Asp52 is proposed to stabilize the transition-state oxocarbonium ion.

Considerable progress has recently been made to both substantiate the two proposed mechanistic pathways outlined above and elucidate their finer details. This evidence is presented in a number of recent excellent reviews [708,711,712], with those by Sinnott being particularly comprehensive [708,711]. The following presents the evidence that the cellulolytic and xylanolytic enzymes follow either a single or double-displacement mechanism involving catalytic acidic (i.e., Glu and/or Asp) amino acid residues.

5.2.1 KINETICS

The presence of both an acid catalyst and a stabilizing anion-nucleophile in β-glycosidases has been inferred from pH activity dependence studies. As indicated in Figures 5.2 and 5.3, the activities of both retaining and inverting enzymes require the presence of at least two essential ionizable groups at their active sites, one remaining protonated (the acid catalyst) while the second is deprotonated (the base catalyst, or stabilizing anion-nucleophile). Consequently, the activity of the enzymes is greatly dependent on pH. As the pH increases, the acid catalyst becomes deprotonated, whereas a decrease of pH will protonate the base catalyst or stabilizing anion-nucleophile. This is reflected in a bell-shaped dependence of activity on pH, and in all cases tested, such behaviour has been noted for the cellulolytic and xylanolytic enzymes.

The apparent p*K* values for the essential ionizable groups on both the free enzyme and enzyme-substrate complexes can be estimated from plots of log k_{cat}/K_m and log k_{cat} versus pH, respectively. The acidic values are typically ascribed to the base catalyst or stabilizing anion-nucleophile, whereas the more basic pK_a values are assigned to the general acid catalysts. However, it is equally possible that these ionizations may not reflect changes in the protonation state of discrete catalytic residues but rather be the consequence of changes to one or more other residues that indirectly influence hydrolytic activity (e.g., binding or stability). With this in mind, the p*K* values obtained from cellulases, β-glucosidases, a cellobiohydrolase, and a xylanase are listed in Table 5.1. Although it was obvious that the ionizable groups with

TABLE 5.1. Apparent p*K* Values for Essential Ionizable Groups in Cellulolytic and Xylanolytic Enzymes. Values Reported Are for the Groups on the Free Enzyme (e) and Enzyme-Substrate Complexes (es).

Enzyme	Family	pK_{e1}	pK_{es1}	pK_{e2}	pK_{es2}	Reference
Cellulase						
Aspergillus niger		3.1	4.2	5.1	5.2	[251]
Schizophyllum commune	5 (A)	3.7	3.8	6.1	6.6	[264]
Cellobiohydrolase						
Cellulomonas fimi	6 (B)	4.1		7.7		[720]
β-Glucosidase						
Agrobacterium faecalis	1	4.9–5.0	3.6–4.8	7.0–7.2	7.6–8.1	[721]
Aspergillus wentii A_3	3				6.1	[823]
Botryodiplodia theobromae		3.50	3.65	6.00	6.75	[713,714]
Helix pomatia		4.5		6.7		[760]
Schizophyllum commune		3.3	3.3	6.6	6.9	[315]
Sweet almonds		4.4		6.7		[719]
Thermoascus aurantiacus			3.19		5.64	[715]
Trichoderma viride	3		3.5		6.8	[824]
Xylanase						
Bacillus sp.	10 (F)	5.2	4.9	6.4	6.9	[741]

the more acidic p*K* values were carboxylates, the identification of the more basic residues proved to be controversial. Early researchers [713–715] naturally invoked the participation of His residues (pK_a in proteins, 5.6–7.0 [716]), but subsequent chemical modification experiments with group specific reagents have generally precluded their direct role in catalysis (see section 5.2.2, Chemical Modification).

In a recent elegant study, the pK_a of an essential ionizable group on a xylanase from *Bacillus circulans* was determined by titration monitoring the FTIR spectrum [717]. A value of pH 6.8 was obtained, and this titration was absent in site-directed mutant enzymes with replacement of either Glu78 or Glu172 with Gln, strongly implicating one of these residues as the catalytic acid.

Further insights into catalytic mechanisms and the nature of the participating residues in cellulolytic enzymes have been gleaned from detailed kinetic studies. Such studies have been primarily conducted with β-glucosidases, likely in view of the relative simplicity, solubility, diversity, and availability of their substrates. Unfortunately, much of the early kinetic analyses of β-glucosidases were performed with a commercial source of almond emulsin [718,719] that contained at least two isozymes with different kinetic properties. Hence, interpretation of these data has to be made with caution. More recently, however, a β-glucosidase from *Agrobacterium faecalis* [720] and the bifunctional xylanase-cellobiohydrolase from *Cellu-*

lomonas fimi [721] were subjected to detailed kinetic investigations. These involved the establishment of Broensted plots that relate rates of hydrolysis with leaving group ability, measurement of primary and secondary kinetic isotope effects, in addition to presteady-state kinetic analyses. The results were consistent with a double-displacement mechanism in which a glycosyl-enzyme intermediate is formed and hydrolysed via oxocarbonium ion-like transition states. Moreover, subtle differences were observed between the structures of the two transition states for the β-glucosidase-catalyzed reactions, with the glycosylation step having more S_N2 character, whereas the deglycosylation step is apparently more S_N1-like [720]. Inverting cellulolytic enzymes have not been subjected to such detailed studies, but kinetic investigations of other inverting glycosidases [722–724] strongly implicates the formation of a transition state with considerable oxocarbonium ion character.

The formation of transition states with substantial oxocarbonium ion character within the active sites of β-glycosidases would necessitate the close proximity of a stabilizing anion. Although the pH activity studies described above can only imply the involvement of one or more essential anions, competitive inhibition studies with various cations give strong evidence for their presence at active sites. D-Glucosylamine is a strong competitive inhibitor of β-glucosidase [725], binding four orders of magnitude more tightly than D-glucose [726]. Likewise, sweet almond [727] and *Aspergillus wentii* [728] β-glucosidases were shown to be more strongly inhibited by basic glucosyl derivatives than by their corresponding neutral analogues, with the additional interaction energy ($\Delta\Delta G°$) provided by the basic or cationic species being on the order of 13 to 23 $kJ \cdot mol^{-1}$. Other inhibition studies on the sweet almond enzyme using phenols and amines confirmed and extended these studies [718]. These inhibition data led each group to conclude that glucosyl cation-like transition states were involved in substrate hydrolysis, and an anionic group (presumably a carboxylate) is present at the active sites of the enzymes.

Further evidence for a catalytic carboxylic acid in the *Schizophyllum commune* β-glucosidase and cellulase is given by the apparent irreversible binding of transition metals to these enzymes with their concomitant inactivation [315,729]. The inactivations were speculated to proceed by chelation involving Asp and Glu residues at the active center in a manner analogous to that observed by both X-ray crystallography and NMR for metal hen egg white lysozyme complexes [730,731].

Trapping experiments have been conducted [732,733] in an attempt to address the question of whether a covalent intermediate is formed during catalysis or the carboxylate anion simply stabilizes the oxocarbonium transition state as originally proposed for lysozyme. The poor (slow) substrates, *p*-nitrophenyl-β-D-2-deoxyglucoside and D-glucal, were used with the *As-*

pergillus wentii β-glucosidase to generate a steady-state concentration of the glycosyl-enzyme intermediates. These were subsequently trapped by denaturation of the enzyme with guanidinium chloride, and one equivalent of substrate was found to remain associated with the enzyme. Protease digestion and peptide mapping allowed the isolation of the substrate-bound peptide, and the binding site was identified as the same Asp residue that binds the affinity label, conduritol-B-epoxide (described below). Although these data would appear to confirm the formation of a covalent adduct between enzyme and substrate, it has been argued that the *binding* may simply involve a strong ion pair intermediate.

5.2.2 CHEMICAL MODIFICATIONS

Group-specific reagents are designed and utilized to form stable, covalent bonds to specific functional side chains of amino acids in proteins. Correlation of enzymatic activity with chemical modification is usually based on various and often tacit assumptions, for example, that the specificity of the reagent employed is absolute, that the conditions under which the reaction is performed do not disturb protein structure, and that the activity changes are not the consequence of either conformational changes or denaturation induced by the modification. In an ideal situation, therefore, the reagents are carefully manipulated to avoid any side reactions that could occur with other residues, and care is taken to assess for such. Nevertheless, the reagents do not discriminate between accessible functional groups on the surface of enzymes and those in active sites. Protection studies employing effective competitive inhibitors may help to ascribe a functional role to a modified residue, but, again, caution must be exercised with the interpretation of results. Notwithstanding these concerns, this approach has the advantage over the widely used, more recent technique of site-directed mutagenesis in that the starting material for the chemical modification experiments is fully active, native enzyme.

Chemical modification of carboxyl groups, primarily with various carbodiimides, has confirmed their essential role in the mechanism of action of both an inverting cellulase [734], and each of the retaining cellulolytic and xylanolytic enzymes as follows: cellulases [264,735–737], cellobiohydrolase [738], β-glucosidases [315,739], xylanases [240,644,740,741], and an exoacting xylobiohydrolase [335]. Treatment of the inverting (Family 6) cellulase from *Thermomonospora fusca* with 1-ethyl-3-[3-(dimethylamino)propyl]carbodiimide led to inactivation of the enzyme, and a kinetic analysis of the reaction indicated that the modification was stoichiometric, i.e., one mole of reagent bound to one mole enzyme for inactivation. Protection from inactivation was provided by both the competitive inhibitor, cellobiose and substrate, (hydroxyethyl)cellulose, implying both the essential nature of the modified residue(s) and its localization within the active site [734]. Similar results have been obtained with retaining enzymes, as exemplified by the carbodiimide modification of the Family 5 cellulase [264], Family 3 β-glucosidase [315], and Family 11 xylanase [644,740] from *Schizophyllum commune*. The studies with the *S. commune* β-glucosidase and xylanase employed a radioactive carbodiimide that permitted quantification of both the total number of accessible residues, and more importantly, those protected by inhibitors three and one, respectively. The use of the chromogenic Woodward's reagent K (*N*-ethyl-5-phenylisoxazolium-3′sulfonate) afforded similar analyses of the reactions with the Family 7 cellobiohydrolase I from *Trichoderma reesei* [738], the Family 9 cellulase D from *Clostridium thermocellum* [735], and a xylanase from an alkalothermophilic *Bacillus* sp. [741]. Differential modification followed by peptide mapping resulted in the identification of catalytically essential Glu residues at the active sites of the *S. commune* xylanase [740], the *C. thermocellum* cellulase D [735], and the *T. reesei* cellobiohydrolase I [738].

A few chemical modification studies have shown that, in addition to Glu and Asp residues, His may play an essential catalytic role (e.g., *C. thermocellum* cellulase D [736] and *T. reesei* β-glucosidase [739]). It is possible that the active sites of some β-glycosidases contain a His residue that either participates in substrate binding or aids catalysis in some other manner. However, it is highly unlikely that His participates as the proton donor in view of the fact that its requirement does not represent a general trend. Indeed, diethylpyrocarbonate (a regeant with specificity for His residues) served to modify all of the His residues present in both the cellulase [264] and xylanase [644] from *S. commune* without significant impairment of activity. The participation of a His residue on the *Aspergillus wentii* [742] and *S. commune* [315] β-glucosidases was likewise discounted.

Although the results obtained with chemical modifications are inherently inconclusive, at least with respect to studies on catalytic mechanisms, the investigations reviewed above strongly implicate the participation of only Asp and Glu residues as the direct catalysts in the single- and double-displacement mechanisms of inverting and retaining enzymes, respectively.

5.2.3 AFFINITY LABELLING AND MECHANISM-BASED INHIBITION

Confirmation of a carboxylate anion-nucleophile at the active sites of β-glycosidases has been obtained with the design and use of a variety of affinity labels and mechanism-based inhibitors. These classes of reagents are far superior to the generic chemical modification reagents because they comprise a sugar moiety or analogue that directs their binding and subsequent reaction to the active sites of the enzymes. To be classified as an affinity label or mechanism-based inhibitor, a compound must fulfil the following criteria: 1) inactivation must be dependent on time, and follow pseudo-first-order kinetics; 2) inhibitions must be irreversible, usually through the formation of a covalent bond; 3) the stoichiometry of the inactivation must be equimolar, that is, one mole of reagent must bind one mole of active site; and 4) a substrate or competitive inhibitor must provide protection from inactivation by the compound. The two classes of reagents are distinquished by the reactivity of the aglycone moieties. Affinity labels typically brandish an electrophilic group that can be attacked by any proximal nucleophile on the enzyme to form stable covalent adducts. Mechanism-based inhibitors, on the other hand, are relatively chemically inert and require the enzyme to perform a mechanism-based activation (usually protonation) before they become reactive (electrophilic). As with chemical modification reagents, the specificity of the affinity labels and mechanism-based inhibitors can be demonstrated by competition experiments using competitive, reversible inhibitors. A variety of reagents have been synthesized to facilitate the study of β-glycosidases, as described in two excellent reviews [712,743], and those used with cellulolytic and xylanolytic enzymes are listed in Table 5.2 and their structures are presented in Figures 5.4 and 5.5.

5.2.3.1 Affinity Labels

Of the affinity labels listed in Table 5.2, *N*-bromoacetyl β-D-glucosylamine [743,744], β-D-glucosyl isothiocyanate [745], conduritol aziridine [746], and cyclophellitol [712,747] have been used sparingly, with studies primarily confined to β-glucosidases. Only one, cyclophellitol, has been used to identify an active site nucleophile [712]. In contrast, the epoxyalkyl derivatives have received much greater attention, and investigations have involved both β-glucosidases and cellulases [748–752]. The epoxide derivatives are presumed to form covalent adducts with active site nucleophiles by the mechanism shown in Figure 5.6. An acidic amino acid residue, either the acid catalyst or another residue in the active-site cleft, protonates the epoxide ring, which then becomes susceptible to nucleophilic attack. Although these epoxyalkyl reagents require activation (protonation) by the enzyme and therefore could be considered as mechanism-based inhibitors, they are still classified as affinity reagents in view of the remoteness of the electrophilic center from the bound sugar, their generally poor binding affinity, and the inherent flexibility of the alkyl chain. Some of these negative features can be minimized because it has been shown that the stereochemistry [753], linkage, and length [751] of the epoxylalkyl part of the inhibitor greatly influence the rate of inactivation. 2′,3′-Epoxypropyl-β-D-glucoside has been used to inactivate the sweet almond and *Aspergillus wentii* β-glucosidases [754], whereas the cellobiose derivative has been successfully

TABLE 5.2. Affinity Labels and Mechanism-Based Inhibitors Used for the Study of Cellulolytic and Xylanolytic Enzymes.

Reagent	Enzyme	Source	Labelled Residue	Reference
Affinity labels				
β-D-Glucosylisothiocyanate	β-Glucosidase	Sweet almond		[745]
N-Bromoacetylglucosylamine	β-Glucosidase	*Agrobacterium faecalis*		[744]
		Aspergillus wentii		[743]
	Xylanase/CBH	*Cellulomonas fimi*		[744]
4′,5′-Epoxypentyl-β-D-glucoside	β-Glucosidase	*Aspergillus wentii*		[754]
		Sweet almond		[745,754]
4′,5′-Epoxypentyl-β-D-cellobioside	Cellulase	*Aspergillus niger*		[750]
		Aspergillus wentii		[750]
		Oxyporus sp.		[750]
		Schizophyllum commune		[748,749,751]
		Trichoderma reesei	Glu329	[752]
Conduritol-B-epoxide	β-Glucosidase	*Aspergillus oryzae*		[757,758]
		Aspergillus wentii	Asp	[754–757]
		Bitter almond	Asp	[761]
		Botryodiplodia theobromae		[759]
		Helix pomatia		[760]
		Sweet almond		[762]
Conduritol aziridine	β-Glucosidase	*Agrobacterium faecalis*		[746]
Cyclophellitol	β-Glucosidase	*Agrobacterium faecalis*	Glu358	[712,747]

TABLE 5.2. (continued).

Reagent	Enzyme	Source	Labelled Residue	Reference
Mechanism-based inhibitors				
Glucosylmethyl-4-(nitrophenyl)triazene	β-Glucosidase	Almond		[722]
	β-Xylosidase	*Bacillus pumilus*		[722]
o,p-(Difluoromethyl)aryl-β-D-glucoside	β-Glucosidase	Almond		[765]
Difluorotolyl-β–D-glucoside	β-Glucosidase	Almond		[765]
2′,4′-Dinitrophenyl-2-deoxy-2-fluoro-β-D-glucoside	Cellobiohydrolase	*Cellulomonas fimi*	Glu274	[721,771,772]
	β-Glucosidase	*Agrobacterium faecalis*	Glu358	[706,766–769]
		Sweet almond		[767]
2′,4′-Dinitrophenyl-2-deoxy-2-fluoro-β-D-cellobioside	Cellulase	*Clostridium thermocellum*	Glu280	[770]
2′,4′-Dinitrophenyl-2-deoxy-2-fluoro-β-D-xylobioside	Xylanase	*Bacillus subtilis*	Glu78	[773]

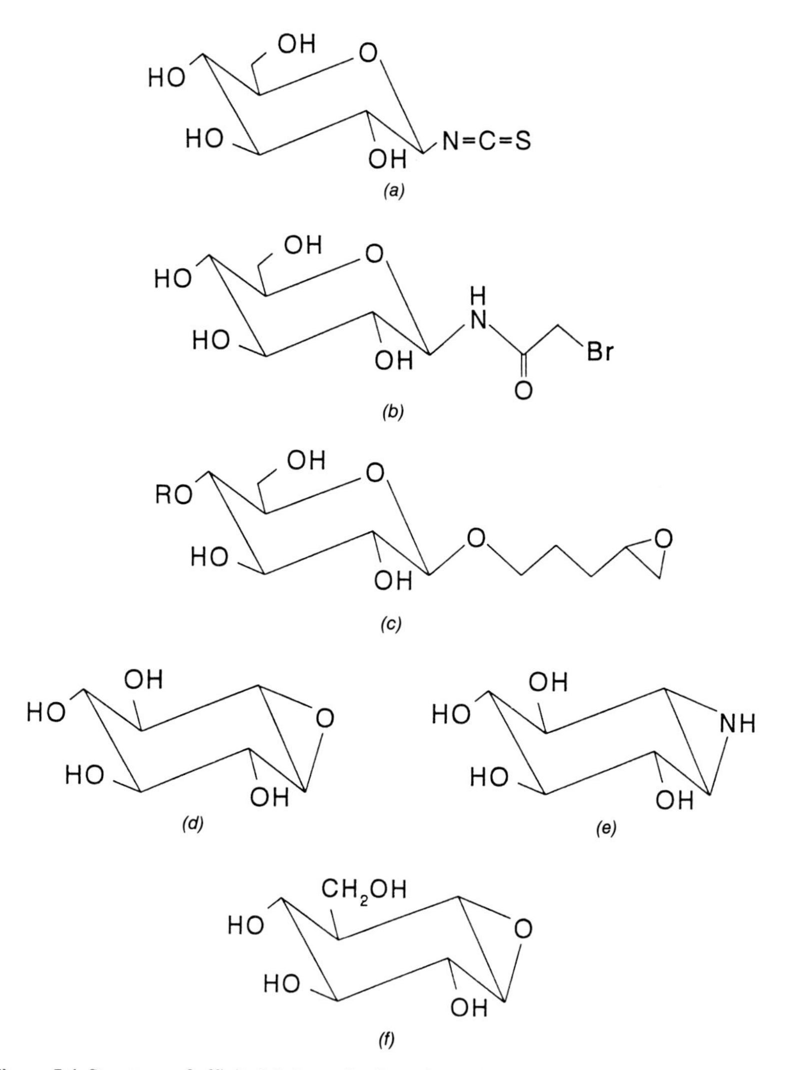

Figure 5.4 Structures of affinity labels used to investigate the mechanism of action of cellulolytic enzymes. (a) β-D-Glucosylisothiocyanate; (b) *N*-bromoacetylglucosylamine (c) 4′,5′-epoxypentyl-β-D-cellobioside and 4′,5′-epoxypentyl-β-D-glucoside when R is β-D-glucose and H, respectively; (d) conduritol-B-epoxide; (e) conduritol aziridine; (f) cyclophellitol.

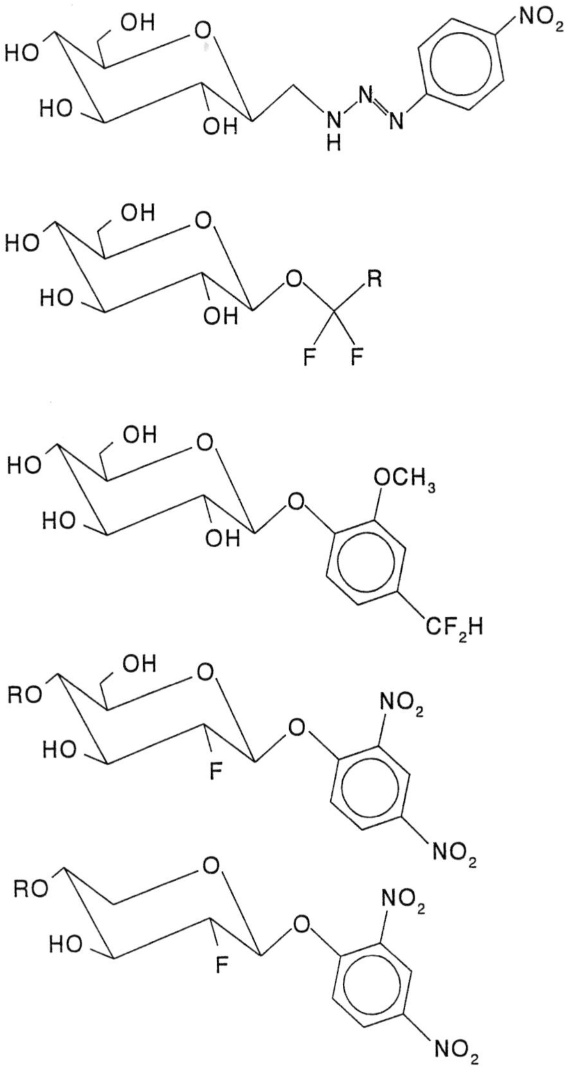

Figure 5.5 Structures of mechanism-based inhibitors used to investigate the mechanism of action of cellulolytic and xylanolytic enzymes. (a) Glucosylmethyl-4-(nitrophenyl)triazene; (b) *o,p*-(difluoromethyl)aryl-β-D-glucoside; (c) difluorotolyl-β-D-glucoside; (f) 2′,4′-dinitrophenyl-2-deoxy-2-fluoro-β-D-cellobioside and 2′,4′-dinitrophenyl-2-deoxy-2-fluoro-β-D-glucoside when R is β-D-glucose and H, respectively; (e) 2′,4′-dinitrophenyl-2-deoxy-2-fluoro-β-D-xylobioside, where R is β-D-xylose.

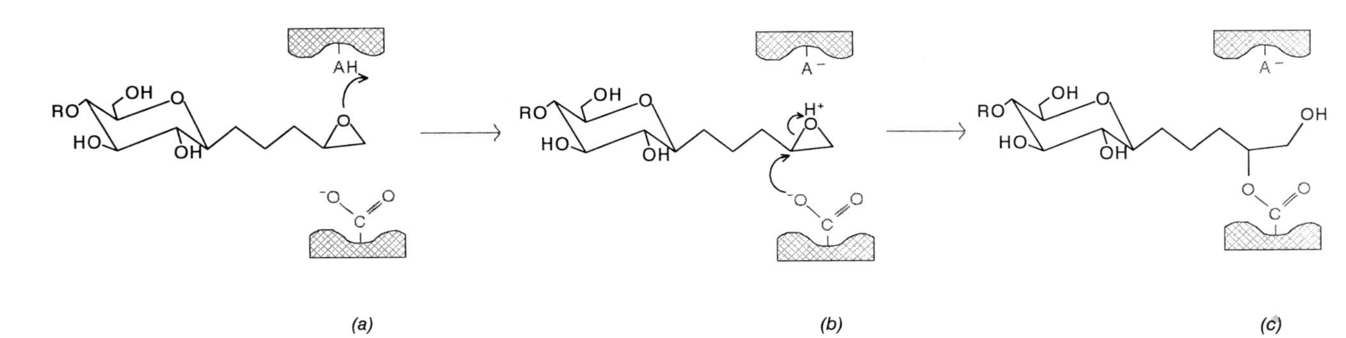

Figure 5.6 Proposed mechanism for the reaction of epoxyalkyl glycosides with β-glycosidases. The acid catalyst protonates the epoxide oxygen (a) which renders the 4′-carbon susceptible to nucleophilic attack by the stabilizing anion-nucleophile (b) producing a covalent adduct (c) via an ester linkage to the active-site nucleophile.

applied to the investigation of active-site residues on cellulases. The highest rates of inactivation of the cellulases from *A. niger, A. wentti, Oxyporus* sp. [750], *S. commune* [748,749,751], and *T. reesei* [752] were obtained with 4′,5′-epoxypentyl-β-D-cellobioside, and the bound nucleophile on the latter enzyme was identified [752].

Reaction of conduritol-B-epoxide and its brominated derivative (following its spontaneous cyclization to the epoxide form) with β-glycosidases follows a mechanism analogous to that of the epoxyalkyl derivatives. These reagents are considered to be a more refined version of the epoxide-based inactivators because the endocyclic epoxide incorporated within a cyclitol ring, which mimics a sugar ring, ensures that the protonation and subsequent nucleophilic attack involve amino acid residues at the binding site. Unfortunately, oligomeric derivatives of these compounds have not been synthesized, thus restricting their use to exoacting enzymes (e.g., β-glucosidases). Nevertheless, they have been used extensively to help delineate the mechanism of action of the β-glucosidases from *A. wentii* [754–757], *A. oryzae* [757,758], *Botryodiplodia theobromae* [759], *Helix pomatia* [760], and bitter [761] and sweet [762] almonds. In those cases examined, the stereochemistry of inactivation was demonstrated by release of the epoxide label with hydroxylamine as (+)-*chiro*-inositol, which would be expected for a retaining glycosidase following a double-displacement mechanism [754,756,762]. Protection studies using substrates and competitive inhibitors [759] and kinetic considerations [756,762] indicated that the conduritol-B-epoxide binds specifically to the catalytic sites and by using tritiated derivatives of the reagent, inactivation of the various β-glucosidases was found to proceed with the binding of one equivalent of affinity label [756,761,762]. For the *A. wentii* [754,755] and bitter almond enzymes [761], the bound nucleophile has been identified as an Asp residue. Finally, inactivation rates for the *A. wentii* [756], *H. pomatia* [760], and sweet almond [762] β-glucosidases show a sigmoidal pH dependence, with values of p*K*s between 5.6 and 7.3. These observations confirm the requirement of an acid catalyst to protonate the relatively inert epoxide prior to nucleophilic attack.

5.2.3.2 Mechanism-Based Inhibitors

In a similar situation to that noted above for the affinity labels, only one of the three classes of mechanism-based inhibitors listed in Table 5.2, the 2′,4′-dinitrophenyl-2-deoxy-2-fluoro-glycosides, has been broadly applied to the study of cellulolytic and xylanolytic enzymes. Upon binding to active site of the enzymes, as directed by the appropriate glycon, each inhibitor is protonated by the acid catalyst to generate a reactive species. With glucosylmethyl-4-(nitrophenyl)triazene, a rapid decomposition follows protonation to generate an arylamine, dinitrogen, and a highly reactive (electrophilic) glucosylmethyl carbenium ion (Figure 5.7). This electrophilic glucose derivative, being generated at the active site of the enzyme, is in position to be attacked by the catalytic nucleophile. The reagent has been used to inactivate the retaining sweet almond β-glucosidase but was found to be ineffective against the inverting xylosidase from *B. pumilus* [763]. Moreover, when used with *Escherichia coli* β-galactosidase, the reagent labelled a noncatalytic Met residue in the active-site cleft in addition to a Glu residue [764].

Enzymatic protonation of difluorotolyl glucoside is believed to be followed by an enzyme-catalyzed cleavage, releasing an aglycone that rapidly eliminates hydrogen fluoride to yield reactive fluorinated quinone methide (Figure 5.8). This compound was used successfully to inactivate the almond β-glucosidase [765], but the fact that the reactive species does not possess any inherent affinity for the binding-site cleft of the enzymes will likely limit its usefulness.

The 2-deoxy-2-fluoroglucosides represent a relatively new class of mechanism-based inhibitors of glycosides, but they have been manipulated and applied to the study of the retaining β-glucosidases [706,766–769], cellulase [770], cellobiohydrolase [721,771,772], and xylanase [773]. This class of inhibitors comprises 2-deoxy-2-fluoro glycosides or xylosides with good leaving groups (either dinitrophenolate or fluoride) that trap a covalent intermediate in the normal catalytic mechanism of the enzymes. The inductive effect of the fluorine at C-2 would destabilize the positively charged carbonium ion transition states, thereby slowing the rates of both glycosyl enzyme formation and hydrolysis. The presence of a reactive leaving group accelerates glycosyl enzyme formation, which results in the accumulation of the glycosyl-enzyme adduct. The enzyme intermediates are stable enough to permit rigorous biochemical analyses. Indeed, the catalytic nucleophiles on the Family 1 *Agrobacterium faecalis* β-glucosidase [768], Family 5 *Clostridium thermocellum* cellulase [770], Family 10 *Cellulomonas fimi* cellobiohydrolase [772], and Family 11 *Bacillus subtilis* xylanase [773] have been unequivocally identified. ^{19}F NMR spectrometry of the adduct formed with the *A. faecalis* β-glucosidase revealed that the 2-deoxy-2-glucose is covalently linked to the enzyme via an α-anomeric linkage, as expected for a retaining glycosidase [767,768]. Furthermore, and perhaps of greater significance, the catalytic competence of the trapped enzyme intermediates formed with the β-glucosidase [768,769], cellulase [770], cellobiohydrolase [772], and xylanase [773] was demonstrated with the addition of an appropriate acceptor molecule to permit transglycosylation reactions and measuring the rates of reactivation. A mechanism for the formation of the trapped enzyme intermediate and subsequent recovery of enzymatic activity by a transglycosylation reaction is presented in Figure 5.9. The pH dependence of both the formation and the hydrolysis (transglycosylation) of the inhibited

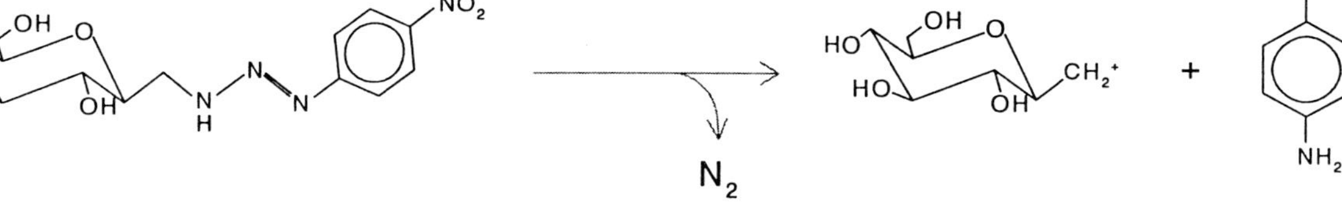

Figure 5.7 Proposed mechanism for the reaction of glucosylmethyl-4-(nitrophenyl)triazene with β-glycosidases.

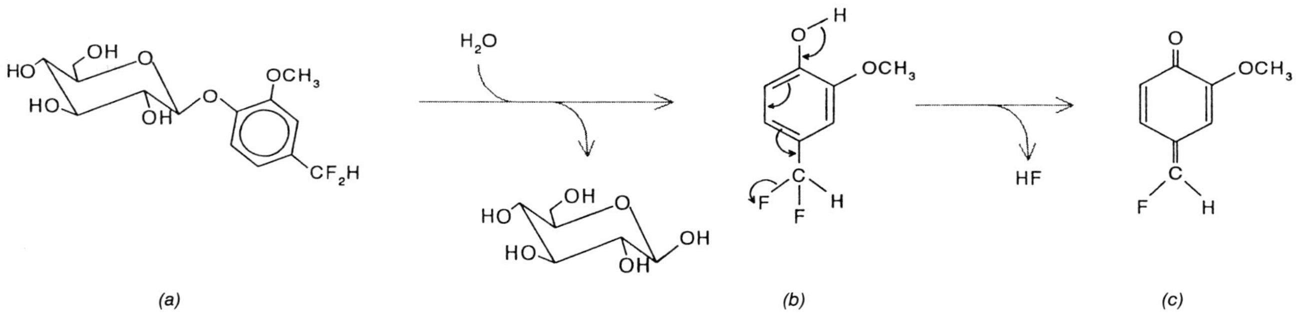

Figure 5.8 Proposed mechanism for the reaction of difluorotolyl-β-D-glucoside with β-glucosidase. Hydrolysis of the inhibitor (a) results in the release of glucose and the aglycone (b) which rapidly eliminates HF to generate the reactive fluorinated quinone methide (c) in the active site of the enzyme.

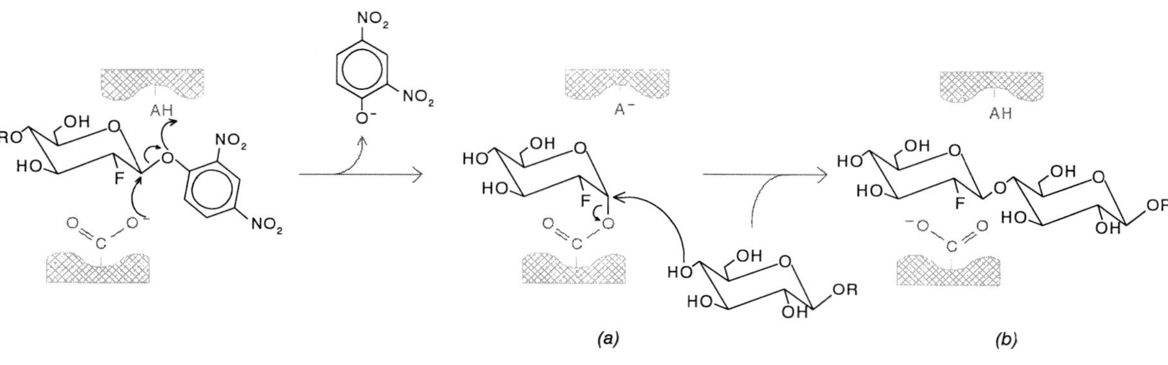

Figure 5.9 Proposed mechanism for the formation of a covalent intermediate between β-glycosidases and 2′,4′-dinitrophenyl-2-deoxy-2-fluoro-β-D-glycoside and its subsequent displacement via a transglycosylation reaction. The enzyme catalyzed hydrolysis of the inhibitor results in the release of dinitrophenol and the concomitant formation of a relatively stable acyl enzyme intermediate (a). In the presence of an appropriate nucleophile, a transglycosylation reaction is promoted to produce a new glycoside (b).

Agrobacterium β-glucosidase was shown to closely resemble those of each step for the normal enzymatic catalysis with substrates, strongly suggesting that the same catalytic groups are involved in both processes.

It is quite apparent that the design and use of the 2-deoxy-fluoro-glycosides has greatly facilitated our understanding of the mechanism of action of the retaining β-glycosidases. Studies with this class of mechanism-based inhibitors have unequivocally identified catalytic nucleophiles, indirectly proven the existence of carbonium ion transition states, and, more importantly, have provided the first direct evidence for the formation of a covalent intermediate to a carboxylate nucleophile on catalytically competent retaining β-glycosidases.

5.2.4 MECHANISM OF CATALYSIS

All the physical, kinetic, and biochemical investigations on the cellulolytic and xylanolytic enzymes reviewed above support the generally accepted single- and double-displacement mechanisms of Koshland [707] for the inverting and retaining β-glycosidases, respectively. The evidence is overwhelming, and there is no doubt that Asp and/or Glu residues participate at the active sites of the respective enzymes as the acid catalyst and base catalyst-nucleophile. Thus, the acid catalyst is thought to protonate the exocyclic oxygen of the substrate to make it a good leaving group, and the transition state with significant oxocarbonium ion character is stabilized by the correctly positioned carboxylate (Exocyclic pathway, Figure 5.10). To accommodate the unfavourable electronic geometry of the cyclic oxocarbonium ion, there is a kinetic requirement for distortion of the sugar ring into a half chair (twist boat) conformation. In view of the energy requirements for such a ring distortion, the enzymes are postulated to induce the conformational change in the ring. With the retaining enzymes, displacement of the exocyclic group by the nucleophilic Asp and Glu residue follows with the formation of a covalent glycosyl-enzyme adduct, which in turn is displaced with water. In the case of the inverting enzymes, no covalent intermediate is achieved, but rather a water molecule displaces the initially formed oxocarbonium ion through general base catalysis involving the active-site carboxylate ion.

Although there is little doubt for the formation of carbonium ion transition states and a covalent enzyme-substrate intermediate in the mechanism of retaining enzymes, the evidence to support the required ring distortion has been controversial. Recently, a proposal for an endocyclic pathway for the formation of the enzyme-substrate adduct was resurrected [774,775]. This latter postulate, which was originally based on kinetic studies on the inhibition of hydroxylated cyclic amines [776], involves the protonation of the ring (endocyclic) oxygen by the acid catalyst, followed by subsequent

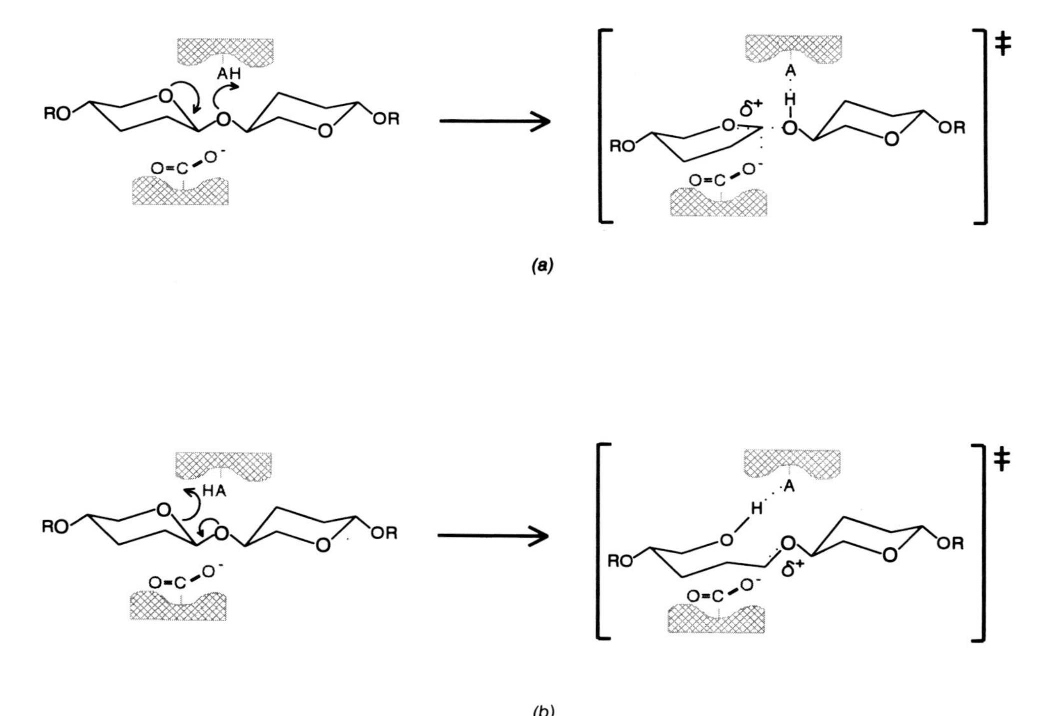

Figure 5.10 Comparison of the proposed reaction pathways for the (a) exocyclic and (b) endocyclic mechanisms of β-glycoside hydrolysis.

nucleophilic attack and displacement (Endocyclic pathway, Figure 5.10). The mechanism requires the rotation of the C1-C2 bond of the carbohydrate substrate rather than invoking substrate distortion for the favourable alignment of electronic orbitals. Naturally, the endocyclic mechanism has been received with some scepticism and has initiated much debate (for example, see Reference [922]). Although it remains to be established what the true mechanism involves, most, if not all, the data support an exocyclic mechanism. Indeed, in a very recent study, the distortion of a bound sugar ring at the catalytic center of a hen egg white lysozyme-oligosaccharide complex has been clearly and unambiguously observed by the X-ray crystallography (1.5 Å resolution) [777].

5.3 STRUCTURE AND FUNCTION RELATIONSHIP OF CATALYTIC SITES

The establishment of a classification scheme based on sequence alignments has been particularly useful for making predictions of amino acid residues that may be responsible for specificity and catalysis. Prior to its development, and at a time when there was a general paucity of detailed structural and mechanistic data for the cellulolytic and xylanolytic glycosidases, a number of predictions of catalytic residues and domains were made solely on the basis of sequence comparisons with other well-characterized enzymes, such as hen egg white lysozyme. For example, an intriguing amino acid sequence homology was noted between the region of lysozyme comprising the catalytic residues Glu35 and Asp52 and segments of a cellulase [265,778], xylanase [339], and three β-glucosidases [779,780] (Figure 5.11). By analogy with hen egg white lysozyme, Asp and Glu residues within the respective enzymes were identified as potential catalytic residues. These bold proposals were made solely on this alignment because very little other homology was found between the different enzymes; indeed, we now know that the three β-glucosidases belong to different families. With the delineation of more sequences and establishment of the classification scheme, it became apparent that the lysozyme-like region was not generally conserved

Hen egg-white lysozyme	34	FESNFNTQAT-N-------RNTDGSTDY
Schizophyllum commune cellulase I	32	NESCAEFGNQ-N-------IPGVKNTDY
Schizophyllum commune β-glucosidase	159	TESSAQVPDI-D-------YSEGLLVDY
Agrobacterium sp. β-glucosidase	310	VEFPATMPAP---------AVSDVKTDI
Candida pelliculosa β-glucosidase	606	IEKVDVPDPV-D------KFTESITVD
Cryptococcus albidus xylanase	39	LESQFDAITPENEMKWEVVEPTEGNFDF

Figure 5.11 Amino acid sequence alignment of a fungal cellulase, both bacterial and fungal β-glucosidases, and a fungal xylanase with the segment of hen egg white lysozyme comprising its catalytic residues. Residues in bold type denote identity with hen egg white lysozyme, whereas the shaded Glu and Asp residues were considered identical to its catalytic residues Glu35 and Asp52.

within the families of the respective enzymes. This would discount any mechanistic significance for the observed homologous regions, which now appear to be no more than an unfortunate coincidence.

Analysis of sequence alignments within a given family for conserved acidic residues that are predicted to be located at active centers has been very useful for both retaining and inverting enzymes. This has been particularly the case when only a few potential residues are conserved, such as with the Family 11 enzymes (see below). In this example, for which the point is taken to the extreme, only two acidic amino acid residues are totally conserved among the twenty-one known sequences and thus, both Glu87 and Glu184 (*S. commune* xylanase A numbering) have been predicted to participate in catalytic roles. Although this prediction is likely to be quite safe, most others are not as conclusive because a number of conserved acidic residues are observed, and not all will be located at the active site or involved in the mechanism of action.

Site-directed mutagenesis of conserved residues has provided the additional insights necessary to further establish their role. This technique has been used to investigate a number of β-glycosidases, inlcuding hen egg white lysozyme [781]. Mutants of lysozyme in which the acid catalyst, Glu35, has been replaced are totally inactivated, whereas mutation of the nucleophile-stabilizing anion, Asp52, results in a mutant enzyme with approximately 5% residual activity [781]. This trend has been observed in all cases where the roles of mutated residues had been previously assigned. However, caution has to be exercised when intrepreting the results of mutation studies. If mutation of a target residue does not significantly alter activity, it is certain that the residue is not catalytically important. Unfortunately, the reverse is not always true because mutation of a residue that is either critical to the structural integrity of the enzyme or directly involved with substrate binding can also lead to substantial loss of activity. These and other hazards that plague both chemical and molecular biological approaches to the study of the structure and function relationships of enzymes are discussed in a review by Schimmel [782], whereas Svensson and Sogaard [783] have comprehensively reviewed the glycosidases subjected to mutational analysis.

The potential of combining site-directed mutagensis with sequence alignments for the investigation of the structure and function relationship of enzymes is probably best illustrated by a recent study of the Family 1 *Agrobacterium* β-glucosidase [784]. Forty-three point mutations were generated at twenty-two different residues in the region surrounding the putative active-site nucleophile. Changes to only five of these residues resulted in any significant alteration in catalytic activity, suggesting that enzymes can accommodate mutations in nonconserved regions, even when drastic changes such as Glu to Lys were made. This study also indicated that sequence alignments alone are not sufficient to identify important residues because replacement of several fully conserved residues induced only modest decreases in catalytic activity.

The following reviews the search for catalytic residues among the various families of cellulolytic and xylanolytic enzymes based on both sequence homology studies and site-directed mutagenesis. This information is integrated with that obtained by X-ray crystallography and the chemical techniques of modification, affinity labelling, and mechanism-based inhibitions to provide a comprehensive representation of what is currently known about the structure and function relationship of the catalytic sites of individual families of cellulolytic and xylanolytic enzymes.

5.3.1 RETAINING ENZYMES

5.3.1.1 Families 1 and 5

Mechanism-based inhibition and trapping experiments have identified Glu358 of the *Agrobacterium faecalis* β-glucosidase as the nucleophilic partner in the double-displacement mechanism of this retaining Family 1 enzyme [706,768]. Not surprisingly, Glu358 represents one of the six totally conserved acidic amino acid residues in the aligned sequences of thirteen of the Family 1 enzymes [520] that include those of bacterial, fungal, and mammalian sources (Figure 5.12). Site-directed mutagenesis of the *Agrobacterium* β-glucosidase gene replacing Glu358 with either Asn or Gln resulted in essentially complete inactivation of the enzyme [784,785], thus confirming the critical role of this residue in catalysis. Twenty-two other residues in the region of Glu358 were also targeted for site-directed mutagenesis [784], but inactivation of the β-glucosidase accompanied only five of these individual point mutations, of which four represented conserved residues. One of these residues, Asp374, belongs to a completely conserved Asp-Arg-Tyr (D-R-Y) motif (Figure 5.12). It was suggested that the conserved Asp residue may participate as the acid catalyst. However, all of the mutant enzymes with replacements at Asp374 did retain some measurable activity, which is inconsistent with this assignment.

Glu358 of the *Agrobacterium* β-glucosidase represents the first residue of a Glu-Asn-Gly motif, which is totally conserved within the Family 1 enzymes (Figure 5.12). An analogous highly conserved Glu-x-Gly motif exists in the Family 5 enzymes (Figure 5.13), where the Glu is completely conserved, and in most cases, x represents an aromatic residue. This Glu of the Family 5 *Clostridium thermocellum* cellulase, Glu280, was found to bind the mechanism-based inhibitor 2′,4′-dinitrophenyl-2-deoxy-2-fluoro-β-D-cellobioside, thus identifying it also as the nucleophile for this family of β-glycosidases [770]. Likewise, the homologous Glu329 residue of the *Trichoderma reesei* cellulase III was identified as the catalytic nucleophile

```
CthBGA     MSKITFPKDFIWGSATAAYQIEGAYNEDGKGESIWDRFSHTPGNIADGH------TGDVACDHYHRYEEDIKIMKEIGIKSYRFSISWPRIFP
CsaBG         MSFPKGFLWGAATASYQIEGAWNEDGKGESIWDRFTHQKRNILYGH------NGDVACDHYHRFEEDVSLMKELGLKAYRFSIAWTRIFP
BpoBGA      MTIFQFPQDFMWGTATAAYQIEGAYQEDGRGLSIWDTFAHTPGKVFNGD------NGDVACDSYHRYEEDIRLMKELGIRTYRFSVSWPRIFP
BpoBGB    MSENTFIFPATFMWGTSTSSYQIEGGTDEGGRTPSIWDTFCQIPGKVIGGD------CGDVACDHFHHFKEDVQLMKQLGFLHYRFSVAWPRIMP
AgrBG   MTDPNTLAARFPGDFLFGVATASFQIEGSTKADGRKPSIWDAFCNMPGHVFGRH------NGDIACDHYNRWEEDLDLIKEMGVEAYRFSLAWPRIIP
LPH            FYHGTFRDDFLWGVSSSAYQIEGAWDADGKGPSIWDNFTHTPGSNVKDNA------TGDIACDSYHQLDADLNMLRALKVKAYRFSISWSRIFP
LPH            FLYGRFPEGFIWSAASAAYQIEGAWRADGKGLSIWDTFSHTPLRVENDA------IGDVACDSYHKIAEDLVTLQNLGVSHYRFSISWSRILP
LcaPGa          MSKQLPQDFVMGGATAAYQVEGATKEDGKGRVLWDDFLDKQGRFKP----------DPAADFYHRYDEDLALAEKYGHQVIRVSIAWSRIFP
SlaPGa          MTKTLPKDFIFGGATAAYQAEGATHTDGKGPVAWDKYLEDNYWYTA----------EPASDFYHKYPVDLELAEEYGVNGIRISIAWSRIFP
SauPGa          MTKTLPEDFIFGGATA-YQAEGATNTDGKGRVAWDTYLEENYWYTA----------EPASDFYNRYPVDLELSEKFGVNGIRISIAWSRIFP
EcoPGu           MKAFPETFLWGGATAANQVEGAWQEDGKGISTSDLQPHGVMGKMEPR-ILGKENIKDVAIDFYHRYPEDIALFAEMGFTCLRISIAWARIFP
SsoBGal1         MLSFPKGFKFGWSQSGFQSEMGTPGSEDPNSDWHVWVHDRENIVSQVVSGDLPENGPGYWGNYKRFHDEAE---KIGLNAVRINVEWSRIFP
SsoBGal2         MYSFPNSFRFGWSQAGFQSEMGTPGSEDPNTDWYLWVHDPENMAAGLVSGDLPENGPGYWGNYKTFHDNAQ---KMGLKIARLNVEWSRIFP

CthBGA     EGTGKL-----------------------------NQKGLDFYKRLTNLLLENGIMPAITLYHWDLPQKLQDK-----------GGWKNRDTTDYFTE
CsaBG      DGFGTV-----------------------------NQKGLEFYDRLINKLVENGIEPVVTLYHWDLPQKLQDI-----------GGWANPEIVNYYFD
BpoBGA     NGDGEV-----------------------------NQEFLDYYHRVVDLLNDNGIEPFCTLYHWDLPQALQDA-----------GGWGNRRTIQAFVQ
BpoBGB     AA-GII-----------------------------NEEGLLFYEHLLDEIELAGLIPMLTLYHWDLPQWIEDE-----------GGWTQRETIQHFKT
AgrBG      DGFGPI-----------------------------NEKGLDFYDRLVDGCKARGIKTYATLYHWDLPLTLMGD-----------GGWASRSTAHAFQR
LPH        TGRNSSI----------------------------NSHGVDYYNRLINGLVASNIFPMVTLFHWDLPQALQDI-----------GGWENPALIDLFDS
LPH        DGTTRYI----------------------------NEAGLNYYVRLIDTLLAASIQPQVTIYHWDLPQTLQDV-----------GGWENETIVQRFKE
LcaPGa     DGAGEV-----------------------------NPRGVAFYHKLFADCAAHHIEPFVTLHHFDTPERLHEA-----------GDWLSQEMLDDFVA
SlaPGa     TGYGEV-----------------------------NEKGVEFYHKLFAECHKRHVEPFVTLHHFDTPEALHSN-----------GDFLNRENIEHFID
SauPGa     NGYGEV-----------------------------NPKGVEYYHKLFAECHKRHVEPFVTLHHFDTPEVLHKD-----------GDFLNRKTIDYFVD
EcoPGu     QGD-EFEP---------------------------NEAGLAFYDRLFDEMAQAGIKPLVTLSHFEMPYGLVKNY----------GGWANRAVIGHFEH
SsoBGal1   RPLPKPEMQTGTDKENSPVISVDLNESKLREMDNYANHEALSHYRHILEDLRHRGFHIVLNMYHWTLPIWLHDPIRVRRGDFTGPTGWLNSRTVYEFAR
SsoBGal2   NPLPRPQNFDES-KQDVTEVEINENELKRL--DEYANKDALNHYREIFKDLKSRGLYFILNMYHWPLPLWLHDPIRVRAGDFTGPSGWLSTRTVYEFAR
```

Figure 5.12 Alignment of deduced amino acid sequences of Family 1 catalytic domains. Residues in bold type denote identity among at least seven of the thirteen sequences, whereas shaded residues are totally conserved. The sequences correspond to the following: CthBGA, *Clostridium thermocellum* β-glucosidase A; CsaBG, *Caldocellum saccharolyticum* β-glucosidase [260]; BpoBGA and BpoBGB, *Bacillus polymyxa* β-glucosidases A and B, respectively [825]; AgrBG, *Agrobacterium* sp. β-glucosidase [826]; LPH, human lactase-phlorizin hydrolase proenzyme [827]; LcaPGa, *Lactobacillus casei* phospho-β-galactosidase [828]; SlaPGa, *Streptococcus lactis* phospho-β-galactosidase [829]; SauPGa, *Staphylococcus aureus* phospho-β-galactosidase [830]; EcoPGu, *Escherichia coli* phospho-β-glucosidase B [831]; SsoBGal1 and SsoBGal2, *Sulfolobus solfataricus* β-galactosidases 1 [832] and 2 [825], respectively. For presentation purposes, a thirteen-amino acid region of nonconserved residues has been removed from between the residues identified by the asterisks (adapted from Reference [520]).

```
CthBGA     YSEVIFKNLGDIVPIWFTHNEPGVVSLLGH-FLGIHAPGIKD-LRTSLEVS-HNLLLSHGKAVKLFREMNIDAQIGIALNLSYHYPASEK------AED
CsaBG      YAMLVINRYKDKVKKWITFNEPYCIAFLGY-FHGIHAPGIKD-FKVAMDVV-HSLMLSHFKVVKAVKENNIDVEVGITLNLTPVYLQTERLGYKVSEIE
BpoBGA     FAETMFREFHGKIQHWLTFNEPWCIAFLSN-MLGVHAPGLTN-LQTAIDVG-HHLLVAHGLSVRRFRELGTSGQIGIAPNVSWAVPYSTS------EED
BpoBGB     YASVIMDRFGERIMWWNTINEPYCASILGY-GTGEHAPGHEN-WREAFTAA-HHILMCHGIASNLHKEKGLTGKIGITLNMEHVDAASER------PED
AgrBG      YAKTVMARLGDRLDAVATFNEPWCAVWLSH-LYGVHAPGERN-MEAALAAM-HHINLAHGFGVEASRHVAPKVPVGLVNAHSAIPASSDG------EAD
LPH        YADFCFQTFGDRVKFWMTFNEPMYLAWLGY-GSGEFPPGVKD-PGWAPYRIAHTVIKAHARVYHTYDEKYRQEQKGVISLSLSTHWAEPKSPGV--PRD
LPH        YADVLFQRLGDKVKFWITLNEPFVIAYQGY-GYGTAAPGVSNRPGTAPYIVGHNLIKAHAEAWHLYNDVYRASQGGVISITISSDWAEPRDPSN--QED
LcaPGa     YAKFCFEEFSE-VKWWITINEPTSMAVQQY-TTGTFPPAESG-RFDKTFQAEHNQMVAHARIVNLYKSMQLGGQIGIVHALQTVYPYSDS------AVD
SlaPGa     YAAFCFEEFPE-VNYWTTFNEIGPIGDGQY-IVGKFPPGIKY-DLAKVFQSHHNMMVSHARAVKLYKDKHYKGEIGVVHALPTKYPYDPEN-----PAD
SauPGa     YAEYCFKEFPE-VKYWTTFNEIGPIGDGQY-IVGKFPPGIKY-DFEKVFQSHHNMMVAHARAVKLFKDGGYKGEIGVVHALPTKYPFDPSN-----PED
EcoPGu     YARTVFTRYQHKVALWLTFNEIN-MSLHAP-FTGVGLAEESG-EAE-VYQAIHHQLVASARAVKACHSLLPEAKIGNIVLGGLVYPLTCQ------PQD
SsoBGal1   FSAYVAWKLDDLASEYATMNEPNVVWGAGYAFPRAGFPPNYL-SFRLSEIAKWNIIQAHARRYDAIKSVS-KKSVGIIYANTSYYPLR--------PQD
SsoBGal2   FSAYIAWKFDDLVDEYSTMNEPNVVGGLGYVGVKSGFPPGYL-SFELSRRHMYNIIQAHARAYDGIKSVS-KKPVGIIYANSSFQPLTDK--------D

                                                                                                     *
CthBGA     IEAAELSFSLAGRWYLDPVLK-GRYPENALKLYKKKGI-------ELSFPEDDLKLISQPIDFIAFNNYSSEFIKYDPSSESGFSPANSILEKF-----
CsaBG      REMVSLSSQLDNQLFLDPVLK-GSYPQKLLDYLVQKDLLDSQKALSM-QQEVKENF---IFPDFPGINYYTRAVRLYDENSSWIFPIRWEHPAGE-----
BpoBGA     KAACARTISLHSDWFLQPITQ-GSYPQFLVDWFAEQGATVP-----IQDGDMDIIG---EPIDMIGINYYSMSVNRFNPEAGFLQSEEINMGLPV-----
BpoBGB     VAAAIRRDGFINRWPAEPLFN-GKYPEDMVEWYGTYLNGLDF----VQPGDMELIQ---QPGDFLGINYYTRSIIRSTNDASLLQVEQVHMEEPV-----
AgrBG      LKAAERAFQFHNGAFFDPVFK-GEYPAEMMEALGDRMPVVEA-------EDLGIISQK---LDWWRLNYYTPMRVADDATPGVEFPATMPAPAVS-----
LPH        VEAADRMLQFSLGWFAHPIFRNGDYPDTMKWKVGNRSELQHLATSRLPSFTEEEKRFIRATADVFCLNTYYSRIVQHKTPRLNPPSYEDDQEMAE-----
LPH        VEAARRYVQFMGGWFAHPIFKNGDYNEVMKTRIRDRSLAAGLNKSRLPEFTESEKRRINGTYDFFGFNHYTTVLAYNLNYATAISSFDADRGVAS-----
LcaPGa     HHAAELQDALENRLYLDGTLA-GEYHQETLALVKEILDANHQPMFQSTPQEMKAIDEAAHQLDFVGVNNYFSKWLRAYHGKSETIHNGDGTKGSSVARLQ
SlaPGa     VRAAELEDIIHNKFILDATYI-GHYSDKTMEGVNHILAENGG-ELDLRDEDFQALDAAKDLNDFLGINYYMSDWMQAFDGETEIIHNGKGEKGSSKYQIK
SauPGa     VRAAELEDIIHNKFILDATYI-GKYSRETMEGVQHILSVNGG-KLNITDEDYAILDAAKDLNDFLGINYYMSDWMRGYDGESEITHNATGDKGGSKYQLK
EcoPGu     MLQAMEENRRWM-FFGDVQAR-GQYPGYM--QRFFRDHNITIEMTESDAEDLK------HTVDFISFSYYMTGCV---------SHDETINK---NAQGN
SsoBGal1   NEAVEIAERLNRWSFFDSIIK-GEITSEGON---------------VREDLR------NRLDWIGVNYYTRTVVTKAESGYLTLPGYGDRCERN-----
SsoBGal2   MEAVEMAENDNRWWFFDAIIR-GEITRGNEK---------------IVRDDLK------GRLDWIGVNYYTRTVVKRTEKGYVSLGGYGHGCERN-----
```

Figure 5.12 (continued) Alignment of deduced amino acid sequences of Family 1 catalytic domains. Residues in bold type denote identity among at least seven of the thirteen sequences, whereas shaded residues are totally conserved. The sequences correspond to the following: CthBGA, *Clostridium thermocellum* β-glucosidase A; CsaBG, *Caldocellum saccharolyticum* β-glucosidase [260]; BpoBGA and BpoBGB, *Bacillus polymyxa* β-glucosidases A and B, respectively [825]; AgrBG, *Agrobacterium* sp. β-glucosidase [826]; LPH, human lactase-phlorizin hydrolase proenzyme [827]; LcaPGa, *Lactobacillus casei* phospho-β-galactosidase [828]; SlaPGa, *Streptococcus lactis* phospho-β-galactosidase [829]; SauPGa, *Staphylococcus aureus* phospho-β-galactosidase [830]; EcoPGu, *Escherichia coli* phospho-β-glucosidase B [831]; SsoBGal1 and SsoBGal2, *Sulfolobus solfataricus* β-galactosidases 1 [832] and 2 [825], respectively. For presentation purposes, a thirteen-amino acid region of nonconserved residues has been removed from between the residues identified by the asterisks (adapted from Reference [520]).

```
          *
CthBGA    TDMGWIIYPEGLYDLLMLLDRDYGKPNI-VISENGAAFKDEIGSNGK----IEDTKRIQYLKDYLTQAHRAIQ-DGVNLKAYYLWSLLDNFEWA-YGYNKRFG
CsaBG     TEMGWEVFPQGLFDLLIWIKESYPQIPI-YITENGAAYNDIVTEDGK----VHDSKRIEYLKQHFEAARKAIE-NGVDLRGYFVWSLMDNFEWA-MGYTKRFG
BpoBGA    TDIGWPVESRGLYEVLH-YLQKYGNIDI-YITENGACINDIVTEDGK----VQDDRRISYMQQHLVQVHRTIH-DGLHVKGYMAWSLLDNFEWA-EGYNMRFG
BpoBGB    TDMGWEIHPESFYKLLTRIEKDFSKGLPILITENGAAMRDEVV-NGQ----IEDTGRHGYIEEHLKACHRFIE-EGGQLKGYFVWSFLDNFEWA-WGYSKRFG
AgrBG     TDIGWEVYAPALHTLVETLYERYDLPEC-YITENGACYNMGV-ENGE----VNDQPRLDYYAEHLGIVADLSR-DGYPMRGYFAWSLMDNFEWA-EGYRMRFG
LPH       TAMNRAA-PWGTRRLLNWIKEEYGDIPI-YITENGVGLTNPNT---------EDTDRIFYHKTYINEALKAYRLDGIDLRGYVAWSLMDNFEWL-NGYTVKFG
LPH       GSFWLKMTPFGFRRILNWLKEEYGDPPI-YVTENGVSQREE-TD-------LNDTARIYYLRTYINEALKAVQ-DKVDLRGYTVWSAMDNFEWA-TGFSERFG
LcaPGa    TDWDWSIYPRGMYDILMRIHNDYPLVPVTYVTENGIGLKESLPENATPDTVIEDPKRIDYVKKYLSAMADAIH-DGANVKGYFIWSLQDQFSWT-NGYSKRYG
SlaPGa    TDWDWIIYPEGLYDQIMRVKNDYPNYKKIYITENGLGYKDEFVDN----TVY-DDGRIDYVKQHLELS-DAIA-DGANVKGYFIWSLMDVFSWS-NGYEKRYG
SauPGa    TDWDWMIYPQGLYDQIMRVVKDYPNYHKIYITENGLGY-DEFIESFK--TVH-DDARIDYVRQHLNVIADAII-DGANVKGYFIWSLMDVFSWS-NGYEKRYG
EcoPGu    SEWGWQIDPVGLRVLLNTLWDRYQKPL--FIVENGLGAKDSVEADGS----IQDDYRIAYLNDHLVQVNEAIA-DGVDIMGYTSWGPIDLVSASHSQMSKRYG
SsoBGal1  SDFGWEFFPEGLYDVLLKYWNRYGLPL--YVMENGIADDA-------------DYQRPYYLVSHIYQVHRALN-EGVDVRGYLHWSLADNYEWS-SGFSMRFG
SsoBGal2  SDFGWEFFPEGLYDVLTKYWNRYHLYM--YVTENGIADDA-------------DYQRPYYLVSHVYQVHRAIN-SGADVRGYLHWSLADNYEWA-SGFSMRFG
```

Figure 5.12 (continued) Alignment of deduced amino acid sequences of Family 1 catalytic domains. Residues in bold type denote identity among at least seven of the thirteen sequences, whereas shaded residues are totally conserved. The sequences correspond to the following: CthBGA, *Clostridium thermocellum* β-glucosidase A; CsaBG, *Caldocellum saccharolyticum* β-glucosidase [260]; BpoBGA and BpoBGB, *Bacillus polymyxa* β-glucosidases A and B, respectively [825]; AgrBG, *Agrobacterium* sp. β-glucosidase [826]; LPH, human lactase-phlorizin hydrolase proenzyme [827]; LcaPGa, *Lactobacillus casei* phospho-β-galactosidase [828]; SlaPGa, *Streptococcus lactis* phospho-β-galactosidase [829]; SauPGa, *Staphylococcus aureus* phospho-β-galactosidase [830]; EcoPGu, *Escherichia coli* phospho-β-glucosidase B [831]; SsoBGal1 and SsoBGal2, *Sulfolobus solfataricus* β-galactosidases 1 [832] and 2 [825], respectively. For presentation purposes, a thirteen-amino acid region of nonconserved residues has been removed from between the residues identified by the asterisks (adapted from Reference [520]).

		58	133	186	265	306
1	BpoCel	SMDDMLDQVKKEGYNLIRLPYSNQLF	AGQRGIQIILDRHRPGSGGQSE	VIGADLHNEPH--GQ	PNRVVYSPHDY	SKQNIAPVLVGEFGGRNVDLS
	CfiCenD	NITQITQQMAQRGINVVRVPVSTQLL	CQKYGIKVFLDVHSAEADNSGH	IVGADIKNEPH----	QDQLVYSPHDY	HDEDIAPLLIGEWGGRLGQD-
	CflCelA	NLEDVTRSMAEHGINIVRVPISTQLL	SEKYGLKVMLDLHSAEADNAGH	LVAMDIKNEPH----	Q--LVYSPHDY	HKQGIAPLLIGEWGGRVGQDE
	CsaCelB	NLKDTLAEIANRGFNLLRVPISAELL	CKEVGLKIMLDIHSIKTDAMGH	IIAFDLKNEPH----	QNKVVYSPHDY	MEENIAPLLIGEWGGHLDGAD
	CthCelB	DIIADIELVADKGINVVRMPIATDLL	FKRVGIKVILDVHSPETDNQGH	IIGFDLKNEPH-TNT	QSQLVYSPHDY	MEEGISPLLLGEWGGMTEGGH
	XcaCelX	NWKDMIVQMQGLGFNAVRLPFCPATL	FNARGMYVLLDHHTPDCAGISE	VLGLDLKNEPH--GA	--RLLLAPHVY	AGTH-A-LLLGEFGGKYGEGD
2	BfiCelA	SAETIKSLRDTWGINVIRLAMTYSDY	ATDNDMYVIIDWHTLNDADPNE	NVIYEICNEPNG---	FDNIMYTYHFY	ALDEGLPVFISEYGLVDADGD
	BspCelA	NYESMKWLRDDWGITVGRAAMYTSS-	AIDLGIYVIIDWHILSDNDPNE	NVIYEIANEPNGS--	DPNVMYAFHFY	ALDQGAAIFVSEWGTSEATGD
	BspCelB	NYDSIKWLRDDWGITVFRAAMYTSS-	AIDLGIYVIIDWHILSDNDPNI	NVIYEIANEPNGH--	DPNVMYAFHFY	ALSRGAAIFVSEWGTSAATGD
	BspCelC	NYDSIKWLRDDWGITVFRAAMYTSS-	AIDLGIYVIIDWHILSDNDPNI	NVIYEIANEPNGH--	DPNVMYAFHFY	ALSRGAAIFVSEWGTSAATGD
	BspEG	NDNAYKALANDWESNMIRLAMYVGE-	AIENDMYVIVDWHVHAPGDPRD	HIIYELANEPSSNNN	DHHTMYTVHFY	ALENGVAVFATEWGTSQANGD
	BspEgl1	NDNAYKALANDWESEMIRLAMYVGE-	AIENDMYVIVDWHVHAPGDPRD	HIIYELANEPSSNNN	DHHTMYTVHFY	ALENGVAVFATEWGTSQANGD
	BspEgl2	NENAFVALSNDWGSNMIRLAMYIGE-	AFEHDMYVIVDWHVHAPGDPRA	YIIWELANEPSPNNN	AENIMYSVHFY	ALDNGVAVFATEWGTSQANGD
	BsuEgl1	NKDSLKWLRDDWGITVFRAAMYTAD-	AKELGIYVIIDWHILNDGNPNQ	NVIYEIANEPNG---	DANVMDALHFY	ALSKGAPIFVTEWGTSDASGN
	BsuEgl2	NKDSLKWLRDDWGITVFRAAMYTAD-	AKELGIYVIIDWHILNDGNPNQ	NVIYEIANEPNG---	DANVMYALHFY	ALSKGAPIFVTEWGTSDASGN
	BsuEG1	NKDSLKWLRDDWGITVFRAAMYTAD-	AKELGIYVIIDWHILNDGNPNQ	NVIYEIANEPNG---	DANVMYALHFY	ALSKGAPIFVTEWGTSDASGN
	BsuEG2	NKDSLKWLRDDWGITVFRAAMYTAD-	AKELGIYVIIDWHILNDGNPNQ	NVIYEIANEPNG---	DANVMYALHFY	ALSKGAPIFVTEWGTSDASGN
	CacEG1	NYDSMKFLRDKWGVNVIRAAMYTNE-	AIDLNMYVIIDWHILSDNNPNT	NVIYEICNEPNG---	YSNIMYTCHFY	AMSKGIAIFVTEWGTSDASGN
	CjoCelA	VIMPLRLSPRDWGSNVIRLAMYVAA-	AIANDMTLFVDWHVLTPGDPNA	HIIYELANEPSPNDP	DNNTAYSFHFY	ALEHGVAVFCSEWGTSEASGN
	EchCelZ	TADTVASLKKDWKSSIVRAAMGVQES	AIANDMYAIIGWH---SHSAEN	NVIYEIYNEPLQ---	AKNIAYTLHFY	ALNNGIALFVTEWGTVNADGN
	SliCelA	DA-SLDRLAQDWKSDLLRVAMYVQE-	AEDRGMYAVIDFHTLTPGDPNY	NVIYEIANEPNG---	ATNIMYAFHFY	A-ATRLPLFVSEFGTVSATAW
	TfuE5	DS-SLDALAYDWKADIIRLSMYIQE-	ATARGLYVIVDWHILTPGDPHY	NVLYEIANEPNG---	ASNIMYAFHFY	A-SELFPVFVTEFGTETYTGD

Figure 5.13 Amino acid sequence alignment of regions of Family 5 catalytic domains. The sequences have been grouped according to subtypes (Subtypes 1–5) as indicated on the left. Numbering is for only the *Bacillus polymyxa* cellulase (BpoCel) [833]. Residues in bold type denote greater than 70% identity within subtypes, whereas shaded residues are totally conserved for all Family 5 β-glycanases. The sequences correspond to the following: CfiCenD, *Cellulomonas fimi* cellulase D [834]; CflCelA, *C. flavigena* open reading frame [835]; CsaCelB, *Caldocellum saccharolyticum* cellulase B (CELB$CALSA); CthCelB, *Clostridium thermocellum* cellulase B (GUNB$CLTM); XcaCelX, *Xanthomonas campestris* cellulase XCA [836]; BfiCelA, *Butyrivibrio fibrosolvens* A46 cellulase A [837]; BspCelA, BspCelB, and BspCelC, *Bacillus* sp. strain N-4 cellulases A (GUNB$BACS4), B (GUN2$BACS4), and C [827], resepectively; BspEG, *Bacillus* sp. KSM-64 cellulase [838]; BspEgl1, *Bacillus* sp. 1139 cellulase 1 (GUN$BACS1); BspEgl2, *Bacillus* sp. KSM-653 cellulase [839]; BsuEgl1, *B. subtilis* BSE616 cellulase [840]; BsuEgl2, *B. subtilis* N-24 cellulase [841]; BsuEG1, *B. subtilis* PAP115 cellulase [842]; BsuEG2, *B. subtilis* DLG cellulase [843]; CacEgl, *Clostridium acetobutylicum* P262 cellulase (GUN$CLOAB); CjoCelA, *C. josui* cellulase A [844]; EchCelZ, *Erwinia chrysanthemi* cellulase Z (GUNZ$ERWCH); SliCelA, *Streptomyces lividans* 66 cellulase A [446]; TfuE5, *Thermomonospora fusca* cellulase 5 [445], CthCelC, *C. thermocellum* cellulase C (GUNC$CLOTM); CspCelC, *Clostridium* sp. F1 cellulase C307 [845]; FusEgl3, *Fibrobacter succinogenes* cellulase 3 (GUN3$FIBSU); RflCelA, *Ruminococcus flavefasciens* cellulase A [846]; BfiEnd1, *Butyrivibrio fibrisolvens* H17c cellulase 1 [415]; BlaCelB, *Bacillus lautus* cellulase B [847]; BruEG1, *Bacteriodes ruminicola* B$_1$4 cellulase [848]; CceCelA, *Clostridium cellulolyticum* cellulase A [849]; CceEngB and CcEngD, *C. cellulovorans* cellulases B [850] and D [851], respectively; CsaManA, *Caldocellum saccharolyticum* mannanase A [852]; CthCelE and CthCelH, *C. thermocellum* cellulases E (GUNE$CLOTM) and H (GUNH$CLOTM), respectively; RalCelA and RalCelB, *Ruminococcus albus* SY3 cellulases A and B, respectively [853]; RalEgl1, *R. albus* F-40 cellulase (GUN1$RUMAL); RflEndA, *R. flavefasciens* cellulase A [854]; PsoEgl, *Pseudomonas solanacearum* cellulase [855]; BspEgl, *Robillarda* sp. Y-20 cellulase [856]; and TreEGII, *Trichoderma reesei* cellulase II (GUN3$TRIRE) (adapted from Reference [770]).

Subtype	Enzyme					
3	CthCelC	ITEKDIETIAEAGFDHVRLPFD---Y	CKKYNLGLVLDMHHAPGYRFQD	HIAFELLNEVVEPDS	DDYIVYNFHFY	REKKKCKLYCGEFGVIAI-AD
	CspCelC	ITEKDIETIAEAGFDHVRLPFD---Y	CKKYNLGLVLDMHHAPGYRFQD	HIAFELLNEVVEPDS	DDYIVYNFHFY	REKKKCKLYCGEFGVIAI-AD
	FsuEgl3	LGESDVKILADNGFKSLRLPIDLDLY	TAKYNMSFVIDYHE---YD-NS	DLFFELLNEPDMSDG	DDNIIYVIHTY	AATNNVPVIINEFGALNLRST
	RflCelA	VKENDIKQIADWGFDHVRLPID---Y	CRKYGLKLVIDLHKTAGFSFDF	NIVFELLNEVTDEAF	DDRVVYNFHCY	-EKYGTTLYCGEYGVIDVVSA
	BfiEnd1	TTKALIDEVAKAGFNTIRIPVSWGQY	CIVNDMYVILNSHHDINSDYC-	HLVFETMNEPRLVGH	--RLILSVHAY	FLSRNIPVVVGETSATNRNTT
	BlaCelB	ITKELIQNIAAQGYKSIRIPVTWDSH	ALDANLYVMINVHHD-SWL---	KLMFESVNEPRFTDG	DKNLIATVHFY	FVAKGIPVVVGEYGLLGFDKN
	BruEG1	TTQDMMTFLMQNGFNAVRIPVTWYEH	AMNAGLYAIVNVHHDTAAGSGA	HLLFEGYNEM--LDG	NNHLIVQVHSY	FTT--IPYIIGEYGTHGESDI
	CceCelA	TTKQMIDAIKQKGFNTVRIPVSWHPH	CIDNKNYVILNTHHDVDKVKG-	HLIFEGMNEPRLVGH	NNKIIVSVHAY	YTSRGIPVIIGECGAVDKNNL
	CceEngB	TTHAMIDKVKAAGFNTLRLPITWDGH	AFDNNMYVIINLHHEDGWLKP-	YLIFETMNEPRPVGA	DSRVIVSLHMY	FVKNGRAVVIGEMGTINKNNL
	CceEngD	TTHAMINKIKEAGFNTLRLPVTWDGH	AFDNDMYVIINLHHENEWLKP-	HLIFETMNEPRPVGA	DSKVIVSLHMY	FVKNGRAVVIGEMGSINKNNT
4	CsaManA	RLDTALRGIRSWGMNSVRVVLSNGYR	RSLGFKAIILEVHDTTGYGEDG	FVIINIGNEPYGNNN	--NLVFSIHMY	FVDKGLPLVIGEFGHQETDGD
	CthCelE	TTKAMIEKVREMGFNAVRVPVTWDTH	VLDCGMYAIINLHHDNTWIIPT	HLLFETMNEPREVGS	DSRVIVSIHAY	FVKNGRAVIIGEFGTIDKNNL
	CthCelH	AMEYYFDDFKAAGYKNVRIPFRWDNH	SLSRGFVTIINSHHD-DWIKED	NLLFEIMNEP-FGNI	------TFHYY	SDRNNIPVYFGEFAVMAYADR
	RalCelA	TTKEMIDAVYNKGFDVIRIPVTWFFH	AYDDGAYVIINSHHEEDW----	HLIFEGLNEPRVKGS	-DHIGFSIHAY	YLDKDIPVIITEYGAVNKDNN
	RalCelB	TNKFMIDMLPEAGFNVLRIPVSWGNH	GIDDGMYVILNTHHEE-W----	HLIFEGLNEPRLRGE	SDKLIISVHAY	FISRDIPVIVGEFGSMNKDNI
	RalEgl1	TNKYMIDMLPEAGFNVLRIPVSWGNH	GIDNGLYVILNTHHEE-W----	HLIFEGLNEPRLRGE	SDKLIISVHAY	FISKGIPVIVGEFGTMNKENT
	RflEndA	ASQELFDAIKAKGFNTVRIPTTWFQH	AYNIGLYVIINLHHEQNWINR-	HLIFECMNEPRAMDT	DDFIAVSIHAY	FINKDIPVVIGEMGTSDFGNT

Figure 5.13 (continued) Amino acid sequence alignment of regions of Family 5 catalytic domains. The sequences have been grouped according to subtypes (Subtypes 1–5) as indicated on the left. Numbering is for only the *Bacillus polymyxa* cellulase (BpoCel) [833]. Residues in bold type denote greater than 70% identity within subtypes, whereas shaded residues are totally conserved for all Family 5 β-glycanases. The sequences correspond to the following: CfiCenD, *Cellulomonas fimi* cellulase D [834]; CflCelA, *C. flavigena* open reading frame [835]; CsaCelB, *Caldocellum saccharolyticum* cellulase B (CELB$CALSA); CthCelB, *Clostridium thermocellum* cellulase B (GUNB$CLTM); XcaCelX, *Xanthomonas campestris* cellulase XCA [836]; BfiCelA, *Butyrivibrio fibrosolvens* A46 cellulase A [837]; BspCelA, BspCelB, and BspCelC, *Bacillus* sp. strain N-4 cellulases A (GUNB$BACS4), B (GUN2$BACS4), and C [827], resepectively; BspEG, *Bacillus* sp. KSM-64 cellulase [838]; BspEgl1, *Bacillus* sp. 1139 cellulase 1 (GUN$BACS1); BspEgl2, *Bacillus* sp. KSM-653 cellulase [839]; BsuEgl1, *B. subtilis* BSE616 cellulase [840]; BsuEgl2, *B. subtilis* N-24 cellulase [841]; BsuEG1, *B. subtilis* PAP115 cellulase [842]; BsuEG2, *B. subtilis* DLG cellulase [843]; CacEgl, *Clostridium acetobutylicum* P262 cellulase (GUN$CLOAB); CjoCelA, *C. josui* cellulase A [844]; EchCelZ, *Erwinia chrysanthemi* cellulase Z (GUNZ$ERWCH); SliCelA, *Streptomyces lividans* 66 cellulase A [446]; TfuE5, *Thermomonospora fusca* cellulase 5 [445], CthCelC, *C. thermocellum* cellulase C (GUNC$CLOTM); CspCelC, *Clostridium* sp. F1 cellulase C307 [845]; FusEgl3, *Fibrobacter succinogenes* cellulase 3 (GUN3$FIBSU); RflCelA, *Ruminococcus flavefasciens* cellulase A [846]; BfiEnd1, *Butyrivibrio fibrisolvens* H17c cellulase 1 [415]; BlaCelB, *Bacillus lautus* cellulase B [847]; BruEG1, *Bacteriodes ruminicola* B14 cellulase [848]; CceCelA, *Clostridium cellulolyticum* cellulase A [849]; CceEngB and CcEngD, *C. cellulovorans* cellulases B [850] and D [851], respectively; CsaManA, *Caldocellum saccharolyticum* mannanase A [852]; CthCelE and CthCelH, *C. thermocellum* cellulases E (GUNE$CLOTM) and H (GUNH$CLOTM), respectively; RalCelA and RalCelB, *Ruminococcus albus* SY3 cellulases A and B, respectively [853]; RalEgl1, *R. albus* F-40 cellulase (GUN1$RUMAL); RflEndA, *R. flavefasciens* cellulase A [854]; PsoEgl, *Pseudomonas solanacearum* cellulase [855]; RspEgl, *Robillarda* sp. Y-20 cellulase [856]; and TreEGII, *Trichoderma reesei* cellulase II (GUN3$TRIRE) (adapted from Reference [770]).

Subtype	Sequence					
5	PsoEG1	SADSVT-YYKNKGMNLVRLPFRWERL	T-ATGQTVLLDPHNYARYY---	-VILFLMNEPNSMPT	--NLVFEVHQY	LRSNGYRGFLGEFGAASNDTC
	RspEG1	SEAAVDVHVDQNHKNKFRVAFKKERN	TVTKGAYAILDPHNYMRYNVLA	----FWAHERATRHG	--NTAIDIHEY	LKENNLKAFITEFGGSNSTSC
	TreEglII	IGQMQH-FVNEDGMTIFRLPVGWQYL	L-SLGAYCIVDIHNYARWN---	-VWFGIMNEPHDVNI	--NLIFDVHKY	LRQNNRQAILTETGGGNVQSC

Figure 5.13 (continued) Amino acid sequence alignment of regions of Family 5 catalytic domains. The sequences have been grouped according to subtypes (Subtypes 1–5) as indicated on the left. Numbering is for only the *Bacillus polymyxa* cellulase (BpoCel) [833]. Residues in bold type denote greater than 70% identity within subtypes, whereas shaded residues are totally conserved for all Family 5 β-glycanases. The sequences correspond to the following: CfiCenD, *Cellulomonas fimi* cellulase D [834]; CflCelA, *C. flavigena* open reading frame [835]; CsaCelB, *Caldocellum saccharolyticum* cellulase B (CELB$CALSA); CthCelB, *Clostridium thermocellum* cellulase B (GUNB$CLTM); XcaCelX, *Xanthomonas campestris* cellulase XCA [836]; BfiCelA, *Butyrivibrio fibrosolvens* A46 cellulase A [837]; BspCelA, BspCelB, and BspCelC, *Bacillus* sp. strain N-4 cellulases A (GUNB$BACS4), B (GUN2$BACS4), and C [827], resepectively; BspEG, *Bacillus* sp. KSM-64 cellulase [838]; BspEgl1, *Bacillus* sp. 1139 cellulase 1 (GUN$BACS1); BspEgl2, *Bacillus* sp. KSM-653 cellulase [839]; BsuEgl1, *B. subtilis* BSE616 cellulase [840]; BsuEgl2, *B. subtilis* N-24 cellulase [841]; BsuEG1, *B. subtilis* PAP115 cellulase [842]; BsuEG2, *B. subtilis* DLG cellulase [843]; CacEgl, *Clostridium acetobutylicum* P262 cellulase (GUN$CLOAB); CjoCelA, *C. josui* cellulase A [844]; EchCelZ, *Erwinia chrysanthemi* cellulase Z (GUNZ$ERWCH); SliCelA, *Streptomyces lividans* 66 cellulase A [446]; TfuE5, *Thermomonospora fusca* cellulase 5 [445], CthCelC, *C. thermocellum* cellulase C (GUNC$CLOTM); CspCelC, *Clostridium* sp. F1 cellulase C307 [845]; FusEgl3, *Fibrobacter succinogenes* cellulase 3 (GUN3$FIBSU); RflCelA, *Ruminococcus flavefasciens* cellulase A [846]; BfiEnd1, *Butyrivibrio fibrisolvens* H17c cellulase 1 [415]; BlaCelB, *Bacillus lautus* cellulase B [847]; BruEG1, *Bacteriodes ruminicola* B₁4 cellulase [848]; CceCelA, *Clostridium cellulolyticum* cellulase A [849]; CceEngB and CcEngD, *C. cellulovorans* cellulases B [850] and D [851], respectively; CsaManA, *Caldocellum saccharolyticum* mannanase A [852]; CthCelE and CthCelH, *C. thermocellum* cellulases E (GUNE$CLOTM) and H (GUNH$CLOTM), respectively; RalCelA and RalCelB, *Ruminococcus albus* SY3 cellulases A and B, respectively [853]; RalEgl1, *R. albus* F-40 cellulase (GUN1$RUMAL); RflEndA, *R. flavefasciens* cellulase A [854]; PsoEgl, *Pseudomonas solanacearum* cellulase [855]; RspEgl, *Robillarda* sp. Y-20 cellulase [856]; and TreEGII, *Trichoderma reesei* cellulase II (GUN3$TRIRE) (adapted from Reference [770]).

through affinity labelling experiments using an epoxyalkyl cellobioside [752].

A highly conserved region involving an Asn-Glu-Pro motif has been observed among the aligned sequences of both Family 1 [520] and Family 5 cellulases [786]. This prompted the prediction that the Glu may be essential for catalytic activity, and the homologous region was extended for both the cellulases and β-glucosidases to include conserved Arg and His residues [520]. Site-directed mutagenesis of the genes coding the Family 5 cellulases from *Bacillus polymyxa*, *B. subtilis* [786], *C. thermocellum* [787], and *Erwinia chrysanthemi* [788] has confirmed the essential role for the Glu residue. Change of Glu194 and Glu169 of the *B. polymyxa* and *B. subtilis* enzymes, respectively, to the isosteric glutamine form resulted in a dramatic, but not complete, loss of catalytic activity (>95% loss of initial activity) [786]. However, mutation of the corresponding Glu140 and Glu133 of the *C. thermocellum* [787] and *E. chrysanthemi* [788] cellulases, respectively, to Ala essentially abolished catalytic activity. Having already established the identity of the catalytic nucleophile, this would suggest that the Glu of the Asn-Glu-Pro motif of the Family 5 cellulases and, although untested, the Family 1 β-glucosidases participate as the acid catalyst in the double-displacement mechanism of action of these retaining enzymes.

With respect to the conserved Arg and His residues, data from the site-directed mutagenesis of the *Clostridium cellulolyticum* cellulase A gene led to the conclusion that the Arg residue either participates in a salt bridge that maintains the structural integrity of the active enzyme or hydrogen bonds to the substrate transition state [789]. The His residue was speculated to be more essential for catalysis, but not as the proton donor because site-directed mutations of both the *C. cellulolyticum* [789] and *C. thermocellum* [736] genes substituting the His residue with a variety of amino acids did not totally abolish catalytic activity.

A three-dimensional structure of a Family 1 or 5 β-glycosidase has yet to be solved, but two preliminary reports concerning the crystallization and preliminary X-ray analyses of the cellulases from *C. cellulolyticum* [790] and *C. thermocellum* [791] have been described. Nevertheless, patient and careful biochemical studies have apparently identified both the acid catalyst and the catalytic nucleophile-stabilizing anion of these enzymes.

5.3.1.2 Family 3

Alignment of the Family 3 enzymes, most of which are β-glucosidases, reveals a number of candidate Asp and Glu residues to serve as the catalytic residues. Unfortunately, very little biochemical analysis has been conducted with these enzymes, but investigations have involved the affinity labelling of the β-glucosidase A_3 from *Aspergillus wentii* with conduritol-B epoxide

[754–756]. The reagent was found to esterify an Asp residue that has since been shown to be conserved throughout the family [792], thus suggesting its function as the catalytic nucleophile. These findings have not been followed up by any further characterization of other Family 3 enzymes, and yet ironically, these experiments represent the first in which an active-site peptide was obtained from a cellulolytic β-glycosidase!

5.3.1.3 Family 7

This small family of β-glycosidases comprises seven cellobiohydrolases and one cellulase, all of fungal origin. As with the Family 3 enzymes, very little is known concerning the structure and function relationship of the Family 7 cellulolytic enzymes. An investigation of the *Trichoderma reesei* cellobiohydrolase I involving chemical modification identified Glu126 as the acid catalyst [738], but replacement of this residue with Gln by site-directed mutagenesis resulted in a mutant enzyme that retained approximately 30% of its catalytic activity relative to wild type [793]. Surprisingly, the replacement of the homologous residue, Glu127, on the *T. reesei* cellulase I did abolish activity [793], and amino acid sequence alignments do reveal that it is totally conserved among the Family 7 enzymes [794]. Nevertheless, the experiments with the cellobiohydrolase suggest that it is unlikely that this residue participates directly in the catalytic mechanism of the enyzmes. It is possible that the loss of catalytic activity associated with the site-directed mutagenesis of the cellulase I was not the direct result of the Glu to Gln replacement but arose from overglycosylation of the expressed enzyme mutant in a yeast host [793].

The catalytic domains of these two enzymes from *T. reesei* have been crystallized [795], and the three-dimensional crystal structure of the cellobiohydrolase I has been determined and refined to 1.8 Å resolution [796]. This catalytic domain consists of two large antiparallel β-sheets that stack to form, a β-sandwich, four short α-helices and a number of loops that connect the β-strands (Figure 5.14). The two β-sheets are highly curved and oppose each other to form a 40-Å-long tunnel that comprises the active site. It is estimated that the active-site tunnel can accommodate seven glucose residues (subsites A–G) and two highly conserved acidic residues, Glu212 and Glu217, are positioned between sites B and C to function as the catalysts. This confirms the interpretation of the site-directed mutagenesis studies described above regarding the nonessential role of Glu126 (Glu126 is, in fact, observed exposed to solvent approximately 20 Å from the active-site tunnel!). Glu212 is in close contact with Asp214 and His228, indicating that it would remain ionized to serve as the catalytic nucleophile. The proximity of Glu217 to the 0–4 of bound cellobiose suggests its role as the acid catalyst [796].

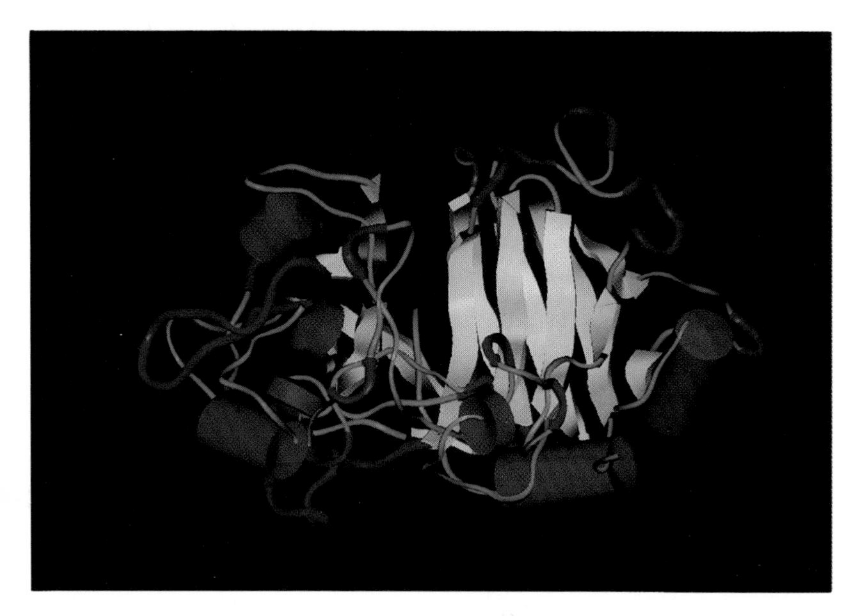

Figure 5.14 Three-dimensional crystal structure (1.81 Å) of cellobiohydrolase I from *Trichoderma reesei.* The β-sheets are depicted in yellow, the α-helices are red, whereas the β-turns and random coil are blue and green, respectively. The figure was generated using the data deposited in the Brookhaven Protein Bank (1cel) and visualized using Insight ver. 2.3.0 software (Biosym Technologies, San Diego).

5.3.1.4 Family 10

Four Asp and two Glu residues are conserved among all twenty of the β-glycosidases assigned to Family 10 (Figure 5.15). This family primarily comprises xylanases, but a cellobiohydrolase and a bifunctional xylanase-cellobiohydrolase are also members.

The three-dimensional structures of the *Streptomyces lividans* xylanase A [797] and the *Cellulomonas fimi* bifunctional xylanase-cellobiohydrolase Cex [798] have independently been solved to 2.6 and 1.8 Å resolution, respectively. Perhaps not too surprising, their structures are remarkably similar (Figure 5.16). Both enzymes form α- and β-barrels that contain elliptical cores of eight parallel β-strands. This folding pattern is unique among the other known structures of β-glycanases. The active sites are located in open clefts situated on the carboxy-terminal ends of the β-barrels. Site-directed mutagenesis of the genes for the xylanases from *Streptomyces lividans* [799] and *Thermoanaerobacterium saccharolyticum* [795] had previously identified the same homologous Glu residues as possible candidates for catalytic roles, and, indeed, they are located at the active site. The

CfiCex	NHVTKVADHFEGK------VASWDVVNEAFA	VDVRITELDIRM
SliXynA	DHINGVMAHYKGK------IVQWDVVNEAFA	VDVAITELDIQ-
CthXynX	THITTVLDHFKTKYGAQNPIIGWDVVNEVLD	VEIQVTELDMNM
CthXynZ	NHITTVMTHYKGK------IVEWDVANECMD	VIVSFTEIDIRI
AkaXynA	NHITTVMQHYKGK------IYAWDVVNEIF-	KEIAVTELDI--
PchXyn	NHITTVMKQYKGK------LYAWDVVNEIF-	EEVAVTELDI--
RflXynA	SMIKNTFAALKSQYPNL-DVYSYDVCNELFL	LEVQITELDITC
BfiXynA	FYVKSVMGHFYSGKTGST-LVYWDVCNETL-	FEVQITELDITN
BfiXynB	SYIHGVLDFVQTNYPGI--IYAWDVVNE-IV	LQIHITELDMH
TBXynA	KHIQTVVGRYKGK------VYAWDVVNEAID	LEIHFTEIDISI
CsaXynA	EHIKTLCERYK------DVVYAWDVVNEAVE	LEIHITELDISV
CsaCelB	QYIYDVVGRYKGK------VYAWDVVNEAID	IEIHITELDMSL
CsaORF4	SYIKQVIEFCQKNYPGV--VYCWDVVNEAIL	LQIHITELNFEI
TsaXylA	THITTVLDHFKTKYGSQNPIIGWDVVNEVLD	VEIQVTELDMNM
BspXynA	NHIKTVVERYK------DDVTSWDVVNEVID	LDNQVTELDMSL
CstXyn	NYIRAVVLRYK------DDIKSWDVVNENIE	LDNIITELDMSI
PflXynA	RHIDTVAAHFAGQ------VKSWDVVNEALF	LKIKITELDVRL
PflXynB	QWIRDYCARYP-------DTAMIDVVNEAVP	KPIYISEYDIGD
CalXynA	NHIDNVIGRYK------DDLAYFDIVNEPL-	-EVPMTELDVRI

Figure 5.15 Alignment of highly conserved regions within deduced amino acid sequences of Family 10 β-glycosidases. Residues in bold type denote identity for at least ten of the nineteen sequences, whereas shaded residues are totally conserved for all Family 10 enzymes. The sequences correspond to (accession numbers for the sequences from either Genbank or SWISS-PROT are provided in parentheses) the following: CfiCex, *Cellulomonas fimi* xylanase/cellobiohydrolase (L11080); SliXynA, *Streptomyces lividans* xylanase A (M54551); CthXynX and CthXynZ, *Clostridium thermocellum* xylanases X (M67438) and Z (M22624), respectively; AkaXynA, *Aspergillus kawachii* xylanase A (D14847); PchXyn, *Penicillium chrysogenum* xylanase (S31307); RflXynA, *Ruminococcus flavefaciens* xylanase A (P29126); BfiXynA and BfiXyB, *Butyrivibrio fibrisolvens* xylanases A (P23551) and B (X61495, S55274), respectively; TBXyn, thermophilic bacterium strain rt8.84 xy-lanase (L18965); CsaXynA, CsaXynB, and CsaORF, *Caldocellum saccharolyticum* xylanases (M34459), cellulase B (A43802, X13602) and open reading frame 4 (M34459), respectively; TsaXylA, *Thermoanaerobacter saccharolyticum* xylanase A (M97882); BspXynA, *Bacillus* sp. C-125 xylanase A (D00087, Po7528); CstXyn, *Clostridium stercorarium* F9 xylanase (D12504); PflXynA and PflXynB, *Pseudomonas fluorescens* subsp. *cellulosa* xylanases A (X15429) and B (P23030), respectively; and CalXynA, *Cryptococcus albidus* xylanase A (JS0734) (adapted from Reference [800]).

carboxyl groups of Glu127 and Glu233 (*C. fimi* numbering) face together across the cleft and are separated by 5.5 Å. Separations of this magnitude are optimal for retaining glycosidases; they permit the protonation of the departing aglycon by the acid catalyst and the concurrent formation of the glycosyl-enzyme intermediate on the nucleophilic carboxylate.

The specific functions of the two active-site Glu residues have been assigned by mechanism-based inhibition studies and site-directed mutagenesis. Inhibition studies with 2′,4′-dinitrophenyl-2-deoxy-2-fluoro-β-D-glucoside led to the identification of Glu233 of the *C. fimi* xylanase-cellobiohydrolase as the active-site nucleophile [772]. Although it would be safe to postulate that the second active-site residue, Glu127, participates as the acid catalyst, an elegant study of a mutant enzyme, in fact, confirmed this role [800]. According to the double-displacement mechanism

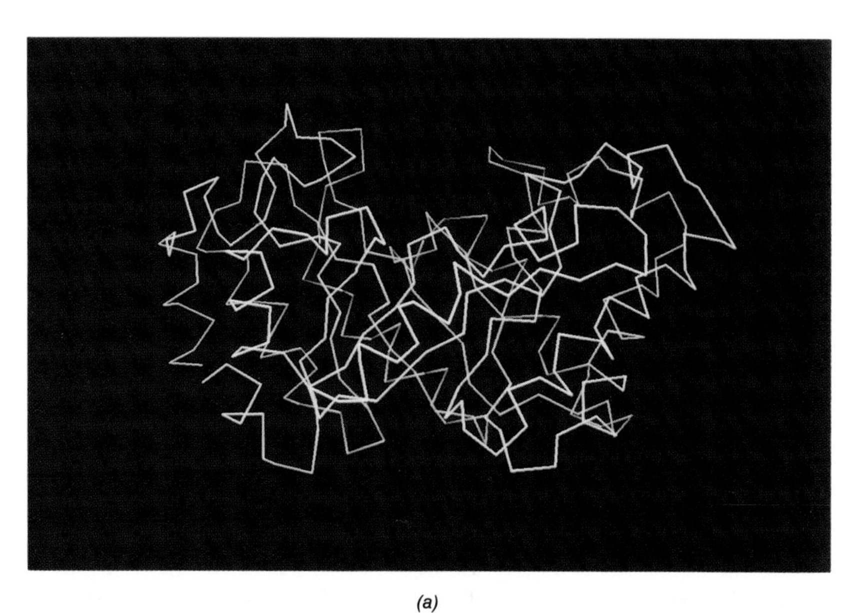

(a)

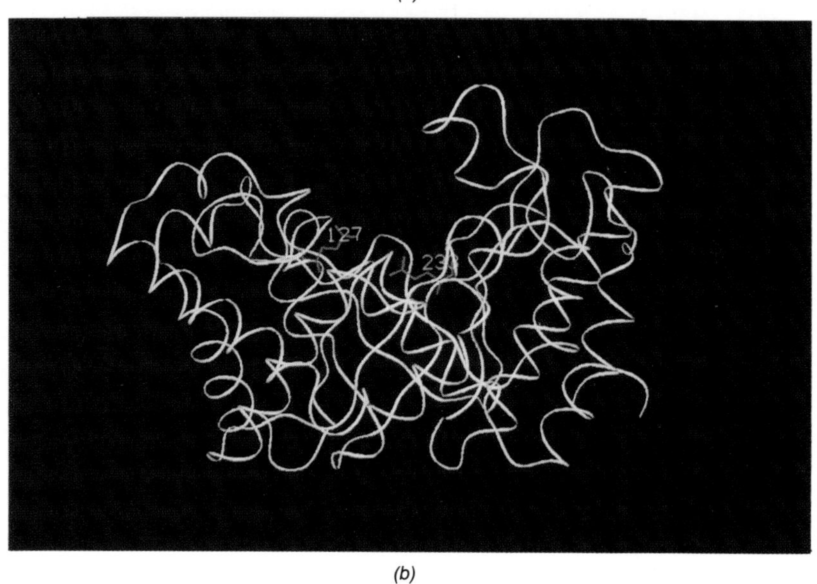

(b)

Figure 5.16 Three-dimensional crystal structures (α-carbon chain) of (a) *Streptomyces lividans* xylanase A and (b) the *Cellulomonas fimi* bifunctional xylanase-cellobiohydrolase Cex. The side chains of the putative catalytic residues, Glu127 and Glu 233, of *C. fimi* Cex are depicted in red. The figures were generated using the data deposited in the Brookhaven Protein Bank (1xas and 2exo, respectively) and visualized using Insight ver. 2.3.0 software (Biosym Technologies, San Diego).

for retaining enzymes as described earlier (Figure 5.3), the acid catalyst participates by first protonating the glycosidic oxygen and then functions as a general base catalyst, removing a proton from water in a concerted manner as the water attacks the anomeric carbon of the covalent glycosyl-enzyme intermediate. Mutation of the acid catalyst would thus affect the rates of both steps in a normal enzymatic reaction, and such was observed with the Glu127Ala mutant of the *C. fimi* xylanase-cellobiohydrolase. However, addition of substrates that do not require protonic activation to the Glu127Ala mutant enzyme led to an accumulation of the covalent glycosyl-enzyme intermediate (because the general base catalyst for deglycosylation of product had been removed). If Glu127 is the acid catalyst, its replacement with Ala would leave a cavity close to the β-face of the substrate that could be filled with small anionic nucleophiles, such as azide. These anions would react in place of water without the need of base catalysis to release the reaction product. With azide serving as the anionic nucleophile, the reaction product would be β-glucosyl azide, and this was indeed recovered [800].

For Glu127 of the *C. fimi* xylanase-cellobiohydrolase to function as the acid-base catalyst, it must remain protonated, and three highly conserved residues, Asn126, Gln203, and Trp84, are thought to be positioned to provide the appropriate environment. The catalytic nucleophile Glu233 is observed to be within hydrogen-bonding distance of His205, which, in turn, interacts with Asp235 [798]. These two latter residues, together with Asn169, which also forms a hydrogen bond to Glu233, are highly conserved among the Family 10 enzymes, and it is likely they maintain the ionization state of the nucleophile.

5.3.1.5 Family 11

This family is comprised of only xylanases, but of both bacterial and fungal origin. A number of amino acid residues appear to be totally conserved among the Family 11 xylanases, including two Glu and eight aromatic residues (Figure 5.17). Chemical modification of the *Schizophyllum commune* xylanase A confirmed the essential role of at least one acidic residue at the active site [644], and it was subsequently identified as Glu87 [740], one of the two completely conserved acidic residues. Crystals of the xylanases from *Bacillus circulans* [801], *B. pumilus* [802], *Trichoderma reesei* (both Xyn I and Xyn II) [803], and *T. harzianum* [804] [805–807] have been prepared and the three-dimensional X-ray crystal structures have been solved at resolutions of 1.8 [801], 2.2 [808,809], 2.0 [810,811], and 1.8 Å, respectively. Secondary structure predictions and hydrophobic cluster analysis had revealed extensive similarities between the Family 11 enzymes, suggesting that their tertiary structures would share many common features [644,812], and this is substantiated by comparison of the recently released

```
An       1      SAGINYVQNYNGNLGD------FTYDESAG-TFSMYWEDGV-SSDFVV--GLGWTTGSSK---------AITY
At       1      SAGINYVQNYNGNLGD------FTYDESAG-TFSMYWEDGV-SSDFVV-GLGGWTTGSSN---------AITY
Bc       1          ASTDYWQNWTD-GGG-IVNAVNGSGGNYSVNWS-N--TGNFV--VGKGWTTGSPFYT--------INY
Bp       1  RTITNNEMGNHSGYDY-ELWKD-YGNT--SMTLNNGGAFSAGWN-N--IGNAL--FRKGKKFDST-RTHHQLGNISINY
Bs       1          ASTDYWQNWTD-GGG-IVNAVNGSGGNYSVNWS-N--TGNFV--VGKGWTTGSPFRT--------INY
Ca      32  KTITSNEIGVNGGYDY-ELWKD-YGNT--SMTLKNGGAFSCQWS-N--IGNAL--FRKGKKFNDT-QTYKQLGNISVNY
Fs(a)   14  VTITSNQTGKIGDIGY-ELWDENGHG-GSATFYSDG-SMDC----NIT-GAKDYLCRAGLSLGS-NKTYKELGGDMIAE
Fs(b)  291  NSSVTGNVG-SSPYHY-EIWYQGG--NNSMTFYDNG-TYKASW--NGT-NDF--LARVGFKYD-EKHTYEELGPI-DAY
Np(a)    5  GNGQTQHKGVADGYSY-SIWLDNTGG-SGSMTLGSGATFKAEWNASVNRGNF--LARRGLDFGSQKKATD-YSYIGLDY
Np(b)  280  GNGQNQHKGVNDGFSY-EIWLDNTGG-NGSMTLGSGATFKAEWNAAVNRGNF--LARRGLDFGSQKKATD-YDYIGLDY
Rf       1  SAADGGTRGNVGGYDY-EMWNQNGGG-QASMNPGAGSFTTCSWS-NIE--NF--LARMGKNYDSQKKNYKAFGNIVLTY
Sc       1   SGTPSSTGTDGGYYY-SWWTD-GAG-DATYQNNGGSYTLTWSGN--NGNLV--GGKGWNPGAASRS--------ISY
Sl(C)    2  TTITTNQTGTDNGM-YYSFWTD-GGG-SVSMTLNGGGSYSTQWT-N--CGNFV--AGKGWANGGR-RT--------VRY
Sl(B)    2  TVVTTNQEGTNNGY-YYSFWTD-SQG-TVSMNMGSGGQYSTSWR-N--TGNFV--AGKGWANGGR-RT--------VQY
Ss       1  TTITTNQTGYD-GM-Y-SFWTD-GGG-SVSMTLNGGGSYSTRWT-N--CGNFV--AGKGWANGGR-RT--------VRY
Th       1   QTIGPGTGYSNGY-YYSYWND-GHA-GVTYTNGGGGSFTVNWS-N--SGNFV--GGKGWQPGTKNKV--------INF
Tr(II)   1   QTIQPGTGYNNGYFY-SYWND-GHG-GVTYTNGPGGQFSVNWS-N--SGNFV--GGKGWQPGTKNKV--------INF
Tr(I)    1        ASINYDQNYQTGGQV------SYSPSNTG--FSVNW--N--QDDFVV--GVGWTTGSSAP---------INF
Tv       1   QTIGPGTGFNNGYFY-SYWND-GHG-GVTYTNGPGGQFSVNWS-N--SGNFV--GGKGWQPGTKNKV--------INF
```

Figure 5.17 Amino acid sequence alignment of Family 11 β-glycosidases (xylanases). Residues in bold type indicate identity among at least ten of the nineteen aligned sequences, whereas shaded residues identify complete conservation. The sequences correspond to the xylanases from the following: An, *Aspergillus niger* awamori [857]; At, *A. tubigensis* [858]; Bc, *B. circulans* [859]; Bp, *B. pumilus* [860]; Bs, *B. subtilis* [861]; Ca, *Clostridium acetobutylicum* [862]; Fs(a) and Fs(b), *Fibrobacter succinogenes* S85—domains A and B, respectively [239]; Np(a) and Np(b), *Neocallimastix patriciarum*—domains A and B, respectively [237]; Rf, *Ruminococcus flavefaciens* [863]; Sl(C) and Sl(B), *Streptomyces lividans* xylanases C and B, respectively [864]; Ss, *Streptomyces* sp. 36a [865]; Th, *Trichoderma harzianum* [866]; Tr(I) and Tr(II), *T. reesei* xylanases I and II, respectively [867]; Tv, *T. viride* [868].

```
An       55  SAEY---SASGSSSYLAVYGWVNY------PQAEYYIVEDYGDYNPCSSAT--SLGTVYSDGSTYQVCTDTRTNEPSIT
At       55  SAEY---SASGSASYLAVYGWVNY------PQAEYYIVEDYGDYNPCSSAT--SLGTVYSDGSTYQVCTDTRTNEPSIT
Bc       54  NAGVW---APNGNGYLTLYGWTRS------PLIEYYVVDSWGTYRP-TGTY--K-GTVKSDGGTYDIYTTTRYNAPSID
Bp       70  NASFN----PSGNSYLCVYGWTQQ------PLAEYYIVDSWGTYRP-TGAY--K-GSFYADGGTYDIYETTRVNQPSII
Bs       54  NAGVW---APNGNGYLTLYGWTRS------PLIEYYVVDSWGTYRP-TGTT--K-GTVKSDGGTYDIYTTTRYNAPSID
Ca       71  DCNYQ----PYGNSYLCVYGWTQS------PLVEYYIVDSWGTWRPPGGTS--K-GTITVDGGIYDIYETTRINQPSIQ
Fs(a)    83  FKLVKSGAQNVGYSYIGIYGWMEGVSGTPSQLVEYYVIDNTLANMPGSWIGNERKGTITVDGGTYIVYRNTRT-GPAIK
Fs(b)   356  YKWSKQ-GSAGGYNYIGIYGWTVD------PLVEYYIVDDW-FNKPGANLLGQRKGEFTVDGDTYEIWQNTRVQQPSIK
Np(a)    79  TATYRQTGSASGNSRLCVYGWFQNRGVQGVPLVEYYIIEDWVDWVPDAQ-G--RMVTI--DGAQYKIFQMDHT-GPTIN
Np(b)   354  AATYKQTASASGNSRLCVYGWFQNRGLNGVPLVEYYIIEDWVDWVPDAQ-G--KMVTI--DGAQYKIFQMDHT-GPTIN
Rf       72  DVEYT----PRGNSYMCVYGWTRN------PLMEYYIVEGWGDWRPPGNDGEVK-GTVSANGNYYDIRKTMRYNQPSID
Sc       64  SGTYQ----PNGNSYLSVYGWTRS------SLIEYYIVESYGSYDPSSAAS-HK-GSVTCNGATYDILSTWRYNAPSID
Sl(C)    62  NGYFN----PVGNGYGCLYGWTSN------PLVEYYIVDNWGSYRP-TGTY--K-GTVTSDGGTYDIYQTTRYNAPSVE
Sl(B)    64  SGSFN----PSGNAYLALYGWTSN------PLVEYYIVDNWGTYRP-TGEY--K-GTVTSDGGTYDIYKTTRVNKPSVE
Ss       62  TGWFN----PSGNGYGCLYGWTSN------PLVEYYIVDNWGSYRP-TGEY--R-GTVHSDGGTYDIYKTTRYNAPSVE
Th       63  SGSYN----PNGNSYLSIYGWSRN------PLIEYYIVENFGTYNPSTGAT--KLGEVTSDGSVYDIYRTQRVNQPSII
Tr(II)   63  SGSYN----PNGNSYLSVYGWSRN------PLIEYYIVENFGTYNPSTGAT--KLGEVTSDGSVYDIYRTQRVNQPSII
Tr(I)    51  GGSF---SVNSGTGLLSVYGWSTN------PLVEYYIMEDNHNY-PAQGT--VK-GTVTSDGATYTIWENTRVNEPSIQ
Tv       63  SGSYN----PNGNSYLSVYGWSRN------PLIEYYIVENFGTYNPSTGAT--KLGEVTSDGSVYDIYRTQRVNQPSII
```

Figure 5.17 (continued) Amino acid sequence alignment of Family 11 β-glycosidases (xylanases). Residues in bold type indicate identity among at least ten of the nineteen aligned sequences, whereas shaded residues identify complete conservation. The sequences correspond to the xylanases from the following: An, *Aspergillus niger* awamori [857]; At, *A. tubigensis* [858]; Bc, *B. circulans* [859]; Bp, *B. pumilus* [860]; Bs, *B. subtilis* [861]; Ca, *Clostridium acetobutylicum* [862]; Fs(a) and Fs(b), *Fibrobacter succinogenes* S85—domains A and B, respectively [239]; Np(a) and Np(b), *Neocallimastix patriciarum*—domains A and B, respectively [237]; Rf, *Ruminococcus flavefaciens* [863]; Sl(C) and Sl(B), *Streptomyces lividans* xylanases C and B, respectively [864]; Ss, *Streptomyces* sp. 36a [865]; Th, *Trichoderma harzianum* [866]; Tr(I) and Tr(II), *T. reesei* xylanases I and II, respectively [867]; Tv, *T. viride* [868].

```
An      122  --GTS-TFTQYFSVRESTRTSG-----------TVTVANHFNFWAQHGFGNSDF-NYQVMA-VEAWSGAGSASVTISS
At      123  --GTS-TFTQYFSVRESTRTSG-----------TVTVANHFNFWAHHGFGNSDF-NYQVVA-VEAWSGAGSASVTISS
Bc      120  --GDRTTFTQYWSVRQSKRPTGSN---------ATITFTNHVNAWKSHGMNLGSNWAYQVMA-TEGYQSSGSSNVTVW
Bp      135  --GIA-TFKQYWSVRQTKRTSG-----------TVSVSAHFRKWESLGMPMGKM--YETAFTVEGYQSSGSANVMTNQLFIG
Bs      120  --GDRTTFTQYWSVRQSKRPTGSN---------ATITFSNHVNAWKSHGMNLGSNWAYQVMA-TEGYQSSGSSNVTVW
Ca      137  --GN-TTFKQYWSVRRTKRTSG-----------TISVSKHFAAWESKGMPLGKMH--ETAFNIEGYQSSGKADVNSMSINIG
Fs(a)   161  NSGNV-TFYQYFSVRTSPRDCG-----------TINISEHMRQWEKMGLTMGK--LYEAKVLGEAGNVNGEVRGGHMDFPHA
Fs(b)   434  --GTQ-TFPQYFSVRKSARSCG-----------HIDITAHMKKWEELGMKMGKM--YEAKVLVEAGGGSGSFDVTYFKMTDK
Np(a)   152  -GGSE-TFKQYFSVRQQKRTSG-----------HITVSDHFKEWAKQGWGIGN--LYEVALNAEGWQSSGIADVTKLDVYTT
Np(b)   427  -GGSE-TFKQYFSVRQQKRTSG-----------HITVSDHFKEWAKQGWGIGN--LYEVALNAEGWQSSGVADVTLLDVYTT
Rf      140  --GTA-TFPQYWSVRQ---TSGSANNQTNYMKGTIDVSKHFDAWSAAGLDMSGT-LYEVSLNIEGYRSNGSANVKSVSV
Sc      131  --GTQ-TFEQFWSVRNPKKAPG------GSISGTVDVQCHFDAWKGLGMNLGSEHNYQIVA-TEGYQSSGTATITV
Sl(C)   127  --GTK-TFQQYWSVRQSKVTSGS---------GTITTGNHFDAWARAGMNMGQFRYYMIMA-TEGYQSSGSSNITVSGGTGG
Sl(B)   129  --GTR-TFDQYWSVRQSKRT-G----------GTITTGNHFDAWARAGMPLGNFSYYMIMA-TEGYQSSGTSSINVGGTGGG
Ss      127  AP-AA--FDQYWSVRQSKVTSGS---------GTITTGNHFDAWARAGMNMGNFRYYMIMA-TEGYQSSGSSTITVSG
Th      130  --GTA-TFYQYWSVRRNHRSSGS-----------VNTANHFNAWASHGLTLGTM-DYQIVA-VEGYFSSGSASITV
Tr(II)  130  --GTA-TFYQYWSVRRNHRSSGS-----------VNTANHFNAWAQQGLTLGTM-DYQIVA-VEGYFSSGSASITVS
Tr(I)   117  --GT-ATFNQYISVRNSPRTSG-----------TVTVQNHFN-WASLGLHLGQMMNYQVVA-VEGWGGSGSASQSVSN
Tv       98  --GTA-TFYQYWSVRRTHRSSGS-----------VNTANHFNAWAQQGLTLGTM-DYQIVA-VEGYFSSGSASITV
```

Figure 5.17 (continued) Amino acid sequence alignment of Family 11 β-glycosidases (xylanases). Residues in bold type indicate identity among at least ten of the nineteen aligned sequences, whereas shaded residues identify complete conservation. The sequences correspond to the xylanases from the following: An, *Aspergillus niger* awamori [857]; At, *A. tubigensis* [858]; Bc, *B. circulans* [859]; Bp, *B. pumilus* [860]; Bs, *B. subtilis* [861]; Ca, *Clostridium acetobutylicum* [862]; Fs(a) and Fs(b), *Fibrobacter succinogenes* S85—domains A and B, respectively [239]; Np(a) and Np(b), *Neocallimastix patriciarum*—domains A and B, respectively [237]; Rf, *Ruminococcus flavefaciens* [863]; Sl(C) and Sl(B), *Streptomyces lividans* xylanases C and B, respectively [864]; Ss, *Streptomyces* sp. 36a [865]; Th, *Trichoderma harzianum* [866]; Tr(I) and Tr(II), *T. reesei* xylanases I and II, respectively [867]; Tv, *T. viride* [868].

structures. Each enzyme is composed of one small α-helix and two β-sheets (primarily antiparallel) that fold in a sandwich-like manner to a shape that has been likened to a right hand (Figure 5.18). The two β-sheets form the fingers, and a short section of a twisted β-sheet together with the α-helix form the palm. An extended loop of residues constitues the thumb, and a loop between two β-strands forms a cord across the cleft on one side.

The active-site cleft lies across the open palm with the two conserved Glu residues suitable disposed opposite each other to participate as the catalysts. Glu93 of the *B. pumilus* xylanase is salt bridged to Arg127 [808], whereas the homologous Glu78 of the *B. circulans* enzyme is within the vicinity of the positive charge contributed by Arg112 [801]. These Arg residues are also totally conserved in the Family 11 enzymes. The proximity of this charged residue will have the effect of lowering the pK_a values of the respective Glu residues, and, therefore, strongly implicates them as the stabilizing anion-nucleophile. Mechanism-based inhibition studies with the *B. subtilis* xylanase have subsequently confirmed that this conserved Glu residue is indeed the catalytic nucleophile [773].

There is no direct evidence for the assignment of the acid catalyst, but the distance between the catalytic nucleophile Glu78 and Glu172 of the *B. circulans* xylanase is 5.4Å [801], which is expected for a retaining enzyme. The roles of the two identified Glu residues in the *B. circulans* [801] and *B. pumilus* [809] xylanases have been further probed by site-directed mutagenesis in combination with both kinetic and X-ray crystallographic analyses. For both enzymes, kinetic analyses showed that enzyme activity is greatly affected by replacements at Glu78/93 and Glu172/182. By monitoring FTIR spectra during a pH titration, the pK_a of an essential amino acid on the *B. circulans* xylanase was determined to be 6.8 [717]. The titration curve was absent with site-directed mutant forms of the enzyme in which either Glu78 or Glu172 was replaced by Gln. On the basis of these data and that obtained from X-ray crystallography, Glu172 was ascribed the pK_a of 6.8 and identified as the catalytic acid. The high pK_a value was thought to arise from the close proximity of the nearby negatively charged side chain of the catalytic nucleophile Glu78.

An analysis of the hydrogen-bonding network in the *T. reesei* xylanase II crystal structure around Glu86 (homologous to Glu78 of *B. circulans* xylanase) indicated that many of the residues are conserved [810]. The cluster of amino acids that directly hydrogen bond to Glu86, viz., Gln136, Tyr77, and Tyr88 (*T. reesei* numbering) are totally conserved (Figure 5.17), as are some of the amino acids involved in a second shell of hydrogen bonding (Trp79 to Tyr77 and Tyr171 to Tyr77). Interestingly, the residues surrounding Glu177, the catalytic acid, are not as well conserved. Hence, it would appear that the preservation of a precise conformation of the catalytic nucleophile is most important for catalysis and that perhaps the greatest

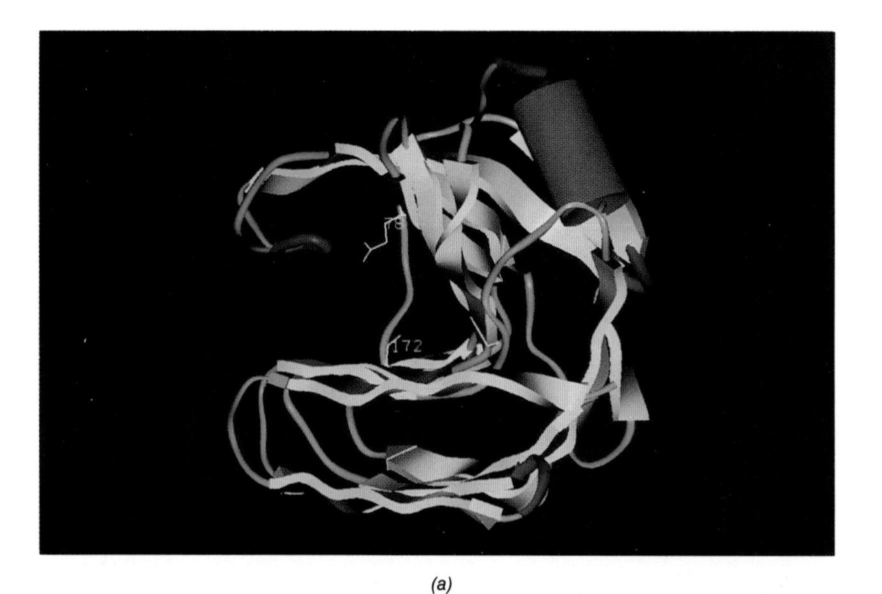

(a)

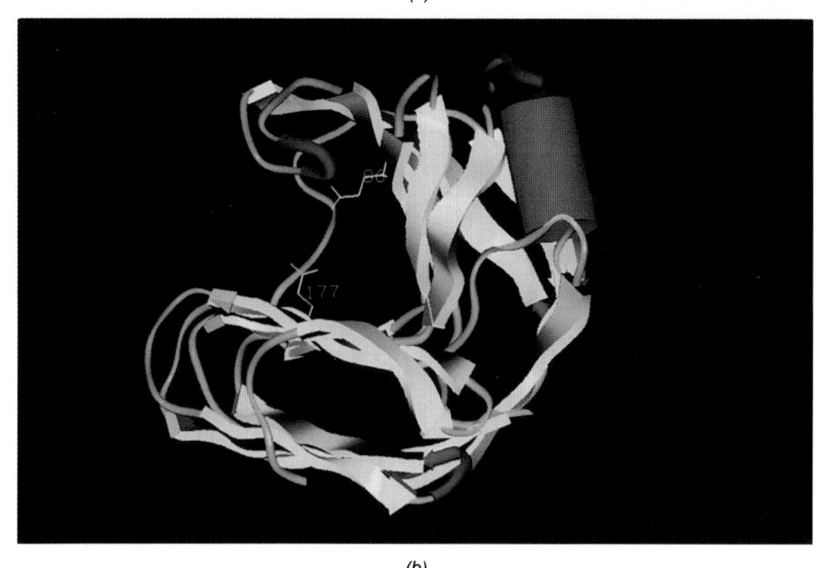

(b)

Figure 5.18 Three-dimensional crystal structures of the xylanases from (a) *Bacillus circulans* and (b) *Trichoderma harzianum* (both at 1.8 Å resolution). The view is from the side of the hand, with the palm below and the fingers curving up and over to form the active-site cleft. The side chains of the pair of putative catalytic residues, Glu78 and Glu172 (*B. circulans*) and Glu86 and Glu177 (*T. harzianum*) are depicted in white. The β-sheets are depicted in yellow, the α-helices are red, whereas the β-turns and random coil are blue and green, respectively. The figures were generated using the data deposited in the Brookhaven Protein Bank (1bcx and 1xnd, respectively) and visualized using Insight ver. 2.3.0 software (Biosym Technologies, San Diego).

influence on the ionization state of the acid catalyst is indeed the close proximity of the negatively charged catalytic nucleophile.

5.3.1.6 Families 12 and 26

At present, Families 12 and 26 comprise only three enzymes each (the cellulases from *Aspergillus aculeatus*, *Erwinia carotovora,* and *Streptomyces rochei* of Family 12; and the *Bacteroides ruminicola* and *Clostridium thermocellum* cellulases and a *Bacillus* sp. β-mannanase of Family 26), thus predictions based on sequence alignments would be superfluous. The *A. aculeatus* cellulase has been crystallized [813], but the three-dimensional structure of the enzyme has not been determined.

5.3.2 INVERTING ENZYMES

5.3.2.1 Family 6

The sequence alignment of the enzymes currently included in Family 6 is presented in Figure 5.19. This relatively small family of six cellulases and a cellobiohydrolase has received very little attention from biochemists, but the three-dimensional crystal structures of two of the enzymes, the *Thermomonospora fusca* cellulase 2 [210] and the *Trichoderma reesei* cellobiohydrolase II [208], have been determined. In addition, the latter enzyme was subjected to small-angle X-ray scattering analysis. X-Ray scattering analyses have shown that these enzymes are tadpole shaped with the catalytic and cellulose-binding domains constituting the head and tail, respectively [814,815]. The three-dimensional structures of the catalytic domains of the *T. fusca* and *T. reesei* enzymes show similar topologies, with many of the conserved amino acids located in the active site regions.

The topology of the catalytic domain of the *T. reesei* cellobiohydrolase II (2.0 Å resolution) is an α- and β-barrel with a central β-barrel consisting of seven parallel strands connected by α-helices [208]. This topology is thus quite different from the typical observed $(\beta\alpha)_8$ parallel (TIM) barrel. The inside of the β-barrel is completely filled by amino acid side chains, but two extended loops at the C-terminal end of the barrel form an enclosed tunnel approximately 20 Å long. Interestingly, a small tube containing water molecules leads from the middle of the tunnel to the surface of the enzyme. Four glucose-binding sites (subsites A–D) exist in the tunnel and four conserved Asp residues, Asp175, Asp221, Asp263, and Asp401, are positioned near subsites B and C. By considering the position of each residue and their hyrogen-bonding patterns, Asp221 was identified as the general acid catalyst for this inverting enzyme, with Asp175 acting to raise its pK_a [208]. Site-directed mutagenesis of the genes coding this enzyme and the *C. fimi* cellulase

```
CfiCbhA    1 APVHVDNPYAGAVQUVNPTWAASVNAAAGRQSADPALAAKMRTVAGQPTAVWMDRISAITGNADGNGLKFHLDNAVAQQKAAGVPLVFNLVIY
CfiCenA   86 --TYSGNPFVGVTPWANAYYASEVSSLAIP-SLTGAMATAAAAVAKVPSFMWLDTLDKTPL-----MEQTLADIRTANKNGGNYAGQ-FVVY
TreCbhII 133 TVTPQPTSGFYVDPTTQGYRAWQAASGTDK----ALLEK----IALTPQAYWVGNWADASH------AQAEVADYTGRAVAAGKTPM--LVVY
MbiCelA    1 -----YDSPFYVDPQSNA-AKWVAANPNDPR-TPVIRDR----IAAVPTGRWFA-NYNPST------VRAEVDAYVGAAAAAGKIPI--MVVY
ShaCel2    1 ADPTTMTNGFYADPDSSA-SRWAAANPGDGR-AAAINAS----IANTPMARWFGSW--SGA------IGTAAGAYAGAADGRDKLPI--LVAY
SspCasA    1 GTTALPSMELY-RAEAGV-HAWLDANPGDHR-APLIAER----IGSQPQAVWFAGAYNPGT------ITQQVAEVTSAAAAAGQLPV--VVPY
TfuEG2     1 -----NDSPFYVNPNMSS-AEWVRNNPNDPR-TPVIRDR----IASVPQGTWFA-HHNPGQ------ITGQVDALMSAAQAAGKIPI--LVVY

CfiCbhA   94 DLPGRD-CFALASNGELPATDAGLARYKSEYIDPIADLLDNPEYESIRIAATIEPDSLPNPTTNISEPACQQAA--PYYROGVKYALDKLHAI-
CfiCenA  160 DLPDRD-CAALASNGEYSIADGGVAKYKN-YIDTIRQIV--VEYSDIRTLLVIEPDSLANLVTNLGTPKCANAQ--SAYLECINYAVTQLNL--
TreCbhII 211 AIPGRD-CGSHSGGGVSE------SEYAR-WVDTVAQGI-----KGNP-IVILEPDALAQLGD------C---SGQGDRVGFLKYAAKSLTLKG
MbiCelA   74 AMPNRD-CGGPSAGGAPN-----HTAYRA-WIDEIAAGL-----RNRPAVIILEPDALPIMTN------CMSPSEQAEVQASMAYAGKKFKAAS
ShaCel2   79 NIYNRDYCGGHSAGGAAS-----PSAYAD-WIARFAGGI-----AARPAVVILEPDSLGDYG-------CMNPAQIDEREAMLTNALVQFNRQA
SspCasA   80 MIPFRD-CGNHSGGGAPS-----FAAYAE-WSGLFAAGL-----GSEPVVVVLEPDAIPLI-D------CLDNQQRAERLAALAGLAEAVTDAN
TfuEG2    74 NAPGRD-CGNHSSGGAPS-----HSAYRS-WIDEFAAGL-----KNRPAYIIVEPDLISLMSS------CMQHVQQ-EVLETMAYAGKALKAGS

CfiCbhA  184 PNVYNYIDIGHSGWLGWDSNAGPSATLFAEVAKSTTAGFASIDGFVSDVANTTPLEEPLLSDSSLTINNTPIRSSKFYEWNFDFDEIDYTAHMH
CfiCenA  256 PNVAMYLDAGHAGWLGWPANQDPAAQLFANVYKNASSPRA-LRGLATNVANYN-------------GWNITSPPSYTQGNAVYNEKLYIHAIG
TreCbhII 280 --ARVYIDAGHAKWL---SVDTAVNRLNQVGF-EYAV------GFALNTSNYQ--------------------------TTADSKAYGQQIS
MbiCelA  150 SQAKVYFDAGHDAWV---PADEMASRLRGADIANSA------DGIALNVSNYR--------------------------YTSGLISYAKSVL
ShaCel2  154 PNTWVYMDAGNPRWA---DAATMARRLHEAGLRQAH-------GFSLNVSNYI--------------------------TTAENTAYGNAVN
SspCasA  154 PEARVYYDVGHSAWH---APAAIAPTLVEAGILEHGA------GIATNISNYR--------------------------TTTDETAYASAVI
TfuEG2   149 SQARIYFDAGHSAWH---SPAQMASWLQQADISNSAH------GIATNTSNYR--------------------------WTADEVAYAKAVL
```

Figure 5.19 Amino acid sequence alignment of Family 6 β-glycosidases. Residues in bold type indicate identity among at least four of the seven aligned sequences, whereas shaded residues identify complete conservation. The sequences correspond to (reference or accession numbers for the sequences from Genbank are provided in parentheses) the following: CfiCbhA and CfiCenA, *Cellulomonas fimi* cellobiohydrolase A (L25809) and cellulase A (M15823), respectively; TreCbhII, *Trichoderma reesei* cellobiohydrolase II (M16190); MbiCelA, *Microbispora bispora* cellulase A [444]; ShaCel1, *Streptomyces halstedii* cellulase 1 (Z12517); SspCasA, *Streptomyces* sp. KSM-9 cellulase A (L03218); and TfuEG2, *Thermomonospora fusca* cellulase 2 (M73321) (adapted from Reference [138]).

```
CfiCbhA   278 RLLVAAGFPSSIGMLVDTSRNGWGGPNRPTSITASTDVNAYVDANRVDRRVHRGAWCNPLGAGIGRFPEATPSGYAASHLDAFVWIKPPGESDG
CfiCenA   335 PLLANHGWSNAF-FITDQGRSGKQPTG----------------------QQQWGDWCNVIGTGFGIRPSANTGD---SLLDSFVWVKPGGECDG
TreCbhII  346 QRLGGK------KFVIDTSRNGNGSNG---------------------------EWCNPRGRALGERPVAVNDG---SGLDALLWVKLPGESDG
MbiCelA   207 SAIGAS----HLRAVIDTSRNGNGPLG--------------------------SEWCDPPGRATGTWSTTDTGD---PAIDAGLWIKPPGEADG
ShaCel2   210 NELAAR-YGYTKPFVVDTSRNGNGSNG---------------------------EWCNPSGRRIGTPTRTGGG------AEMLLWIKTPGESDG
SspCasA   211 AELGGG-----LGAVVDTSRNGNGPLG--------------------------SEWCDPPGRLVGNNPTVNPGV---PGVDAFLWIKLPGELDG
TfuEG2    208 SAIGNP----SLRAVIDTSRNGNGPAG--------------------------NEWCDPSGRAIGTPSTTNTGD---PMIDAFLWIKLPGEADG

CfiCbhA   372 ASTDIPNDQGKRFDRMCDPTFVSPKLNNQLTGATPNAPLAGQWFEEQFVTLVKNAYPVIG
CfiCenA   403 TSDSSAP----RFDSHCALP-----------DALQPAPQAGAWFQAYFVQLLTNANPSFL
TreCbhII  394 A---------------CNGG-----------------PAAGQWWQEIALEMARNARW
MbiCelA   268 ----------------CIAT-------------------PGVFVPDRAYELAMNAAPPTY
ShaCel2   270 N---------------CGVG---------------SGSTAGQFLPEVAYKMIYGY
SspCasA   270 ----------------CDGP-------------------AGSFSPAKAYELAGG
TfuEG2    268 ----------------CIAG-------------------AGQFVPQAAYEMAIAAGGT
```

Figure 5.19 (continued) Amino acid sequence alignment of Family 6 β-glycosidases. Residues in bold type indicate identity among at least four of the seven aligned sequences, whereas shaded residues identify complete conservation. The sequences correspond to (reference or accession numbers for the sequences from Genbank are provided in parentheses) the following: CfiCbhA and CfiCenA, *Cellulomonas fimi* cellobiohydrolase A (L25809) and cellulase A (M15823), respectively; TreCbhII, *Trichoderma reesei* cellobiohydrolase II (M16190); MbiCelA, *Microbispora bispora* cellulase A [444]; ShaCel1, *Streptomyces halstedii* cellulase 1 (Z12517); SspCasA, *Streptomyces* sp. KSM-9 cellulase A (L03218); and TfuEG2, *Thermomonospora fusca* cellulase 2 (M73321) (adapted from Reference [138]).

A confirmed the proposed role of Asp221 [208,816]. Replacement of Asp216 in the *C. fimi* cellulase (homologous to Asp175) only moderately affected k_{cat} values for a variety of substrates, but replacement of Asp252 (homologous to Asp221) effectively abolished substrates requiring general acid catalysis [816]. Asp263 or Asp401 were thought to serve as the active-site base, but no specific assignment was made.

The crystal structure (1.8 Å resolution) of the *T. fusca* cellulase [210] (Figure 5.20) retains many of the structural features observed with the *T. reesei* cellobiohydrolase. Thus, its overall topology is an atypical α- and β-barrel, consisting of eight β-strands and ten α-helices. However, as expected for an endoacting cellulase, the two extended loops do not close to form a tunnel but rather create a cleft, 11 Å deep at its deepest point and 10 Å wide, which extends the full length of the enzyme. Of the four Asp residues present in the active-site tunnel of the *T. reesei* cellobiohydrolase, only three, Asp117, Asp156, and Asp265 (corresponding to Asp221, Asp263, and Asp401 of the cellobiohydrolase), are suitably positioned in the cellulase cleft. Asp117 and Asp265 are located closest to the proposed cleavage site, and they are appropriately postioned across the cleft opposite each other,

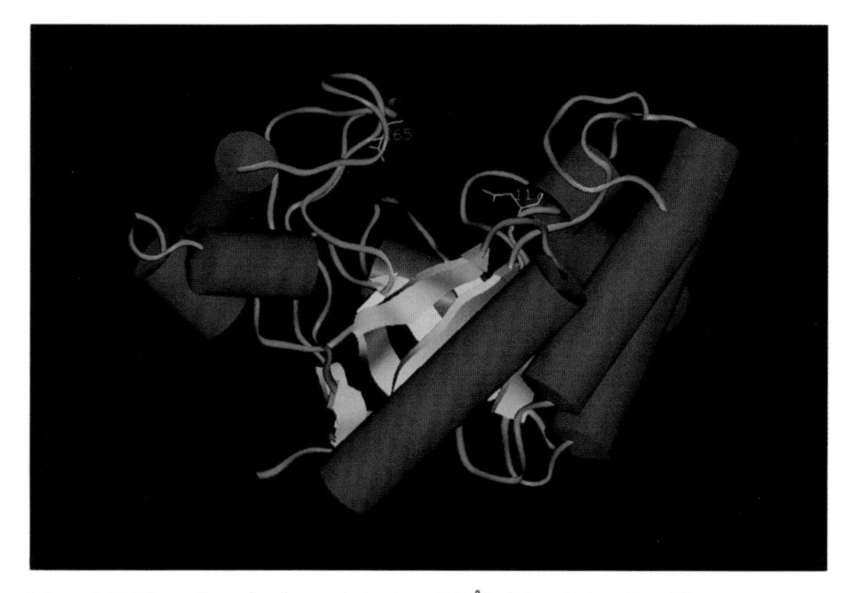

Figure 5.20 Three-dimensional crystal structure (1.8 Å) of the cellulase from *Thermomonospora fusca.* The side chains of the putative catalytic residues Asp117 and Asp265 are clearly seen opposing each other across the active-site cleft of the enzyme. The β-sheets are depicted in yellow, whereas the α-helices and the random coils are red and green, respectively. The figure was generated using the data deposited in the Brookhaven Protein Bank (1tml) and visualized using Insight ver. 2.3.0 software (Biosym Technologies, San Diego).

separated by a distance of approximately 11 Å [210]. This separation is expected for an inverting enzyme because a molecule of water has to be positioned between the general base catalyst and glycosidic oxygen. The environment of Asp265 suggested that it would exist predominantly in its ionized state; it forms a salt bridge to Arg221 and weakly interacts with Lys259 and Tyr191. Asp117, like the homologous Asp221 of the *T. reesei* cellobiohydrolase, was proposed to be the general acid catalyst [210].

5.3.2.2 Family 9

Alignment of the sequences of the cellulases comprising Family 9 reveals four totally conserved acidic residues (Figure 5.21). One of these residues, a Glu, comprises a conserved Asn-Glu-X motif (where X is a hydrophobic residue), which closely resembles the Asn-Glu-Pro motif observed in Family 1 and 5 β-glycosidases. With these latter enzymes, the Glu of this motif is proposed to be the catalytic acid residue. Chemical modification studies using the chromogenic reagent Woodward's reagent K identified this Glu (Glu555) as an essential residue in the Family 9 cellulase from *Clostridium thermocellum* [735], and all subsequent investigations on Family 9 enzymes pertained to this enzyme.

Site-directed mutagenesis of its gene replacing each of the four conserved Asp and Glu residues supported the proposal that Glu555 is essential for catalysis [817]. The enzyme has been crystallized [818], and its three-dimensional structure solved and refined to 2.3 Å resolution [819]. It is comprised of two domains, an N-terminal β-barrel and a larger, mostly α-helical domain (Figure 5.22). The β-barrel contains seven antiparallel β-strands, forming two β-sheets packed against each other. The large domain has been coined a twisted α-barrel that is comprised of twelve major helices, six inner helices oriented in the same direction and the outer six oriented in the opposite sense.

Cocrystallization of the inhibitor *o*-iodobenzyl-β-D-thiocellobioside with the cellulase localized the active site to a cleft at the surface, formed by three loops connecting six of the helices. Glu555 is indeed observed to be positioned appropriately adjacent to the inhibitor to serve as the acid catalyst. Asp201, as predicted by site-directed mutagenesis studies [817], is thought to act as the general base, removing the proton from water in the single-displacement mechanism of action [819].

5.3.2.3 Family 45

Family 45 glycosidases comprise only two cellulases, those from *Pseudomonas fluorescens* and *Humicola insolens*. Again, sequence alignments for the purpose of identifying essential residues would be meaningless. How-

Subtype 1

```
CthEGD    191 DSTKGWHDAGDYNKY   502 SYVTGLGINPPMNPHDR   544 WVDIQDSYQTNEIAINWNAALIYALA
PflEGA    119 NVTKGWYDAGDHGKY   457 SFITGLGTNTVAQPHHR   571 WLDSIDAWSTNEITINWNAPLAWVLG
CfiCenC   496 DVSGGWYDAGDHGKY   818 SYVTGWGEVASHQQHSR   880 YVDEIQAWSTNELTVNWNSALSWVAS
BfiCed1   138 DVTGGWHDAGDYGRY   455 SYVTGNGEKAFKNPHLR   513 YIDHIDLYSLNEITIYWNSPLVFALS
```

Subtype 2

```
PamEG      72 DLVGGYYDAGDNLKF   397 SYMVGFGERYPQHVHHR   460 FSDDRNNYQQSEPATYINAPLVGALA
CfiCenB    81 DLTGGWYDAGDHVKF   396 SYVVGFGANPPTAPHHR   447 YTDSRQDYVANEVATDYNAGFTSALA
DdiEG      75 NLSGGYFDAGDGVKF   377 SFVVGMGPNYPINPHHR   426 YTDDRTDYISNEVATDYNAGFVGALA
CstEGZ     74 DLTGGWYDAGDHVKF   386 SYVVGFGVNPPKRPHHR   436 YTDDINNYINNEVACDYNAGFVGALA
CthEGF     75 DLTGGWYDAGDHVKF   386 SFVVGFGKNPPRNPHHR   436 YVDRLDDYQCNEVANDYNAGFVGALA
```

Figure 5.21 Alignment of highly conserved regions within deduced amino acid sequences of Family 9 β-glycosidases. The sequences have been grouped according to subtypes as indicated, and the numbering indicates the position of the first amino acid in the respective sequences. Residues in bold type denote identity within at least five of the nine sequences, whereas shaded residues are totally conserved for all Family 9 β-glycanases. The sequences correspond to the following: CthEGD, *Clostridium thermocellum* cellulase D [515]; PflEGA, *Pseudomonas fluorescens* subsp. *cellulosa* cellulase A [420]; CfiCenC, *Cellulomonas fimi* cellulase C [869]; BfiCed1, *Butyrivibrio fibrisolvens* H17c cellobiohydrolase I [870]; PamEG, *Persea americana* cellulase [871]; CfiCenB, *C. fimi* cellulase B [417]; DdiEG, *Dictyostelium discoideum* cellulase [872]; CstCelZ, *Clostridium stercorarium* cellulase Z [873]; and CthEGF, *C. thermocellum* cellulase F [504] (adapted from Reference [817]).

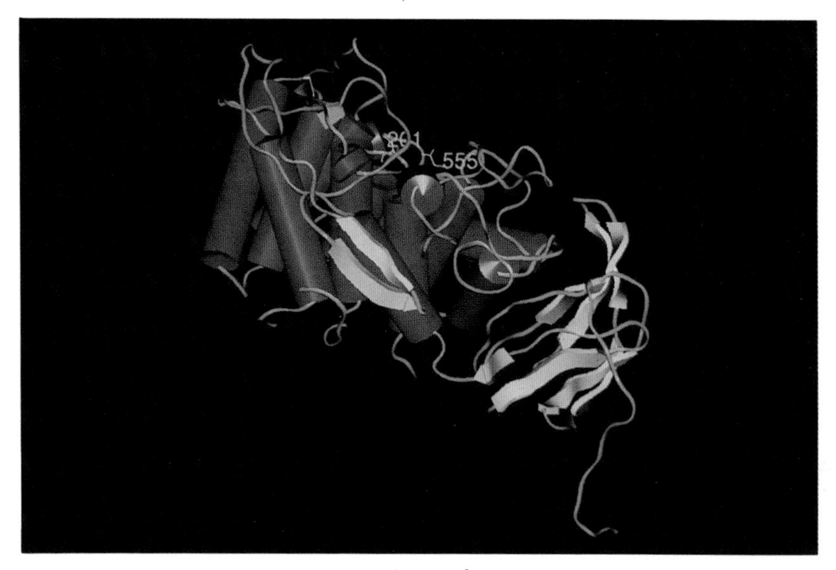

Figure 5.22 Three-dimensional crystal structure (2.3 Å) of the cellulase from *Clostridium thermocellum*. The cellulase is comprised of an N-terminal β-barrel (yellow) comprised of seven antiparallel β-strands, and the twisted α-barrel (red). The side chains of the putative catalytic residues associated with the larger C-terminal domain, Asp201 and Glu555, are depicted are also shown. The figure was generated using the data deposited in the Brookhaven Protein Bank (1clc) and visualized using Insight ver. 2.3.0 software (Biosym Technologies, San Diego).

ever, the catalytic domain of the *H. insolens* cellulase V has been crystallized [820], and its structure has been determined at a resolution of 1.6 Å [821].

The catalytic core of the *H. insolens* cellulase V is a six-stranded β-barrel, and hence represents yet another type of protein topology for the cellulolytic and xylanolytic enzymes. The β-barrel is composed of both parallel and antiparalleled β-strands, making it unique among the known protein structures. A deep groove runs across the surface of the molecule, which is large enough to accommodate seven subsites for substrate binding. Site-directed mutagenesis has identified Asp10 and Asp121 as essential residues, and they are located on either side of the major groove. The C_{α} atoms of these two residues are separated by 11.5 Å, consistent with the dimensions between the catalytic residues of other inverting β-glycosidases. Cellobiose was observed to bind between the two Asp residues so that a water molecule bound to Asp10 could nucleophilically attack the substrate [821]. This would identify Asp10 as the general acid-base catalyst and Asp121 as the stabilizing anion in the single-displacement mechanism of action.

5.3.2.4 Family 46

This is a newly created family of cellulases and cellobiohydrolases [206]

```
Bsp EGK    91  YNDWKKKYLKNDLSSLPGGYYVKGEITGNPDGFRPLGTSEGQGYGMIITVLMAGHDSNAQTIYDGLKFTARAFKSSINPN
Cce EGC    67  WEQWKSAHITSN----GARGYKRVQRDATTN-YDTV--SEGLGYGL----LLSV-YFGEQQLGDDLYRYVKVFL--NSNG
Cth EGA    63  WEDWKSKRITSN----GAGGYKRVQRDASTN-TDTV--SEGMGYGL----LLAV-CFNEQALFDDLYRYVKSHF--NGNG
Cud EG     27  WERTJARFNNODAR-------------IIDTANGNVSHTEGQGFAM----LLAV-ANNDRPAFDKLWQWTDSTLRDKSNG
Ech EGY    27  WEIYKSRFMTTDGR-------------IQDTGNKNVSHTEGQGFAM----LMAV-HYDDRIAFDNLWNWTQSHLRNTTSG
Bci LIC    66  WDSWSAYLKTAGT---GKYYVK-----YQSNG-DTV--SEAHGYGMLATVIMAGYDGNAQTYFDGLYQYYKAHPSANNSK

Bsp EGK   171  LMGWVVADDKKAQGH---FDSATDGDLDIAYSLLLAHKQWGSSGKINYLKEAQNMITKGIKASNV---TKNNGLNLGD-W
Cce EGC   133  LMSWRIDSS-GNIMGKDSIGAATDADEDIAVSLVFAHKKWGTSGGFNYQTEAKNYINNIYNKMVE---PGYTVIKAGDTW
Cth EGA   129  LMHWHIDANNNVTSHDGGDGAATDADEDIALALIFADKLWGSSGAINYGQEARTLINNLYNHCVE---HGSYVLKPGDRW
Cud EG     89  LFYWRYNPVAPDPIAD--KNNATDGDTLIAWALLRAQKQWQ---DKRYATASDAITASLLKYTVVTFAGRQVMLPGVKGF
Ech EGY    89  LFYWRYDPSAANPVVD--KNNASDGDVLIAWALLKAGNKWQ---DNRYLQASDSIQKAIIASNIIQFAGRTVMLPGAYGF
Bci LIC   136  LMAWKQNSSFQNIEG---ADSATDGDMDIAYSLLLADKQWGSSGSINYLQAGKDIINAIMQSDVN---QSQWTLRLGD-W

Bsp EGK   244  GDKSTF--DTRPSDWMMSHLRAFYEFTGDKTWLNVIDNLYNTYTNFTNKYSPKTGLISDFVVKNPP---QPAPKDFLDES
Cce EGC   209  G--GSN--VTNPSYFAPAWYRIFADFTGNSGWINVANKCYEIA---DKARNSNTGLVPDWCTANGT----PASGQGFDFY
Cth EGA   206  G--GSS--VTNPSYGAPAWYKVYAQYTGDTRWNQVADKCYQIVEE-VKKYNNGTGLVPDWCTASGT----PASGQSYDYK
Cud EG    164  NRNDHL--NLNPSYFIFPAWRAFAERTHLTAWRTLQSDGQA--LLGQMGWGKSH-LPSDWVALRADGKMLPAKEWPPRMS
Ech EGY   164  NKNSYV--ILNPSYFLFPAWRDFANRSHLQVWRQLIDDSLS--LVGEMRFGQVG-LPTDWAALNADGSMAPATAWPSRFS
Bci LIC   209  ATDNTFKNATRPSDFMLNHLKAFQAATGDARWANVIDKTYTIINSLYSGYSSSTGLLPDFVVLSGS-TYKPASADFLEGA
```

Figure 5.23 Amino acid sequence alignment of Family 8 β-glycosidases. Residues in bold type indicate identity among at least four of the six aligned sequences, whereas shaded residues identify complete conservation. The sequences correspond to (reference or accession numbers for the sequences from SWISS-PROT are provided in parentheses) the following: Bsp EGK, *Bacillus* sp. KSM-330 cellulase K (P29019); Cce EGC, *Clostridium cellulolyticum* cellulase C [874]; Cth EGA, *C. thermocellum* cellulase A [511]; Cud EG, *Cellulomonas uda* cellulase [875]; Ech EGY, *Erwinia chrysanthemi* cellulase Y (P27032); and Bci LIC, *Bacillus circulans* WL-12 lichenase [876].

for which very little is known. Indeed, the activity of one of the proteins comprising this family, a 70-kDa polypeptide called P70 that constitutes a subunit of the *Clostridium cellulovorans* cellulosome remains unknown, whereas the assignments of CelA from *Caldocellum saccharolyticum* and CelCCF from *C. cellulolyticum* as cellulases was only recently made [109]. With very limited available information concerning these enzymes, a discussion of their active-site structure would be premature.

5.3.3 ENZYMES WITH UNCHARACTERIZED STEREOCHEMISTRY

5.3.3.1 Family 8

Six β-glycosidases have been classified into Family 8, five bacterial cellulases and a bacterial lichenase. Alignment of their amino acid sequences (Figure 5.23) reveals that five acidic residues are completely conserved, four Asp and one Glu residue [822]. Interestingly, however, neither His nor Arg residues appeared to be conserved in this family of enzymes. In keeping with the fact that the anomeric configuration of the reaction products catalyzed by these enzymes remains unknown, very little information is available concerning the structure and function relationship of their catalytic sites. In fact, only one investigation has been conducted. The five acidic residues in the cellulase from *Bacillus* sp. KSM-330 were replaced by site-directed mutagenesis with the corresponding Gln and Asn residues. Each mutant enzyme was expressed and purified. Kinetic analysis of each revealed that the k_{cat} values for three of the mutants (Glu130Gln, Asp191Asn, and Asp193Asn) were less than 5% of the value for the wild-type enzyme. Hence, two of these residues likely participate in the catalytic mechanism of action of the enyzmes, although the two catalytic roles were boldly ascribed to Glu130 and Asp191 [822].

CHAPTER 6

Structure and Function Relationship

THE substrate-binding regions of all carbohydrases have been rationalized on the basis of a subsite model [877]. In this model, the binding region of the enzyme is comprised of a sequential array of modules (subsites) that are formed by a number of amino acid residues that interact with the monomeric units of substrate through hydrogen bonding and both hydrophobic (stacking) and ionic interactions. Thus, a subsite is a particular region within a binding cleft or pocket that is complementary to an individual substrate residue. The interaction between a subsite and a substrate residue will give rise to a decrease in free energy (viz., increase in affinity) and contribute to the binding of the substrate molecule as a whole [878,879]. The specificity of the carbohydrase is thus dependent on the following: the number of subsites and their individual contribution to the overall affinity for substrate(s); the size and structural properties of each binding subsite; and the location of the catalytic site in relation to the binding subsites. Subtle differences in the composition of the amino acid residues within a subsite may have profound effects on the enzymatic properties of a β-glycosidase, and with the establishment of the mechanism of hydrolysis, attention is now being focused on the mode of action of the enzymes with the view to control and possibly alter substrate specificities.

6.1 SUBSTRATE-BINDING SUBSITES

The affinities of individual subsites can be measured and mapped, providing a graphical representation of the binding region of carbohydrases. The technique of subsite mapping was pioneered by Hiromi and co-workers [878–880] and later extended to more quantitative applications by Thoma and Allen [881]. A series of oligomeric substrates ranging in chain length from the smallest hydrolysed by the enzyme of study to one that is larger than the binding region is required. With these oligomers, the maximum rate

(V_{max}) and Michaelis constant (K_M) of the enzymatic hydrolysis of each is determined, together with the inhibition constants for any smaller nonhydrolysed reaction products. Because the location of the catalytic site within the subsite array is not known from the product distribution produced by endohydrolases, such as cellulases and xylanases, a series of reducing end-labelled oligomers must be used to establish the frequency of cleavage of each bond in the substrate. From these bond-cleavage frequencies and the kinetic parameters, binding energies can be calculated for each predicted subsite and thereby indicate the role of each in the substrate binding and hydrolysis.

Few β-glycosidases have been investigated with respect to their cleavage patterns and binding-site affinities, perhaps because of the unavailability of defined oligomeric substrates and the complicated, time-consuming data handling involved in product analysis. This is especially likely for the study of xylanases, and yet, ironically, this class of enzyme represents the best characterized among the cellulolytic and xylanolytic hydrolases. For the analysis of bond-cleavage frequencies, the reducing sugar of the oligosaccharides may be labelled with a chromogenic group, such as 2-chloro-4-nitrophenol [630] and 4-methylumbelliferol [630,882,883], or reduced to the corresponding alditol (for example, References [270,631,636,882,884]). Separation of reaction products following enzymatic hydrolysis is best achieved by either paper chromatography [885–888], thin-layer chromatography (for example, References [636,884]), or high-performance liquid chromatography (HPLC) (for example, References [270,630,631,882, 883,889,890]). In most cases, detection of the alditols is facilitated by their synthesis as radioactive derivatives using sodium borotritiide. However, the development of pulsed amperometric detection coupled with anion exchange chromatography at alkaline pH has provided an alternative to the radioactive methods. Complete resolution of both reduced and native oligosaccharides is acheived by this HPLC methodology, enabling unambiguous assessment of both rate of substrate loss and rates of product formation in addition to the quantitative analysis of predominant reaction products. This technique has been applied to the analysis of the mode of action of xylanase A from *Schizophyllum commune* [270,631], and the bond-cleavage frequencies obtained by the action of this enzyme on homoxylo-oligosaccharides are presented in Figure 6.1. Also included in this figure are examples of the bond-cleavage frequencies together with the kinetic parameters obtained with a *Trichoderma reesei* β-glucosidase [884] and cellobiohydrolase II [883], and a cellulase from *Penicillium pinophilum* [882]. With each type of enzyme, increasing the chain length of the substrate results in an increase in affinity (decreasing values of K_M) and, with the exception of β-glucosidases, an increase in turnover number (k_{cat}). The influence of the degree of polymerization of substrate on the kinetic parameters of a given

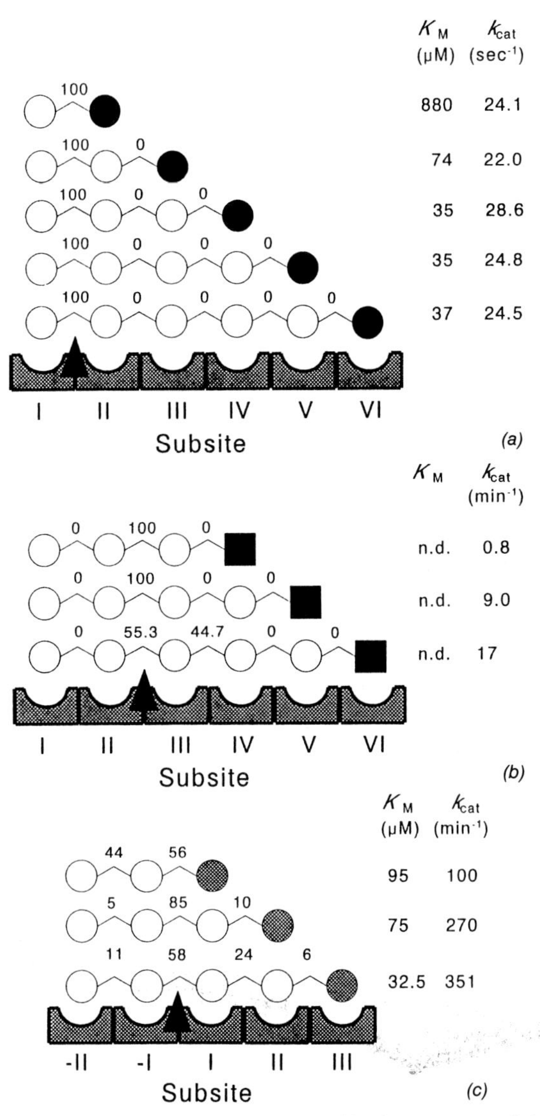

Figure 6.1 Bond-cleavage frequencies and steady-state kinetic parameters of the hydrolysis of oligosaccharides by (a) and (b) the β-glucosidase [884] and cellobiohydrolase II [883] from *Trichoderma reesei*, respectively, (c) cellulase IV from *Penicillium pinophilum* [882], and (d) xylanase A from *Schizophyllum commune* [631]. Frequencies are expressed as percentages of total cleavage events. (O) Glucosyl residues; (◇) xylosyl residues; (●, ♦) alditols; (⊗) ^{3}H-labelled glucosyl residues; (■) methylumbelliferyl glucosides. The arrow head denotes the catalytic site located between the binding subsites.

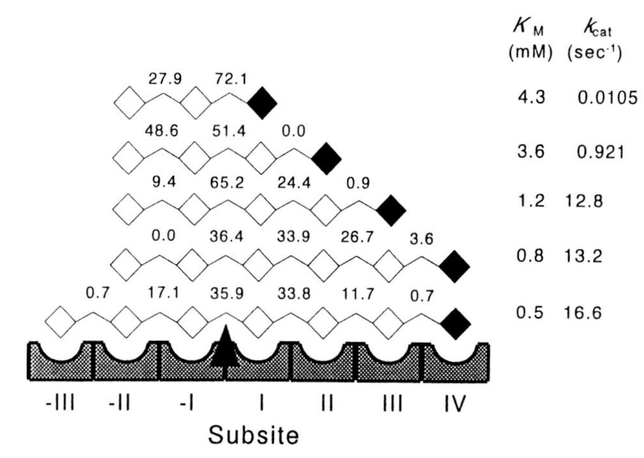

Figure 6.1 (continued) Bond-cleavage frequencies and steady-state kinetic parameters of the hydrolysis of oligosaccharides by (a) and (b) the β-glucosidase [884], and cellobiohydrolase II [883] from *Trichoderma reesei*, respectively, (c) cellulase IV from *Penicillium pinophilum* [882], and (d) xylanase A from *Schizophyllum commune* [631]. Frequencies are expressed as percentages of total cleavage events. (O) Glucosyl residues; (◇) xylosyl residues; (●, ◆) alditols; (⊗) ^{3}H-labelled glucosyl residues; (■) methylumbelliferyl glucosides. The arrow head denotes the catalytic site located between the binding subsites.

enzyme is easily observed when presented graphically, as illustrated in Figure 6.2 for the *S. commune* xylanase A hydrolysis of xylo-oligosaccharides [270,631]. An estimation of the number of binding subsites in the enzyme is readily indicated by the point at which no further increase in affinity or catalytic efficiency (asymptote) is observed. For the example provided, a minimum of five affinity subsites would be predicted for the xylanase [270,631], but this estimate is not wholly conclusive because a bend or break in a curve of Michaelis parameters plotted against chain length may underestimate subsite numbers [881,891].

Analysis of bond-cleavage frequencies provides further information on the dimensions of the binding-site regions. As to be expected for endoacting enzymes, the highest frequencies of bond cleavage catalyzed by the cellulase and xylanase are found internally within the oligosaccharide substrates (Figure 6.1). This strong preference for internal linkages, which becomes more apparent with longer oligosaccharide substrates, indicates that the catalytic site is located asymmetrically within the binding subsites of the enzyme. Another feature of hydrolysis shared by many endoacting glycosidases [877,888,891–894], as exemplified by *S. commune* xylanase A [270,631], is that the frequency of enzyme attack is not constant but shifts with increasing size of oligomer. However, assuming that K_M as well as V_{max} and bond-cleavage frequencies are directly related to the free energy rein the process of monomer-subsite interaction [891], it was concluded that xylanase A has at least seven subsites capable of binding xylose residues [270,631]. The exoacting enzymes β-glucosidase and cellobiohydrolase release predominantly, if not exclusively, glucose and cellobiose, respectively, from the nonreducing ends of the substrates, and the catalytic site is appropriately positioned.

The cellobiohydrolases I and II from *Trichoderma reesei* share many common features, such as similar binding properties to affinity columns and high activity toward crystalline cellulose, but their modes of action appear to differ considerably [883]. Cellobiohydrolase I is capable of hydrolysing chromogenic cellobiosides, whereas cellobiohydrolase II is inactive against these smaller substrates. Both enzymes hydrolyse cellotriosides but compared with cellobiohydrolase II, cellobiohydrolase I is at least ten times more efficient. Reaction patterns toward longer substrates were found to be totally different for the two enzymes. For example, the chromogenic cellotrioside is the main product released by the action of cellobiohydrolase II on chromogenic cellopentaside, but none was observed in the product mixture produced by cellobiohydrolase I. These and other properties led to the proposal that the two enzymes are capable of interacting with internal cellobiosyl units within cellulose [883], thereby accounting for their high activity toward insoluble substrates. Moreover, their different specificities may at least partially explain the well-documented synergistic action (exo-exo) of these two cellobiohydrolases.

Although the assignment of the site of cleavage is facile for β-glucosidases

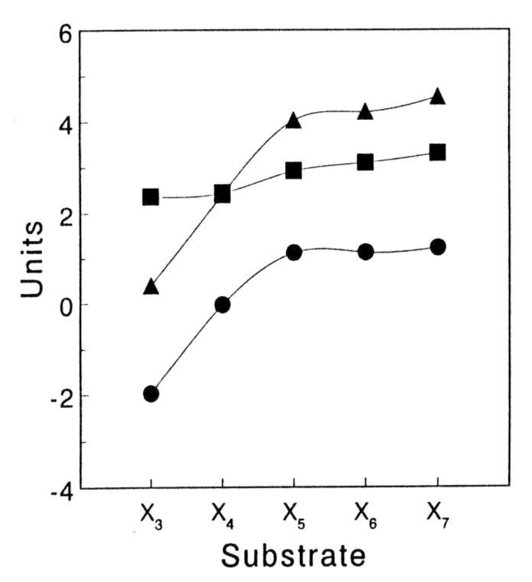

Figure 6.2 Influence of the degree of polymerization of substrate on the kinetic parameters for the *S. commune* xylanase A-catalyzed hydrolysis of xylo-oligosaccharides at pH 5.7 and 25°C. (●) log k_{cat}; (■) pK_M; (▲) log $k_{cat} \cdot K_M^{-1}$ (adapted from Reference [270]).

and cellobiohydrolases, it is not always the case for cellulases and xylanases. Complications arise from the choice of substrates because in most cases, bond-cleavage frequencies are determined using a form of unnatural substrate and differences have been observed between those used for investigations. This has been clearly shown in a meticulous study of five cellulases from *Penicillium pinophilum* [882]. The hydrolysis pattern using cello-oligosaccharides as substrate was found to be rarely the same as that obtained with either chromogenic or reduced substrates (Figure 6.3). For example, neither cellotriose, cellotetraose, nor cellopentaose were hydrolysed by cellulase I, and yet both the 4-methylumbelliferyl and reduced derivatives of cellopentaose were readily cleaved by the enzyme. In contrast, cellulases III and IV degraded cellotriose to cellobiose and glucose, whereas the cellotriose derivatives did not serve as substrates for either enzyme. The study of xylanases is further complicated by the heterogeneous nature of its substrate. To obtain relatively unambiguous substrate cleavage profiles for the majority of the enzymes of this class, care should be taken to use homoxylo-oligosaccharides as substrates (i.e., neither side-chain carbohydrates nor other modifications should be present). However, as indicated in Chapter 2, some xylanases have an affinity for specific side-chain modifications within heteroxylans, rendering such studies as described above extremely difficult, with the availability of appropriate substrates posing the most serious barrier.

Thermodynamic binding arrays may be calculated by the methods of Hiromi [878,879] or Thoma and Allen [881] using both kinetic data and bond-cleavage frequencies. A comparison of subsite-binding affinities calculated in this manner for various cellulolytic and xylanolytic enzymes is illustrated in Figure 6.4. For the endoacting xylanases, the sum of affinities of subsites directly spanning the catalytic site constitutes a region of positive free energy that may distort the chair form of a bound xylosyl residue into a half-chair, producing the enzyme-induced strain that is proposed to lower the activation energy barrier to form the transition state. With the method of Hiromi [878,879], binding affinities of subsites are estimated from the dependence of $V_{\max}$ and K_M values on the chain length of substrates. These values would be directly proportional to the association constant of the single productive complex formed between enzyme and substrate, assuming that the hydrolytic rate coefficients for all bonds within the substrate are independent of chain length. The interaction energy for these two subsites is determined as a sum because all productive complexes have to fill both subsites. The effects of a subsite of low affinity near the active site would be overcome upon binding of substrates with a degree of polymerization greater than four because of the positive binding affinities associated with distant subsites. The strong endocharacter of these binding arrays is most apparent with these rapidly hydrolysed substrates and is reflected in the reduced likelihood of a complex being formed the closer one end of the

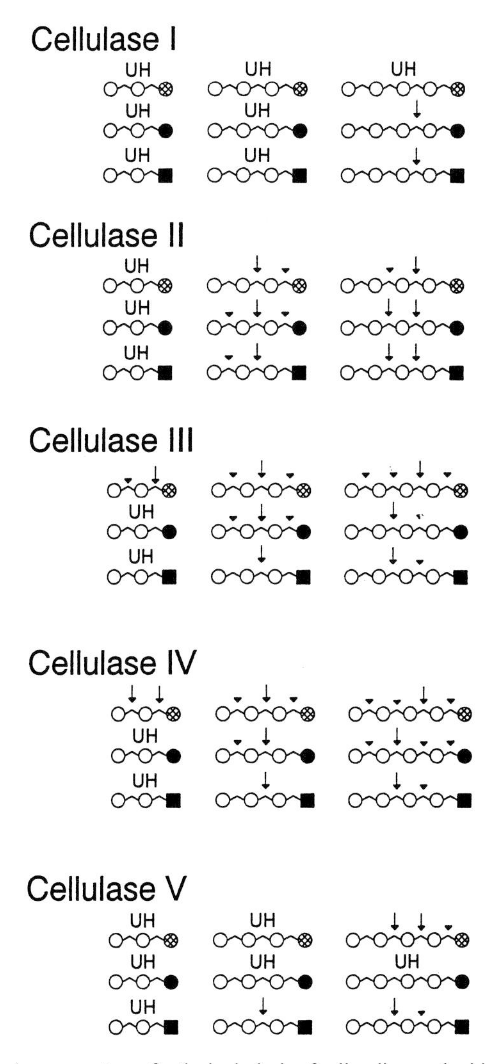

Figure 6.3 Bond-cleavage patterns for the hydrolysis of cello-oligosaccharides by cellulases from *Penicillium pinophilum*. (○) Glucosyl residues; (●) alditols; (⊗) ^{3}H-labelled glucosyl residues; (■) methylumbelliferyl glucosides. The bonds cleaved preferentially and nonpreferentially by the enzymes are denoted by arrows and arrow heads, respectively, whereas UH denotes unhydrolysed bonds.

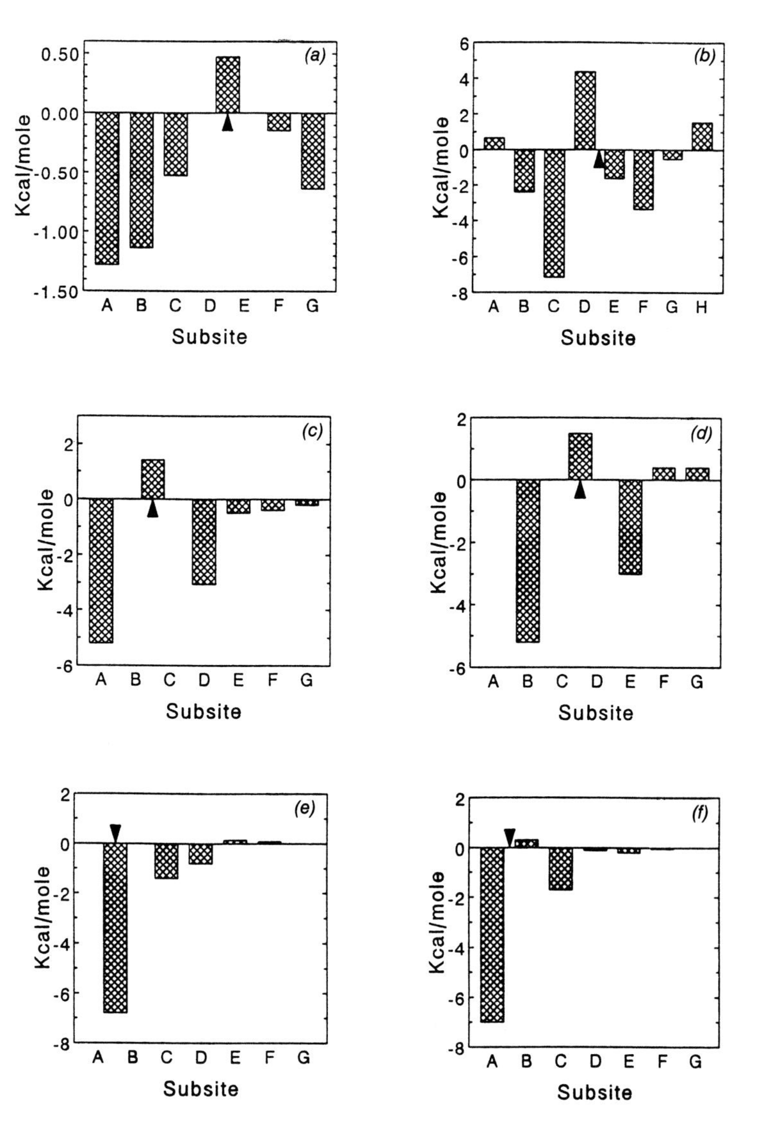

Figure 6.4 Subsite affinity maps for (a) xylanase A from *Schizophyllum commune* [631], (b) and (c) the xylanase [892] and acidic xylanase [885,888] from *Aspergillus niger*, (d) the xylanase from *Cryptococcus albidus* [886], (e) and (f) the β-glucosidase from *Trichoderma reesei.* Subsite maps (a–e) were originally calculated from Michaelis constants whereas inhibition constants were used to calculate affinity map (f). The arrow denotes the position of the catalytic groups. Because individual affinities of the subsites at the catalytic center cannot be evaluated using the method based on Michaelis parameters, the sum of affinities of the two subsites is presented.

oligomer comes to the repelling effects of subsites at the catalytic site. Indeed, as suggested by Biely [895], this would account for the inability of most xylanases to hydrolyse xylobiose [33].

The substrate-binding site of xylanase A from *S. commune* appears to be composed of seven subsites with the outer subsites providing the greatest contribution to binding affinity [Figure 6.4(a)]. This model readily shows why xylopentaose is the shortest oligomer that is rapidly hydrolysed [270,631], because the strong affinities of subsites B or G would be within reach for productive binding. The stronger binding at subsite C compared to subsite 6, and its greater ability to counteract the repulsion generated by subsite D would suggest that the more productive binding mode of xylopentaose would involve subsites C to G, and this is reflected in the bond-cleavage frequencies [Figure 6.1(a)]. The subsite-binding cleft of the *Aspergillus niger* xylanase, another Family 11 enzyme, is likely comprised of six subsites with the catalytic site in the middle, between subsites C and D [Figure 6.4(b)] [892]. Similarly, the acidic xylanase from *A. niger* has a binding site consisting of seven subsites [Figure 6.4(c)], with the catalytic groups located between subsites B and C [885,888]. Although there are obvious differences in both the subsite maps and bond-cleavage frequencies between the two *A. niger* enzymes, both display strong endo character and resemble xylanase A from *S. commune* in their mode of action.

An attempt has been made to correlate the specificity mapping of β-glycosidases with their classification into families [896]. With larger chromogenic substrates, a characteristic specificity pattern was observed. Thus, Family 6, 8, and 9 enzymes tend to catalyze the hydrolysis of internal linkages within the middle of oligosaccharide substrates, whereas the activities of Family 5, 7, and 10 β-glycosidases are more confined to the reducing ends. This study involved only a few representatives of each family of cellulolytic enzyme, and given the dearth of available information, it is still too premature to establish the validity of the correlation. However, perusal of the available data does provide some support for the hypothesis. For example, the *Cryptococcus albidus* xylanase, a Family 10 β-glycosidase, appears to be comprised of only four subsites with the catalytic site localized in the middle [Figure 6.4(d)]. Subsites B and E, which neighbour the repulsion subsites C and D of the catalytic site, provide all of the affinity for substrate binding, and this model is consistent with the fact that xylotetraose is the smallest substrate that is hydrolysed rapidly to xylobiose by the enzyme [33,886,887]. As a consequence of the asymmetric distribution of negative affinities in the binding site, the enzyme displays a strong preference for hydrolysis near the reducing end of substrates [886], a feature wholly consistent with other Family 10 enzymes [896]. Likewise, the mode of action of xylanase A from *Streptomyces lividans*, another Family 10 β-glycosidase, is quite distinct from the two Family 11 enzymes, xylanase

B and xylanase C, produced by the same bacterium [636]. Xylanase A was observed to have a much higher catalytic efficiency compared with the other two on short xylo-oligosaccharides, and the predicted subsite structure was correspondingly smaller and differently organized relative to those of xylanases B and C. Xylanase A of *S. lividans* thus more closely resembles the other Family 10 enzymes, whereas xylanases B and C share similar binding features of the Family 11 β-glycosidases.

An interesting phenomenon that has been observed with cellobiohydrolase II from *T. reesei* [883] and one that may apply to other cellulolytic and xylanolytic enzymes, is a form of positive cooperativity in binding. The binding site of this enzyme is thought to contain four subsites, and glucose was observed to bind predominantly in subsite 1. Chromogenic cellobioside appeared to bind at either subsites A and B or B and C, but in the presence of glucose, an enhanced binding of the cellobioside was observed. This suggested a positive cooperativity in the binding of the different substrates at contiguous subsites [883].

Studies with the *Trichoderma reesei* β-glucosidase provide an example of the differences observed in the subsite affinity maps obtained by the two methods. The data presented in Figure 6.4 (e) were obtained by manipulation of the Michaelis parameters according to the method of Hiromi. As typical of exoacting enzymes, substrate binds in only one productive mode, which ensures the release of monomeric products. The results indicated that the active site of the β-glucosidase comprises four subsites [884], but the map is incomplete because the individual contributions of subsites A and B cannot be determined. The alternative method for evaluating subsite-binding affinities is based on calculation of the contribution of glycosyl residues to the free energy of binding from the ratio of inhibition constants for a series of substrates (in this case, cello-oligosaccharides). As seen in Figure 6.4 (f), this method provides free energy data for each subsite, and for the *T. reesei* β-glucosidase, it is evident that subsite A of the four subsites contributes the greatest proportion of binding energy [148]. This finding is somewhat surprising because it suggests that once hydrolysed from the nonreducing end of an oligosaccharide, glucose may remain bound to subsite A and thereby inhibit further catalysis. However, the k_{cat} value for this enzyme, approximately 24 sec^{-1} [884], is similar to those of other cellulolytic enzymes, indicating that other factors come into play. For example, it is conceivable that the released glucose is displaced from subsite A as the bound cello-oligosaccharide shifts (translocates) along the binding site, with the repulsive contribution at subsites E and F providing the driving force.

The differences noted above indicate that care must be taken when comparing subsite affinity maps. In addition to the different methods available for generating such maps, other factors have to be considered. Calculations can be complicated with enzymes that readily catalyze transglycosylation reactions, and the composition of reaction products from the hydrolysis of oligosaccharides by endoacting enzymes is usually strongly dependent on substrate concentration [895]. Moreover, there are many equivalent ways in which a given amount of energy may be distributed between subsites, with each obeying the constraints imposed by the experimental data [892]. Quite often, this flexibility arises from the margins of error that can be relatively large. This is exemplified by an analysis of the results obtained from the subsite mapping of an *A. niger* xylanase [892]. Comparison of experimentally obtained values of K_M and V_{max} with theoretical values calculated from determined binding energies revealed significant differences, especially for the shorter substrates. Calculated values of K_M varied from approximately one-half the observed value for xylooctaose to two to three orders of magnitude for xylotetraose and xylotriose, respectively [892]. Another point that should be considered is the fact that calculations of binding energies based on the interaction of short oligosaccharides do not account for any effects that arise from interactions between extended substrates and regions extending beyond the active site of the enzymes. Such dynamic interactions may induce conformational changes in the enzyme that may be transmitted to the active center of the enzyme to alter catalytic properties [891]. Nevertheless, although the subsite model does not fully account for the experimental data, it does provide a preliminary view of the binding-site regions of the β-glycosidases.

6.2 IDENTIFICATION OF BINDING RESIDUES

Aromatic amino acids have been shown to play a prominent role in the binding of carbohydrate ligands to binding proteins and carbohydrases, including the cellulolytic and xylanolytic enzymes. In addition to the strong hydrogen-bonding characteristics of the phenolic hydroxyl of tyrosyl residues, both tyrosine and tryptophan residues participate in stacking interactions with the hydrophobic patches of carbohydrates. The hydrophobic patch of a sugar results from the disposition of the equatorial and axial hydroxyls to one side of the pyranose ring of a sugar monomer, and this may align with the aromatic ring face. The presence and positioning of aromatic amino acid residues within binding sites thus contributes to selectivity of fit of the substrate to the enzyme [206,897,898].

6.2.1 SPECTROSCOPY

The presence of essential aromatic amino acid residues in the binding-site pockets and clefts of the cellulolytic and xylanolytic enzymes has been demonstrated by both physical and chemical techniques. Both difference ultraviolet absorbance and fluorescence spectroscopies provide a nondestructive means for probing the involvement of aromatic residues at the binding sites of enzymes. The interaction of a ligand with an exposed aromatic residue perturbs its spectral properties, and signature spectra are observed for each of Phe, Tyr, and Trp residues. Application of spectroscopic studies can thus provide information concerning the presence, number, and to some extent, function of the aromatic amino acid residues in an enzyme. Surprisingly, however, these techniques have been largely neglected by researchers investigating cellulolytic and xylanolytic enzymes, perhaps in part because of the requirement of relatively high concentrations of enzyme, at least for difference absorbance measurements.

Difference ultraviolet absorbance spectroscopy has been applied to the study of the cellulases from *Penicillium notatum* [899] and *Schizophyllum commune* [900,901], cellobiohydrolases I from *Trichoderma reesei* [890,902], and the *S. commune* xylanase [812]. Binding of cello-oligosaccharide substrates to the cellulases and cellobiohydrolases produced characteristic difference spectra (Figure 6.5) with maxima at 292, 285, and 279 nm indicating that the environment of a Trp residue(s) located in the binding cleft becomes more hydrophobic upon binding of substrate. With the *S. commune* cellulase, binding of the competitive inhibitor cellobiose induced a similar absorbance perturbation, but its intensity was not as great. The extent of ligand perturbation was shown to be dependent on concentration, and values of the dissociation constants, K_d, for cellobiose (22 mM and cellodextrins (3.5 mM) were calculated. Similar kinetic parameters were also obtained by a fluorometric titration of the cellulase with cellobiose, which when complexed with the enzyme was found to quench its intrinsic Trp fluorescence [901]. From the difference absorbance molar extinction coefficients, one and two Trp residues were estimated to be perturbed by the binding of cellobiose and cellodextrins, respectively. These residues, together with a Tyr residue, were also protected from solvent perturbation. Based on these data, it was concluded that at least two Trp residues were located in separate subsites but close to the catalytic site and participated in the binding of substrate to the enzyme [900]. Binding constants were obtained in a similar manner from the interaction of cellobiose and cellotriose with Trp residue(s) in *T. reesei* cellobiohydrolase I [890]. With this enzyme, a second binding site for cellobiose was observed, and binding of ligand to this site is thought to provide stabilization of the core catalytic region to thermal denaturation [902].

Ultraviolet spectroscopic studies of *S. commune* xylanase A [812] provided evidence for the participation of one or more Tyr residues in the binding of ligands (Figure 6.5). The slight shift in difference absorbance maxima indicated either the formation of new hydrogen bonds or the disruption of the intrinsic hydrogen-bonding network involving the Tyr residue(s). Because the xylo-oligomers used in these experiments would

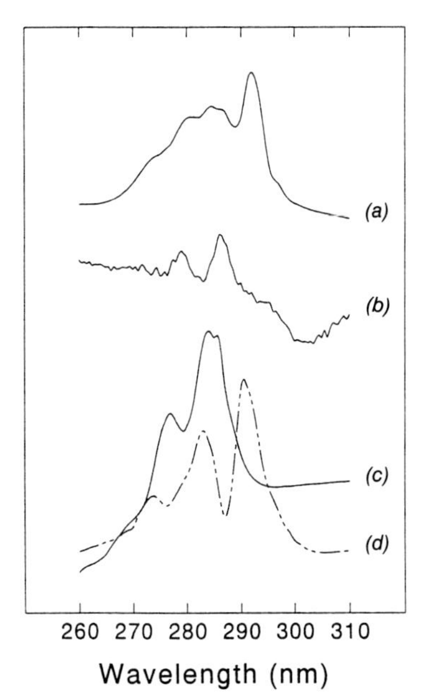

Figure 6.5 Difference ultraviolet absorbance spectra. The spectra presented are of (a) cellulase [900] and (b) xylanase A [812] from *Schizophyllum commune* induced by cellodextrins and xylo-oligosaccharides, respectively, (c) and (d) the model compounds *N*-acetyl-L-tyrosine ethyl ester and *N*-acetyl-L-tryptophan ethyl ester, respectively, perturbed by dioxane.

have been rapidly hydrolysed to xylobiose and xylotriose, the perturbed Tyr residue was thought to be located in subsites A, B, or C of Figure 6.4(a). However, the apparent subsite-binding energies would predict that xylose preferentially binds subsite A, and this weak inhibitor was observed to neither interact directly with the Tyr residue(s) nor induce sufficient conformational changes to perturb it. Thus, the perturbed Tyr residue was postulated to be localized to subsites B or C [812].

6.2.2 CHEMICAL MODIFICATION

The chemical modification reagent *N*-bromosuccinimide oxidizes the indole ring of Trp residues to oxindole, causing significant changes to the properties of the residue within enzymes. This and another reagent, 2-hydroxy-5-nitrobenzyl bromide, have been used to demonstrate the presence of essential Trp residues in cellulases from *Schizophyllum commune* [901], *Aspergillus niger* [737], and *Penicillium notatum* [899], the *Trichoderma reesei* cellobiohydrolase I [902], the xylanases from *Bacillus* sp. [903], *Fibrobacter succinogenes* [240], *Irpex lacteus* [343], *S. commune* [812,904], *Streptomyces* sp. [905], and *Trametes hirsuta* [355], and the xylosidase from

Aspergillus sydowii [335]. In some cases, oxidations were performed in the presence of substrates or inhibitors, which provided protection from inactivation, indicating that the essential Trp residue(s) is located in the binding-site regions [737,812,901,903,904]. Quantification of the reactions indicated that one and two Trp residues were essential for the activity of the cellulases from *A. niger* [737] and *S. commune* [901], respectively. With respect to the latter enzyme, further investigations involving both difference absorbance and fluorescence spectroscopies of oxidized derivatives revealed that one Trp residue directly participates in the binding of substrate, whereas the other was postulated to maintain the structural integrity of a catalytically competent enzyme [901]. However, it is conceivable that both Trp residues are required for binding, as predicted by difference spectroscopic studies described above [900] and that oxidation of either results in loss of catalytic activity. This would imply that the one apparent residue protected from oxidation by ligands is, in fact, a reflection of the fractional oxidation of two different residues (i.e., oxidation of one Trp residue in a fraction of the enzyme molecules and a second Trp residue in the remainder of the enzyme molecules). With the *S. commune* xylanase, one Trp residue in the binding site appears to be protected by xylo-oligosaccharides from oxidation [812,904].

In a similar manner, chemical modifications with *N*-acetylimidazole and tetranitromethane have shown Tyr residues to be essential for substrate binding to almond β-glucosidase [906] and the xylanases from *T. hirsuta* [355] and *S. commune* [812]. On the basis of *N*-acetylimidazole modifications of the almond β-glucosidase, three Tyr residues were postulated to comprise the active site and participate in the binding of glucosides [906]. Reaction of *S. commune* xylanase A with a 100-fold molar excess of tetranitromethane led to the nitration of approximately 3 Tyr residues and the concomitant elimination of catalytic activity, whereas *N*-acetylimidazole in great excess had little effect. The differences in both the chemical properties of the two Tyr-specific reagents and their effects on this xylanase are consistent with modified Tyr residues being located in the relatively hydrophobic environment of a substrate-binding cleft. Tetranitromethane is a nonpolar reagent and typically reacts with buried residues, whereas *N*-acetylimidazole is quite polar and has been shown to preferentially modify exposed surface Tyr residues [907]. As with the difference spectroscopic investigations described above, xylobiose and xylotriose would have been the predominant protective ligands remaining in the reaction mixtures during the protection studies involving tetranitromethane. Hence, the Tyr residue nitrated by the chemical reagent was predicted to be the residue located in subsites 2 or 3 and perturbed by ligands.

Through differential labelling with tetranitromethane employing xylopentaose as protective ligand and peptide mapping studies, the catalytically essential Tyr residue of xylanase A was identified as Tyr97 [812]. The nitration of a Tyr would have several consequences, including the effects of the bulky nitro group being placed *ortho* to the phenolic hydroxyl function. In addition, the substituent nitro group would push electrons into the benzene ring (inductive effect) and thereby lower the pK_a of the phenolic hydroxyl from approximately 10.3 to 7.3 [908]. Thus, the nitrated Tyr97 would be partially ionized at physiological pH, and this is likely to critically disrupt the function of a binding residue within the confines of the active center of the enzyme.

6.2.3 X-RAY CRYSTALLOGRAPHY

The localization of active-site clefts or pockets together with the identification of amino acid residues that directly interact with ligands can be readily made if crystals are prepared of enzyme-substrate or inhibitor complexes. This is not a trivial process because the binding of substrate or inhibitor to enzymes leads to conformational changes that quite often result in the cracking of crystals. If the crystal survives the binding process or cocrystallization experiments were successful, substrate residues may only be observed in electron density maps if they are tightly bound to the crystallized enzyme. Moreover, enzymatic hydrolysis of substrate within crystallized enzymes will result in the production of a variety of reaction products leading to crystallographic disorder and hence undiscernible electron densities. As an alternative, identification of catalytic residues permits the docking of substrate through computer-modelling techniques. As with studies designed to identify catalytic residues, site-directed mutagenesis in combination with X-ray crystallography has proven to be very successful in probing the amino acid residues critical for binding of substrate.

The three-dimensional structure of three different enzyme-inhibitor complexes has been determined for the Family 6 cellobiohydrolase from *Trichoderma reesei* [208]. Four binding subsites for glucosyl residues were observed within the active-site tunnel of this exoacting enzyme, thus confirming earlier results obtained by subsite mapping experiments [883]. Only sugar residues in subsites A and C were detected in the electron density maps, but enough information was gleaned from these investigations to allow the accurate modelling of a cellotetraose ligand into the active site. The glucosyl residues bound in subsites A, C, and D pack against Trp135, Trp367, and Trp269, respectively, whereas Tyr169 is positioned in subsite B. It is also possible that Trp272 and Trp364 interact with the glucosyl residue at subsite D. The other face of the sugar residues is exposed to a variety of hydrophilic charged residues (Asp137, Asp175, Asp221, His266, Lys395, Glu399, Asp401, Asp412, His414) and a few hydrophobic residues (Ala178, Ala304, Asn305, Ala427, Gly428). The restricted volume of the active-site tunnel was thought to preclude extensive conformational rearrangement of the cellulose substrate following the release of a cellobiose product from its nonreducing end and its subsequent translocation for the next hydrolytic event. Based on the modelled three-dimensional structure of this enzyme-substrate complex together with bond cleavage frequency data [883,890], a model was proposed to account for the mode of action of this cellobiohydrolase [208]. Once an extended cellulose chain with its zig-zag pattern of glucosidic linkages enters the active-site tunnel, productive binding is directed by amino acid residues at subsites A and B, which ensures the correct orientation of the bond to be cleaved at the active site. Thus, with the first hydrolytic event half of the molecules, by chance, will have their 2-3 glucosidic linkage in a conformation susceptible to catalytic hydrolysis, whereas with the other half, the 3-4 glucosidic linkage will be correctly positioned for cleavage. Threading the substrate onto the next susceptible linkage ensures production of only cellobiose. This model would explain the cleavage patterns observed with cellopentaside substrates that yielded products arising from hydrolysis at linkages 2-3 or 3-4 [883,890]. Binding of the competitive inhibitor cellobiose to another Family 6 cellulase from *Thermomonospora fusca* was observed by difference map analysis (at 2.2 Å) of isomorphous cocrystals of the enzyme complex [210]. Not surprisingly, the binding cleft of this cellulase is also thought to be composed of four subsites (A–D), and cellobiose was observed to bind subsites A and B. Like the *T. reesei* cellobiohydrolase, the binding cleft is lined with three Trp residues and an array of polar amino acids. Two of the Trp residues, Trp41 and Trp162, are completely conserved in all Family 6 enzymes. These residues are exposed at the base of the cleft, and Trp41 was observed to stack against the hydrophobic surface (patch) of the nonreducing glucosyl residue of cellobiose. Because of its conservation among Family 6 enzymes despite unusual main-chain and side-chain torsion angles, Trp162 was predicted to participate in the binding of longer substrates. Residues Lys259 and Glu263 appear to form hydrogen bonds with hydroxyl groups of the nonreducing glucosyl residue, whereas Asp265 and Lys259 interact with individual hydroxyl moieties on the reducing-end residue. The hydroxyl group of Ser189 also appears to participate in the binding of the nonreducing end but indirectly through a water molecule that binds upon the binding of ligand. Although cellobiose was estimated to bind at approximately 50% occupancy, with the nonreducing glucosyl ring binding more tightly than the reducing glucosyl ring, glucose gave no evidence of binding.

Alignment of the three-dimensional structures of the two Family 6 enzymes revealed that their central β-strand regions are structurally highly conserved and correspond in both placement and length, whereas more variation was seen in helical regions [210]. An obvious difference between the two enzyme structures is the topology of the active sites. Two extended loops are present in the cellobiohydrolase, which cover the active-site cleft to form its tunnel. Based on amino acid sequence comparisons, these loops had been predicted to be missing in the cellulase to provide a more open active-site cleft [208]. The cellulase cleft is indeed more exposed, which would be expected for an endoacting enzyme, but one of the extended loops is still present. This loop in the cellulase crystal structure is pulled back away from the active-site cleft, but the possibility that in solution this loop has more flexibility to close down and contribute to substrate binding was not excluded [210]. Other differences noted between the two enzymes concern the varying affinities for specific ligands. For example, cellobiose does not appear to bind in equivalent subsites of the enzymes, and although glucose has been shown to bind the cellobiohydrolase with high affinity [883], it does not appear to complex at all with the cellulase [210]. These data thus reveal that although β-glycosidases of the same family may share many common features, including identical catalytic residues, subtle variations in the amino acid composition within binding clefts profoundly alter activity profiles.

Examination of the three-dimensional structure of the Family 7 cellobiohydrolase I from *T. reesei* suggests that it has seven binding subsites (A–G) [796] and not three as previously suggested by earlier subsite mapping studies [890]. This observation does, however, corroborate more recent subsite mapping experiments with this enzyme that indicated the presence of six subsites [889]. As with cellobiohydrolase II, these subsites are located in a long (40 Å) tunnel. Cocrystallization of the inhibitor *O*-iodobenzyl-1-thio-β-D-cellobioside with cellobiohydrolase I permitted the further characterization of this active-site tunnel [796]. The inhibitor was found to bind at one end of the tunnel with the aglycone moiety at subsite A and a glucosyl residue in subsite B. The second glucosyl residue of the cellobioside was not observed in the electron density maps, perhaps because of its release through the hydrolytic activity of the enzyme. Nevertheless, the first glucosyl residue stacks against the indole ring of Trp376 at subsite B, and a second Trp, Trp367, is positioned in subsite C to stack against the α-face of a glucosyl residue. All the hydroxyl groups of the first glucosyl residue appear to form hydrogen bonds with other amino acids lining the subsite or bound water molecules. In an atypical situation for β-glycosidases, the aglycone in subsite A was found to stack against the charged (ionized), planar guanido group of Arg251. Two other Trp residues, Trp38 and Trp40, are located at the other end of the tunnel, comprising subsites E and G, respectively.

A model based on differences observed between the structures of the active-site tunnels of the *T. reesei* cellobiohydrolases I and II has been proposed to account for the synergistic action of the two enzymes acting on crystalline cellulose [796]. Acting alone, cellobiohydrolase II binds to the tip of cellulose fibrils [909], where the enzyme is proposed to continuously attach, hydrolyse cellobiose from the nonreducing ends of a cellulose chain, and then either translocate to the next cellobiosyl unit or separate from the substrate. The rapid and irreversible sorption of cellobiohydrolase I to crystalline cellulose is postulated to arise from the insertion of a cellulose chain into the long active-site tunnel that continues to be threaded along the binding subsites following release of cellobiose from its nonreducing end. Such activity along the complete length of a cellulose chain would disrupt the hydrogen-bonding network to neighbouring chains and thus render them more susceptible to attack by cellobiohydrolase II [796].

The three-dimensional structure of the Family 9 cellulase D from *Clostridium thermocellum* complexed with the inhibitor *O*-iodobenzyl-1-thio-β-cellobioside has been solved and permitted the modelling of cellohexaose into the six subsites, labelled A–F, of active-site cleft [819]. The glucosyl residue at subsite E is stacked between Phe276 and Tyr551, whereas the indole ring of Trp457 is thought to stack against the glucosyl residue at subsite D. Tyr456 may participate in substrate binding at subsites B or C, whereas Trp400 located on the opposite side of the cleft protrudes into these sites and likely contributes to binding. The putative catalytic residues, Glu555 and Asp201, are appropriately positioned at the glycosidic bond between subsites D and E. Among the polar residues thought to bind substrate include the two highly conserved, basic residues His516 and Lys518. Chemical modification experiments employing diethylcarbonate, a reagent with specificity for His residues, and site-directed mutagenesis had earlier suggested an important functional role for His516 [736].

An X-ray structure of a Family 10 enzyme-substrate complex is not currently available. However, identifying the conserved amino acid residues of the Family 10 glycosidases within the three-dimensional structure of the *C. fimi* xylanase-cellobiohydrolase revealed that most are located at or near the identified catalytic site [798] (Figure 6.6). It is likely that many of these conserved residues, other than those already assigned to assisting the catalytic Glu residues, participate in the binding of substrate. Among these conserved residues are three Trp residues that are positioned very closely to the catalytic residues Glu127 and Glu233. On the basis of a study involving site-directed mutagenesis and subsequent kinetic analyses, one of these conserved residues in the *S. lividans* xylanase, Asn173 (Asn172 in the *C. fimi* enzyme), was predicted to form a hydrogen bond to the xylosyl residue bound in subsite F of the binding-site cleft [910]. This prediction was found to be wholly consistent with the three-dimensional structures of both the *S. lividans* and *C. fimi* enzymes [797,798] and, moreover, permits the orientation of substrate within the active-site cleft for modelling studies.

The crystal structure of the Family 11 *Bacillus circulans* xylanase-substrate complex has been determined to 1.8 Å resolution using a genetically engineered, catalytically incompetent mutant (Glu172Cys) [801]. The binding of the substrate xylotetraose does not significantly alter the structure of the enzyme (the two structures superimpose with an RMS deviation of 0.137

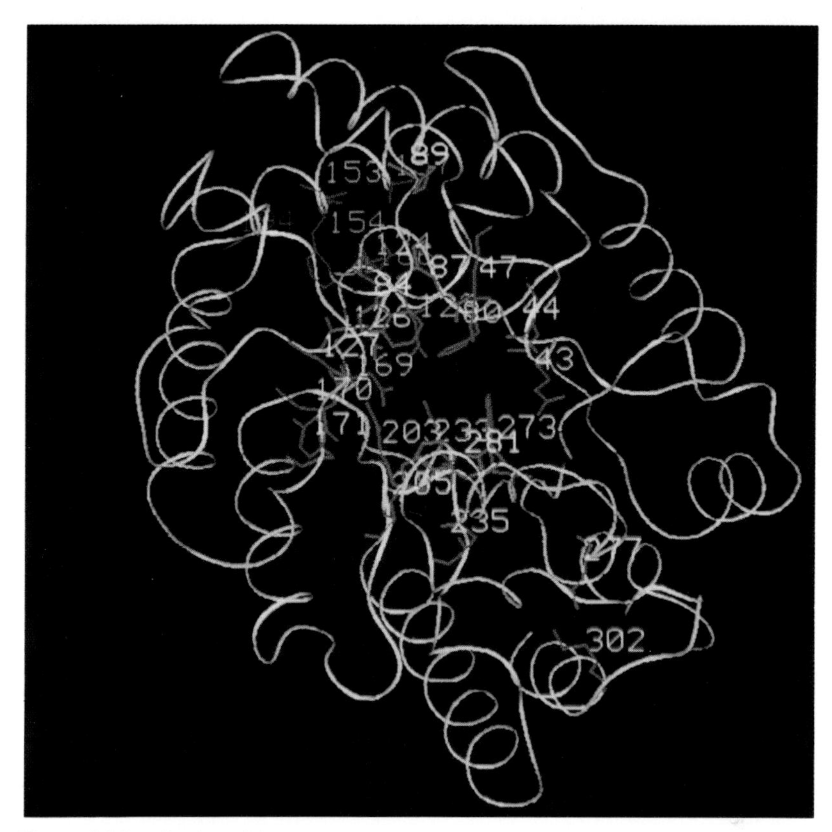

Figure 6.6 Localization of the side chains of the conserved amino acid residues of Family 10 β-glycosidases within the three-dimensional structure (α-carbon) of Cex from *Cellulomonas fimi* (adapted from Reference [803]) using Insight ver. 2.3.0 software (Biosym Technologies, San Diego) and the atomic coordinates (2exo) obtained from the Brookhaven Protein Data Bank).

Å for all main-chain atoms), and the amino acid residues that make close contacts to the visible xylose residues in the electron density map are listed in Table 6.1. As revealed in Figure 5.17, each of these contact residues, which include one Trp and three Tyr residues, are either highly or totally conserved. In addition to its binding role, Tyr69 is thought to hydrogen bond to the putative stabilizing anion-nucleophile Glu78 and play a role in positioning this residue for catalysis [801]. This function was also ascribed to the homologous residue, Tyr77, on the *T. reesei* xylanase II [810,811].

On the basis of difference ultraviolet absorbance and chemical modification experiments with *S. commune* xylanase A (described above), a different conserved aromatic residue, Tyr97, was identified as essential to binding. This residue was predicted to be located in either subsite B or subsite C of the binding-site cleft [812]. The contact residues listed in Table 6.1 are for

TABLE 6.1. Summary of Close Contacts between *B. circulans* Xylanase and the Substrate Xylotetraose.

Substrate Ring	Substrate Atom	Protein Atom	Distance (Å)
Xylose 1	O3	OH Tyr166	2.82
	O2	OH Tyr69	2.88
	C5	CE2 Trp9	3.37
	C5	NE1 Trp9	3.10
Xylose 2	C1	OE1 Glu78	3.37
	O1	OE1 Glu172	2.15
	O2	OE2 Glu78	3.00
	C1	OH Tyr80	3.57
	O2	NE Arg112	2.97
	O3	NH2 Arg112	3.15
	O3	O Pro116	2.56

Source: Compiled from Reference [801].

the *B. circulans* xylanase complexed with xylotetraose, which was observed to span the catalytic site and thus bind in subsites B, C, D, and E. Hence, it is conceivable that Tyr88 of the *B. circulans* xylanase, which is located in the crystal structure on the edge of the active-site cleft [801], comprises subsite A and as predicted for the xylanase A Tyr97, makes binding contacts with substrates of greater degrees of polymerization. Indeed, preliminary studies of enzyme-substrate complexes with the *T. reesei* Xyn II suggest that the homologous Tyr96 packs against a xylose ring of xylan in subsite A [811].

Soaking experiments with various ligands and crystals of both XynI [811] and XynII [810] from *T. reesei* failed to provide high-resolution structures of the complexes. The lack of success with this technique was suggested to be due to the poor binding affinity of the xylo-oligomers. However, weak electron density was observed for one xylosyl residue of a xylan substrate at the active-site of XynII [811]. Using the approximate position of this residue, modelled xylo-oligomers were docked to the active-site clefts of both XynI and XynII [811]. Examination of the subsite architecture in their respective active-site clefts revealed distinctions. The molecular and functional properties of these two *T. reesei* xylanases differ, especially with respect to their mode of action and pH optima (pH 3–4 and 5–5.5, respectively). The binding cleft of XynI appears to comprise three subsites, whereas that of XynII contains five, which may account for the differences in the activity profiles of the two xylanases [811]. Trp18 of XynII, which is homologous to Trp9 of the *B. circulans* xylanase, is thought to stack against the xylosyl residue at subsite E. In XynI this residue is replaced by Tyr9, but it may function similarly to bind the xylosyl residue at subsite C. Two other

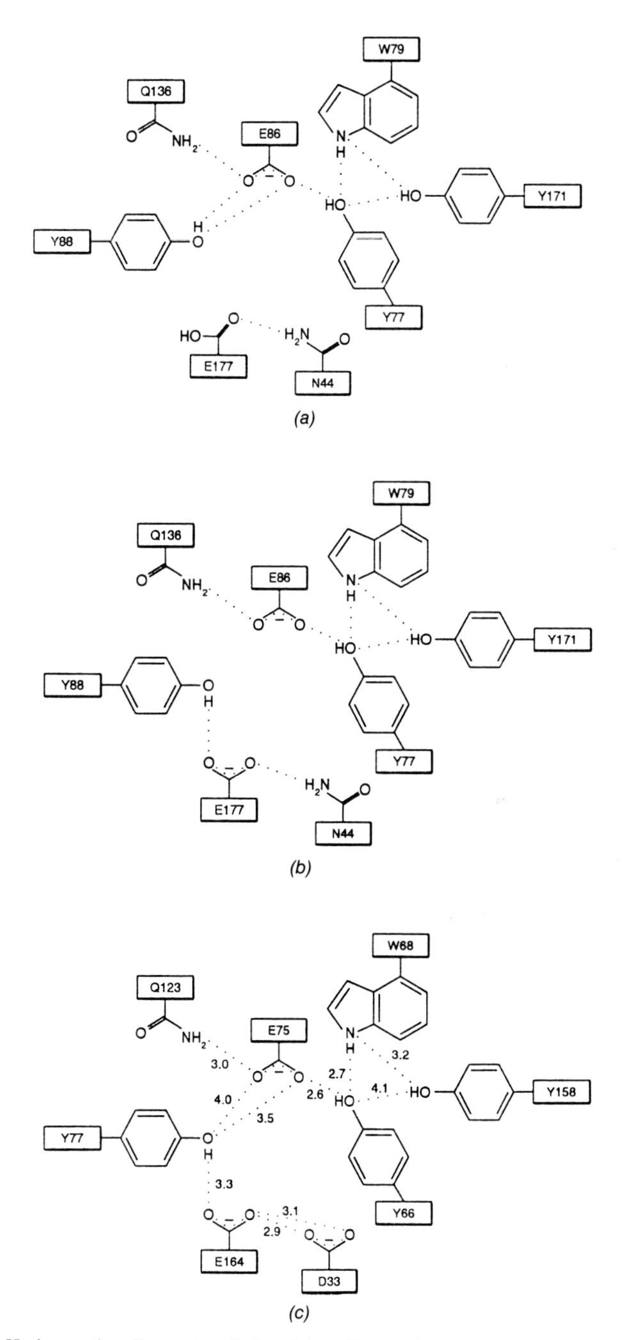

Figure 6.7 Hydrogen-bonding network involving the catalytic residues of xylanases from *Trichoderma reesei*: (a), (b) xylanase II at pH 4.5 and 6.5, respectively, and (c) xylanase I at pH 4.5 (adapted from Reference [811]).

aromatic residues of XynII that are conserved among Family 11 β-glycosidases, Tyr96 and Tyr179, are predicted to participate in stacking interactions at subsites A and B, respectively. The homologous residue Tyr179 in XynI, Trp166, may contribute to binding at subsite B of this enzyme.

The different pH optima associated with these enzymes has been postulated to arise from the observed subtle variations in positioning of the network of residues surrounding the catalytic acids and nucleophiles [811]. The constellation of amino acid residues surrounding the catalytic nucleophiles of XynI and XynII, Glu75 and Glu86, respectively, were observed to be highly conserved in the Family 11 xylanases. These clusters include the homologous residues to Tyr residues (Tyr69 and Tyr80) in the *B. circulans* xylanase identified as binding ligand. These residues may also contribute to the binding of substrate to the *T. reesei* xylanases, but it is thought that they may also help to position and fine-tune the properties of the catalytic nucleophiles. The hydrogen-bonding network involving both the catalytic nucleophiles and the acid-base catalysts for XynI and XynII are depicted in Figure 6.7. The hydrogen-bonding network in XynII at pH 4.5 changes with adjustment of the pH to 6.5 to a pattern closely resembling that of XynI at pH 4.5. It is postulated that this latter conformation represents the catalytically competent active-site and thus accounts for the differences in pH optima of the two xylanases; the pH optima of XynI and XynII are 3.0–4.0 and 5.0–5.5, respectively [811].

Seven subsites are evident in the X-ray crystal structure of cellulase V from *Humicola insolens* [821]. Cocrystallization of this Family 45 β-glycosidase with cellobiose revealed it binds to subsites D and E. Details of this complex have not been described, but two Tyr residues, Tyr8 and Tyr147, composing these subsites appear to be in position to interact with the competitive inhibitor.

6.3 MODIFICATION OF CATALYTIC PROPERTIES BY RATIONAL DESIGN

Manipulation of the genes coding for the cellulolytic and xylanolytic enzymes by the powerful technique of site-directed mutagenesis permits the production of mutant enzymes with new desired properties. These enzymes could possess enhanced physical and/or catalytic properties, such as increased thermostability, broader or narrower substrate profiles, greater catalytic efficiencies, and altered transglycosylation activities. The development of these *designer* enzymes is greatly facilitated by knowledge of the structure and function relationship of an enzyme obtained through the approaches described above. With the great strides recently made in this direction, the rational design of altered enzymes represents the future of research involving the cellulolytic and xylanolytic enzymes.

The enzymatic properties of xylanase A from *Streptomyces lividans* have been altered through the replacement of an amino acid residue located within its active-site cleft [910]. Examination of the three-dimensional structure of this enzyme indicates that Asn173 is located at subsite F, and its replacement with Asp profoundly affected the mode of action of the enzyme. The catalytic residues are positioned between subsites C and D, and the binding of xylopentaose thus resulted in the equal hydrolysis of the second and third glycosidic bonds from the nonreducing end of the substrate. Production of the Asn173Asp mutant forced this substrate back one xylosyl residue with binding occurring at only subsites A–E. Thus, hydrolysis catalyzed by this mutant enzyme occurred preferentially at only the second glycosidic linkage. This alteration to the structure of subsite E also affected the transglycosylation reactions catalyzed by the Asn173Asn mutant. Instead of producing xylohexaose, xyloheptaose, xylooctaose, and larger xylo-oligosaccharides from xylopentaose, as observed with the wild-type enzyme, the mutant produced only a small amount of xyloheptaose and xylooctaose [910]. This significant alteration in the action pattern of the enzyme caused by the replacement of a single residue at its binding site indicates the feasibility for developing novel enzymes with desired activity profiles.

The enzymatic properties of cellobiohydrolase I from *Trichoderma reesei* have also been altered, but not by amino acid replacement. Chemical modification of the enzyme with pentaamine ruthenium (III) resulted in the formation of an enzyme adduct that was found to have a two- to three-fold increase in specific activity toward both soluble and insoluble substrates [911]. This finding alone is interesting, but what makes the study intriguing is the fact that the covalently modified enzyme was endowed with additional catalytic properties. The modified enzyme was able to catalyze a hydrogen peroxide-dependent oxidation of veratryl alcohol, which is a substrate for lignin peroxidase [912]. What contribution the protein backbone (to which the transition metal is bound) provides to this new catalytic activity remains to be established, but the possibility of creating cellulolytic enzymes with additional ligninase activity is an exciting one.

Enzymes with increased thermostability are also of great benefit for various industrial processes. The role of disulfide bonds in maintaining the conformation of proteins, including the *T. reesei* cellulase I [913], and providing stability to denaturing conditions has been well established. Oxidation of the intrinsic disulfide bonds (cystine) in, for example, a thermostable xylanase from *Humicola lanuginosa* [914] and the cellulase from *Thermomonospora fusca* [734] led to considerable loss of thermostability. Although it is simple to oxidize cystine bonds and thereby decrease enzyme stability, the formation of stabilizing cross-links while maintaining catalytic activity is not facile. Early attempts at increasing thermostability of cellulolytic and xylanolytic enzymes involved chemical cross-linking, involving reagents such as glutaraldehyde. For example, treatment of a cellulase from *T. reesei* and a β-glucosidase from *Aspergillus niger* with glutaraldehyde with a view to increase thermostability was met with limited success [915]. Approximately 25% of the β-glucosidase activity was lost upon glutaraldehyde modification, but the enzyme was considerably more thermostable. In contrast, no increase in the heat stability of the *T. reesei* cellulase was observed when it was treated with glutaraldehyde [916]. A major problem with this chemical modification approach is one of control; cross-links will only form between preexisting residues that neighbour each other. Moreover, neutralization of the charges associated with amino acid residues involved in the chemical cross-links, especially those in the vicinity of the active-site, often affects catalytic activities.

Genetic engineering provides a tenable alternative to the chemical modification approach by introducing authentic disulfide bonds at specific sites while maintaining most of the stabilizing hydrogen-bonding networks and salt bridges intact. This approach was recently applied for the thermal stabilization of a xylanase from *Bacillus circulans* [638]. With knowledge of its three-dimensional structure, site-directed mutagenesis of the gene coding for this xylanase permitted the introduction of Cys residues where favourable geometry for a disulfide bridge existed. Computer modelling involving additional residues also enabled the creation of a suitable site for a new Cys residue. In most cases, the introduced disulfide bonds were buried within the enzyme, rendering them relatively insensitive to oxidation under physiological conditions. Each of the eight engineered mutant xylanases displayed increased thermostability, but only two retained full catalytic activity at both 50 and 60°C compared with wild-type enzyme. However, two other mutants were considerably more active than the wild-type enzyme at 60°C, a temperature at which the wild-type enzyme loses complete activity within thirty minutes.

6.4 CONCLUDING REMARKS

Considerable effort has been made to delineate the mechanism of action of cellulolytic and xylanolytic enzymes over the past forty-five years since Reese and coworkers first demonstrated their presence in fungi [917]. During this period, analytical methods have been refined while other more powerful ones have been developed, with the result that extremely rapid progress can now be made. Indeed, in only the past three years, the three-dimensional structures of the catalytic domains of ten cellulolytic and xylanolytic enzymes have been determined and more are in press. Information gleaned from these investigations, combined with that obtained by molecular modelling, genetic replacements, affinity labelling, and other chemical approaches, places the bioengineer in the position to efficiently and effectively exploit these enzymes. The pioneering studies described above exemplify the potential for the production of designer cellulolytic and xylanolytic enzymes and should provide the impetus for continued research in this direction. It is pertinent, however, to remember that such manipulations are not trivial and that each case has to be treated separately. For example, increased thermostability was afforded the *Bacillus circulans* xylanase by the introduction of a disulfide linkage [638]. The same group of researchers who conducted this latter study also revealed that the xylanase from *Schizophyllum commune* has a disulfide linkage but is, in fact, *less* thermostable than another Family 11 xylanase from *Bacillus subtilis,* which is devoid of such a stabilizing factor [918]. One should also consider whether there is a real need to create a thermostable enzyme when nature already provides. As noted in Chapter 2, some enzymes that have been recently isolated and characterized have been shown to have optimal temperatures close to, and in one case, exceeding 100°C [178]!

Attention has been directed toward the thermostable enzymes in recent years, and the list of those isolated, purified, and characterized continues to grow. In addition, we can expect more important discoveries concerning the bifunctional enzymes and the multienzyme complexes. With the establishment of the mechanism of action for β-glycosidases in general, investigators may now focus their effort and resources toward a better understanding of the interactions that direct the specificity of a given enzyme. For example, what primary and secondary structural features are responsible for the different activity profiles of the two highly conserved domains of the bifunctional xylanase from *Fibrobacter succinogenes* or between cellobiohydrolases I and II of *T. reesei?* Can the information gleaned from such studies be exploited to produce tailored enzymes with desired activities? Another promising area of study involves the manipulation of the cellulosome and its scaffoldin protein. Consideration has already been given to the possibility of producing heterocellulosomes by substituting resident cellulosomal components with enzymes more suitable for a desired process [496]. The fusion of a genetic element encoding a dockerin sequence to a β-glucosidase gene and its subsequent expression may permit the production of a cellulosome that would not be susceptible to product inhibition by cellobiose. This is just one simple potential change that could be made; and as stated by Ed Bayer [496], the biotechnologists may likely be limited only by their imaginations.

Families of β-Glycosidases: Classification of Cellulolytic and Xylanolytic Enzymes

THE table is compiled from those prepared by Henrissat [204], Henrissat and Bairoch [205], and Shen et al. [206] and also includes newer additions that are individually referenced. Abbreviations used conform to those of the original tables and they are as follows: AIDU, α-L-iduronase; ARAF, α-arabinofuranosidase; BGAL, β-galactosidase; BGLU, β-glucosidase; BMAN, β-mannanase; BXYL, β-xylosidase; CBH, cellobiohydrolase; CED, cellodextrinase; EG, cellulase; EXG, exo-1,3-β-glucanase; LIC, lichenase; LPH, lactase phlorizin hydrolase; ORF, open reading frame; PBGAL, 6-phospho-β-galactosidase; PBGLU, 6-phospho-β-glucosidase; SGSP, spore-germination specific protein, XYN, xylanase.

Family	Enzyme	Source	EC Number	Reference
1	BGLU	*Agrobacterium* sp.	3.2.1.21	
	BGLU A	*Bacillus polymyxa*	3.2.1.21	
	BGLU B	*Bacillus polymyxa*	3.2.1.21	
	BGLU A	*Caldocellum saccharolyticum*	3.2.1.21	
	BGLU A	*Clostridium thermocellum*	3.2.1.21	
	BGLU	*Erwinia chrysanthemi*	3.2.1.21	
	BGLU Q60	*Hordeum vulgare* (Barley)	3.2.1.21	[655]
	BGLU	*Manihot esculenta*	3.2.1.21	
	BGLU B	*Microbispora bispora*	3.2.1.21	

Family	Enzyme	Source	EC Number	Reference
	BGLU 3	*Streptomyces* sp.	3.2.1.21	[330]
	BGLU 1	*Trifolium repens*	3.2.1.21	
	BGLU 2	*Trifolium repens*	3.2.1.21	
	BGAL	*Sulfolobus solfataricus*	3.2.1.23	
	BGAL S	*Sulfolobus solfataricus*	3.2.1.23	
	PBGAL	*Lactobacillus casei*	3.2.1.85	
	PBGAL	*Staphylococcus aureus*	3.2.1.85	
	PBGAL	*Streptococcus lactis*	3.2.1.85	
	PBGLU	*Escherichia coli*	3.2.1.86	
	PBGLU 2	*Escherichia coli*	3.2.1.86	
	LPH	Human	3.2.1.62/108	
	LPH	Rabbit	3.2.1.62/108	
	LPH	Rat	3.2.1.62/108	
3	BGLU	*Agrobacterium tumefaciens*	3.2.1.21	
	BGLU 3	*Aspergillus wentii*	3.2.1.21	
	BGLU A	*Butyrivibrio fibrisolvens*	3.2.1.21	
	BGLU B	*Clostridium thermocellum*	3.2.1.21	
	BGLU	*Hansenula anomala*	3.2.1.21	
	BGLU	*Kluyveromyces fragilis*	3.2.1.21	
	BGLU	*Ruminococcus albus*	3.2.1.21	
	BGLU 1	*Saccharomyces fibuligera*	3.2.1.21	
	BGLU 2	*Saccharomyces fibuligera*	3.2.1.21	
	BGLU	*Schizophyllum commune*	3.2.1.21	
	BGLU	*Trichoderma reesei*	3.2.1.21	
	CED D	*Pseudomonas fluorescens*	3.2.1.74	
5 (A)	EG	*Bacillus* sp. strain 1139	3.2.1.4	

Family	Enzyme	Source	EC Number	Reference
	EG	*Bacillus* sp. strain KSM-64	3.2.1.4	
	EG	*Bacillus* sp. strain KSM-635	3.2.1.4	
	EG A	*Bacillus* sp. strain N-4	3.2.1.4	
	EG B	*Bacillus* sp. strain N-4	3.2.1.4	
	EG C	*Bacillus* sp. strain N-4	3.2.1.4	
	EG B	*Bacillus lautus*	3.2.1.4	
	EG C	*Bacillus lautus*	3.2.1.4	
	EG	*Bacillus polymyxa*	3.2.1.4	
	EG	*Bacillus subtilis* N-24	3.2.1.4	
	EG	*Bacillus subtilis* BSE616	3.2.1.4	
	EG	*Bacillus subtilis* PAP 115	3.2.1.4	[842]
	EG	*Bacillus subtilis* DL G	3.2.1.4	[843]
	EG	*Bacteroides ruminicola*	3.2.1.4	[848]
	EG 1	*Butyrivibrio fibrisolvens*	3.2.1.4	
	EG A	*Butyrivibrio fibrisolvens*	3.2.1.4	
	EG/CBH	*Caldocellum saccharolyticum*	3.2.1.4	
	EG D	*Cellulomonas fimi*	3.2.1.4	[834]
	EG	*Clostridium acetobutylicum*	3.2.1.4	
	EG A	*Clostridium cellulolyticum*	3.2.1.4	
	EG D	*Clostridium cellulolyticum*	3.2.1.4	
	EG B	*Clostridium cellulovorans*	3.2.1.4	
	EG D	*Clostridium cellulovorans*	3.2.1.4	

Family	Enzyme	Source	EC Number	Reference
	EG A	*Clostridium josui*	3.2.1.4	
	EG B	*Clostridium thermocellum*	3.2.1.4	
	EG C	*Clostridium thermocellum*	3.2.1.4	
	EG E	*Clostridium thermocellum*	3.2.1.4	
	EG H	*Clostridium thermocellum*	3.2.1.4	
	EG C307	*Clostridium* sp. F1	3.2.1.4	
	EG 1	*Cryptococcus flavus*	3.2.1.4	
	EG Z	*Erwinia chrysanthemi*	3.2.1.4	
	EG 3	*Fibrobacter succinogenes*	3.2.1.4	
	EG A	*Prevotella ruminicola*	3.2.1.4	
	EG	*Prevotella ruminicola*	3.2.1.4	
	EG	*Pseudomonas solanacearum*	3.2.1.4	
	EG 1	*Robillarda* sp. Y-20	3.2.1.4	
	EG 1	*Ruminococcus albus*	3.2.1.4	
	EG A	*Ruminococcus albus*	3.2.1.4	
	EG B	*Ruminococcus albus*	3.2.1.4	
	EG A	*Ruminococcus flavefaciens* 17	3.2.1.4	
	EG A	*Ruminococcus flavefaciens* FD1	3.2.1.4	
	EG I	*Schizophyllum commune*	3.2.1.4	
	EG	*Streptomyces lividans*	3.2.1.4	
	EG 5	*Thermomonospora fusca*	3.2.1.4	
	EG II	*Trichoderma reesei*	3.2.1.4	
	EG III	*Trichoderma reesei*	3.2.1.4	
	EG	*Xanthomonas campestris*	3.2.1.4	
	CED C	*Pseudomonas fluorescens*	3.2.1.74	

Family	Enzyme	Source	EC Number	Reference
	EXG	*Candida albicans*	3.2.1.58	
	EXG	*Saccharomyces cerevisiae*	3.2.1.58	
	BMAN A	*Caldocellum saccharolyticum*	3.2.1.78	
6 (B)	EG A	*Cellulomonas fimi*	3.2.1.4	
	EG A	*Microbispora bispora*	3.2.1.4	
	EG A	*Streptomyces halstedii*	3.2.1.4	
	EG A	*Streptomyces* sp. KSM-9	3.2.1.4	
	EG 2	*Thermomonospora fusca*	3.2.1.4	
	CBH A	*Cellulomonas fimi*	3.2.1.91	[138]
	CBH II	*Phanerochaete chrysosporium*	3.2.1.91	[919]
	CBH II	*Trichoderma reesei*	3.2.1.91	
7 (C)	EG 1	*Trichoderma reesei*	3.2.1.4	
	CBH I	*Humicola grisea*	3.2.1.91	
	CBH I	*Phanerochaete chrysosporium*	3.2.1.91	
	CBH I-1	*Phanerochaete chrysosporium*	3.2.1.91	
	CBH I-2	*Phanerochaete chrysosporium*	3.2.1.91	
	CBH I	*Trichoderma reesei*	3.2.1.91	
	CBH I	*Trichoderma viride*	3.2.1.91	
8 (D)	EG	*Bacillus* sp. KSM-330	3.2.1.4	
	EG	*Cellulomonas uda*	3.2.1.4	
	EG C	*Clostridium cellulolyticum*	3.2.1.4	
	EG A	*Clostridium thermocellum*	3.2.1.4	
	EG Y	*Erwinia chrysanthemi*	3.2.1.4	
	LIC	*Bacillus circulans* WL-12	3.2.1.73	

Family	Enzyme	Source	EC Number	Reference
9 (E)	EG 1	*Butyrivibrio fibrisolvens*	3.2.1.4	
	EG B	*Cellulomonas fimi*	3.2.1.4	
	EG C	*Cellulomonas fimi*	3.2.1.4	
	EG C	*Clostridium cellulovorans*	3.2.1.4	
	EG Z	*Clostridium stercorarium*	3.2.1.4	
	EG D	*Clostridium thermocellum*	3.2.1.4	
	EG F	*Clostridium thermocellum*	3.2.1.4	
	EG	*Dictyostelium discoideum*	3.2.1.4	[872]
	EG A	*Fibrobacter succinogenes*	3.2.1.4	
	EG	*Persea americana*	3.2.1.4	
	EG 1	*Persea americana*	3.2.1.4	
	EG 2	*Persea americana*	3.2.1.4	
	EG	*Phaseolus vulgaris*	3.2.1.4	
	EG A	*Pseudomonas fluorescens*	3.2.1.4	
	EG 4	*Thermomonospora fusca*	3.2.1.4	
	SGSP	*Dictyostelium discoideum*		
10 (F)	XYN A	*Aspergillus kawachii*	3.2.1.8/4	
	XYN A	*Bacillus* sp. C-125	3.2.1.8	
	XYN A	*Butyrivibrio fibrisolvens*	3.2.1.8	
	XYN B	*Butyrivibrio fibrisolvens*	3.2.1.8	
	XYN A	*Caldocellum saccharolyticum*	3.2.1.8	
	XYN	*Clostridium stercorarium*	3.2.1.8	[800]
	XYN X	*Clostridium thermocellum*	3.2.1.8	
	XYN Z	*Clostridium thermocellum*	3.2.1.8	

Family	Enzyme	Source	EC Number	Reference
	XYN	*Cryptococcus albidus*	3.2.1.8	
	XYN B	*Neocallimastix patriciarum*	3.2.1.8	[230]
	XYN	*Penicillium chrysogenum*	3.2.1.8	
	XYN A	*Pseudomonas fluorescens*	3.2.1.8	
	XYN B	*Pseudomonas fluorescens*	3.2.1.8	
	XYN A	*Ruminococcus flavefaciens*	3.2.1.8	
	XYN A	*Streptomyces lividans*	3.2.1.8	
	XYN I	*Streptomyces thermoviolaceus*	3.2.1.8	
	XYN A	*Thermoanaerobacter saccharolyticum*	3.2.1.8	[800]
	XYN	*Thermoascus aurantiacus*	3.2.1.8	
	EG/CBH	*Caldocellum saccharolyticum*	3.2.1.91	
	CBH	*Cellulomonas fimi*	3.2.1.91	
	ORF4	*Caldocellum saccharolyticum*		
11 (G)	XYN A	*Aspergillus niger*	3.2.1.8	
	XYN C	*Aspergillus kawachii*	3.2.1.8	
	XYN A	*Aspergillus tubigensis*	3.2.1.8	
	XYN A	*Aureobasidium pullulans*	3.2.1.8	[920]
	XYN A	*Bacillus circulans*	3.2.1.8	
	XYN A	*Bacillus pumilus*	3.2.1.8	
	XYN A	*Bacillus subtilis*	3.2.1.8	
	XYN	*Chainia* sp.	3.2.1.8	
	XYN B	*Clostridium acetobutylicum*	3.2.1.8	
	XYN C(A)	*Fibrobacter succinogenes*	3.2.1.8	[239]
	XYN C(B)	*Fibrobacter succinogenes*	3.2.1.8	[239]

Family	Enzyme	Source	EC Number	Reference
	XYN A(A)	*Neocallimastix patriciarum*	3.2.1.8	
	XYN A(B)	*Neocallimastix patriciarum*	3.2.1.8	
	XYN 2	*Nocardiopsis dassonvillei*	3.2.1.8	
	XYN A	*Ruminococcus flavefaciens*	3.2.1.8	
	XYN	*Schizophyllum commune*	3.2.1.8	
	XYN B	*Streptomyces lividans*	3.2.1.8	
	XYN C	*Streptomyces lividans*	3.2.1.8	
	XYN	*Streptomyces* sp. 36a	3.2.1.8	
	XYN II	*Streptomyces thermoviolaceus*	3.2.1.8	
	XYN A	*Thermomonospora fusca*	3.2.1.8	[405]
	XYN	*Trichoderma harzanium*	3.2.1.8	
	XYN 1	*Trichoderma reesei*	3.2.1.8	
	XYN 2	*Trichoderma reesei*	3.2.1.8	
	XYN	*Trichoderma viride*	3.2.1.8	
12 (H)	EG	*Aspergillus aculeatus*	3.2.1.4	
	EG S	*Erwinia carotovora*	3.2.1.4	
	EG S	*Streptomyces rochei*	3.2.1.4	[921]
26 (I)	EG	*Bacteroides ruminicola*	3.2.1.4	
	EG H	*Clostridium thermocellum*	3.2.1.4	
	BMAN	*Bacillus* sp.	3.2.1.78	
39	BXYL B	*Caldocellum saccharolyticum*	3.2.1.37	
	BXYL	*Thermoanaerobacter* sp. B6A	3.2.1.37	
	AIDU	Dog	3.2.1.76	
40	BGLU RolB	*Agrobacterium rhizogenes*	3.2.1.x	

Family	Enzyme	Source	EC Number	Reference
	BGLU RolB	*Nicotinia glauca*	3.2.1.x	
41	BGLU RolC	*Agrobacterium rhizogenes*	3.2.1.x	
	BGLU RolC	*Nicotinia glauca*	3.2.1.x	
43	BXYL	*Bacillus pumilus*	3.2.1.37	
	BXYL/ARAF	*Butyrivibrio fibrisolvens*	3.2.1.37/55	
44 (J)	EG A	*Bacillus lautus*	3.2.1.4	
	BMAN/EG	*Caldocellum saccharolyticum*	3.2.1.4	
45 (K)	EG B	*Pseudomonas fluorescens*	3.2.1.4	
	EG V	*Humicola insolens*	3.2.1.4	
	EG V	*Trichoderma reesei*	3.2.1.4	[425]
46 (L)	EG A (2)	*Clostridium saccharolyticum*	3.2.1.4	
	EG E	*Cellulomonas fimi*	3.2.1.4	
	EG CCF	*Clostridium cellulolyticum*	3.2.1.4	
	CBH S8	*Clostridium thermocellum*	3.2.1.91	
	CBH II	*Clostridium stercorarium*	3.2.1.91	
	ORF P70	*Clostridium cellulovorans*		
	ORF Ss	*Clostridium thermocellum*	3.2.1.x	

Commercial Sources of Substrates for Cellulases and Heteroxylanases

Substrate	Supplier
Celluloses	
Avicel	FMC, Merck
Crystalline cellulose	Fluka, ICN, Sigma, Whatman
Cotton linters	Fluka
Spruce cellulose	Fluka
Chemically modified celluloses	
O-Carboxymethyl cellulose	Fluka, Hercules, ICN, Sigma
O-Hydroxyethyl cellulose	Fluka, Serva
Cello-oligosaccharides	
Cellobiose	Biosynth, Dextra, Fluka, ICN, Seikagaku, Sigma, V-Labs
Cellotriose	Dextra, Fluka, Seikagaku, Sigma, V-Labs
Cellotetraose	Dextra, Fluka, Seikagaku, Sigma, V-Labs
Cellopentaose	Dextra, Seikagaku, Sigma, V-Labs
Cellohexaose	Dextra, Seikagaku, V-Labs
Cello-oligosaccharides (mixture)	Sigma
Chromogenic cellulose	
Remazol-Brilliant Blue (AZCl)-carboxymethyl cellulose	Megazyme

Substrate	Supplier
Chromogenic cellulose (continued)	
Cross-linked Remazol-Brilliant Blue (AZCl)-HE cellulose	Megazyme
Ostazin Brilliant Red (OBR)-hydroxyethyl cellulose	Fluka, ICN, Sigma
Cellulose-Azure	Sigma
Chromogenic cello-oligosaccharides	
5-Bromo-4-chloro-3-indoxyl-β-D-cellobioside	Biosynth
5-Bromo-4-chloro-3-indoxyl-β-D-glucopyranoside	Biosynth
6-Chloro-3-indoxyl-β-D-glucopyranoside	Biosynth
3-Indoxyl-β-D-glucopyranoside	Biosynth
4-Methylumbelliferyl-β-D-cellotrioside	Sigma
4-Methylumbelliferyl-β-D-cellobioside	Biosynth, Sigma
4-Methylumbelliferyl-β-D-glucoside	Biosynth, Fluka, Sigma
1-Napthyl-β-D-glucopyranoside	Biosynth
p-Nitrophenyl-β-glucoside	Biosynth, Fluka, ICN, Sigma
m-Nitrophenyl-β-glucoside	Biosynth
o-Nitrophenyl-β-glucoside	Biosynth, Sigma
p-Nitrophenyl-β-cellobioside	Biosynth, Fluka, ICN, Sigma
o-Nitrophenyl-β-cellobioside	Sigma
p-Nitrophenyl-β-cellotrioside	Sigma
p-Nitrophenyl-β-cellotetraoside	Sigma
p-Nitrophenyl-β-cellopentaoside	Sigma
Xylans	
Arabinogalactoglucuronoxylan	Megazyme
Arabinoxylan (from rye flour; D-xylose:L-arabinose, 1:1)	Megazyme
Arabinoxylan (from wheat flour; D-xylose:L-arabinose, 3:2)	Megazyme
Beechwood xylan	Serva
Birchwood xylan	Sigma
Larchwood 4-*O*-methyl glucuronxylan	Fluka, ICN, Sigma
Oat spelt arabinoglucuronoxylan	Sigma

Substrate	Supplier
Xylo-oligosaccharides	
Xylobiose	Megazyme, Sigma
Xylotriose	Megazyme, Sigma
Xylotetraose	Megazyme
Xylopentaose	Megazyme
Aldotriouronic acid (MeGlcA-α-1,2-Xyl-β-1,4-Xyl)	Megazyme
Aldotetrauronic acid (MeGlcA-α-1,2-Xyl-β-1,4-Xyl-β-1,4-Xyl)	Megazyme
Chromogenic xylans	
Remazol-Brilliant Blue (AZCl)-xylan	
Birchwood	Megazyme
Oat spelts	Megazyme
Remazol-Brilliant Blue (AZCl)-arabinoxylan (wheat)	Megazyme
Cross-linked Remazol-Brilliant Blue (AZCl) xylan	
Birchwood	Megazyme
Oat spelts	Megazyme
Cross-linked Remazol-Brilliant Blue (AZCl) xyloglucan	Megazyme
Remazol-Brilliant Blue xylan (RBBX)	Fluka, Megazyme, Sigma
Chromogenic xylo-oligosaccharides	
5-Bromo-4-chloro-3-indoxy-β-D-xylopyranoside	Biosynth
4-Methylumbelliferyl-β-D-xylopyranoside	Biosynth
p-Nitrophenyl-β-xylopyranoside	Biosynth, Fluka, Serva, Sigma
o-Nitrophenyl-β-xylopyranoside	Sigma
Others	
Arabinogalactan	Serva
Arabino-oligosaccharides (arabinobiose-arabinooctaose)	Megazyme
4-Methylumbelliferyl-acetate	Sigma
p-Nitrophenylacetate	Sigma
p-Nitrophenylarabinofuranoside	Sigma
Pectic galactan	
Lupin	Megazyme
Potato	Megazyme

Biosynth International, Inc.
1665 West Quincy Avenue, Suite 155
Naperville, Illinois
60540
Phone: (800) 270-2436
Fax: (800) 276-2436

Dextra Laboratories
The Innovation Centre
The University
P.O. Box 68
Reading, RG6 2BX, UK

FMC Corporation
Food and Pharmaceutical Products Division
2000 Market Street
Philadelphia, Pennsylvania
19103
Phone: (215) 299-6000

ICN Pharmaceuticals Inc.
3300 Hyland Avenue
Costa Mesa, California
92626
Phone: (714) 545-0113

E. Merck
Frankfurter Strasse 250
Postfach 4119
D-64293 Darmstadt
Germany
Phone: 06151 72 0
Fax: 06151 72 2000

Serva Feinbiochemica GmbH & Co.
Carl-Benz Strasse 7
D-69115
Heidelberg, Germany
Phone: 06221 502 0
Fax: 06221 502 113

V-Labs, Inc.
423 North Threard Street
Covington, Louisiana
70433
Phone: (504) 893-0533
Fax: (504) 893-0533

Fluka Chemie AG
Industriestrasse 25
CH-9470
Buchs, Switzerland
Phone: 081 755 25 11
Fax: 081 756 54 49

Hercules Incorporated
Hercules Plaza
Wilmington, Delaware
19894
Phone: (800) 247-4372

Megazyme
2/11 Ponderosa Parade
Warrlewood (Sydney)
N.S.W. 2102, Australia
Phone: (612) 979-8163
Fax: (612) 979-8272

Seikagaku America Incorporated
2502 Urbanna Pike
Suite 100
Ijamsville, Maryland
21754
Phone: (301) 874-3656
Fax: (301) 874-3677

Sigma Chemical Co.
P.O. Box 14508
St. Louis, Missouri
63178
Phone: (314) 771-5750
Fax: (314) 771-5757

Whatman Incorporated
9 Bridewell Place
Clifton, New Jersey
07014
Phone: (201) 773-5800
Fax: (201) 472-6949

References

1 Young, R. A. and R. M. Rowell. 1986. *Cellulose. Structure, modification and hydrolysis*. New York: John Wiley and Sons.

2 Carpita, N. C. and D. M. Gibeaut. 1993. "Structural models of primary cell walls in flowering plants: Consistency of molecular structure with the physical properties of the walls during growth," *Plant J.*, 4:1–30.

3 Fincher, G. B. and B. A. Stone. 1986. "Cell walls and their components in cereal grain technology," *Adv. Cereal Sci. Technol.*, 8:207–295.

4 Meinert, M. C. and D. P. Delmer. 1977. "Changes in biochemical composition of the cell wall of the cotton fiber during development," Plant Physiol., 59: 1088–1097.

5 Payen, M. 1839. "Rapport sur un Memoire de M. Payen, relatif a la composition de la matiere ligneuse," *Compte. Rendu Acad. Sci. Paris*, 8:51–53.

6 Schulze, E. 1891. "Zur kenntnis der chimeschen zusammensetzung der pflanzlichen zellmembranen," *Ber. Dtsch. Chem. Ges.*, 24:2277–2287.

7 Ring, S. G. and R. R. Selvendron. 1981. "An arabinogalactoglucan from the cell wall of *Solanum tuberosum*," *Phytochemistry*, 20:2511–2519.

8 Jermyn, M. A. and F. A. Isherwood. 1956. "Changes in the cell wall of the pear during ripening," *Biochem. J.*, 64:123–132.

9 Atalla, R. H. and D. L. Van der Hart. 1984. "Native cellulose: A composite of two distinct cyrstalline forms," *Science*, 223:283–285.

10 Marx-figini, M. 1982. "The control of molecular weight and molecular-weight distribution in the biogenesis of cellulose," in *Cellulose and other natural polymer systems*, R.M. Brown, Jr., ed. New York: Plenum, pp. 243–271.

11 Timpa, J. D. 1991. "Application of universal calibration in gel permeation chromatography for molecular weight determinations of plant cell wall polymers: Cotton fibers," *J. Agr. Food Chem.*, 39:270–275.

12 Frey, A. 1928. *Die Micellartheorie von Carl Nageli*. Leipzig: Akademische verlagsgesellschaft M.B.H.

13 Kyu, K. L., K. Ratanakhanokchai, and M. Tanticharoen. 1994. "Induction of xylanase in *Bacillus circulans* B6," *Bioresource Technol.*, 48:163.

14 Gardner, K. H. and J. Blackwell. 1974. "The hydrogen bonding in native cellulose," *Biochim. Biophys. Acta*, 343:232–237.

15 Preston, R. D. 1979. "Polysaccharide conformation and cell wall function," *Annu. Rev. Plant Physiol.*, 30:55–78.

16 Pizzi, A. 1985. "The structure of cellulose by conformational analysis. 2. The cellulose polymer chain," *J. Macromol. Sci. Chem.*, A22:105–137.

17 Pizzi, A. and A. Eaton. 1985. "The structure of cellulose by conformational analysis. 3. Crystalline and amorphous structure of cellulose I," *J. Macromol. Sci. Chem.*, A22:139–168.

18 Van der Hart, D. L. and R. H. Atalla. 1984. "Studies of microstructure in native celluloses using solid-state ^{13}C-NMR," *Macromolecules*, 17:1465–1472.

19 Sarko, A. 1978. "What is the crystalline structure of cellulose?" *Tappi*, 61:59–63.

20 French, A. 1978. "Modification of polysaccharides containing uronic acid residues," *Carbohydr. Res.*, 61:81–87.

21 Meyer, K. H. and L. Misch. 1937. "The constitution of the crystalline part of cellulose. VI. The positions of the atoms in the new space model of cellulose," *Helv. Chem. Acta*, 20:232–244.

22 Nieduszynski, I. A. and E. D. T. Atkins. 1970. "Preliminary investigations of algal cellulose. I. X-Ray intensity data," *Biochim. Biophys. Acta*, 222:109–118.

23 Delmer, D. P. and B. A. Stone. 1988. "Biosynthesis of plant cell walls," in *The biochemistry of plants. Vol. 14. Carbohydrates*, J. Preiss, ed. San Diego: Academic Press, pp. 372–420.

24 Atalla, R. H. 1987. *The structure of cellulose. Characterization of the solid states. ACS Symposium Series, Vol. 340.* Washington, D.C.: American Chemical Society.

25 Henrissat, B. 1985. "Structure et réactivité enzymatique de la cellulase," Ph.D. diss., University de Genoble.

26 Wood, T. M. 1988. "Preparation of crystalline, amorphous, and dyed cellulase substrates," *Methods Enzymol.*, 160:19–25.

27 Marx-figini, M. 1966. "Comparison of the biosynthesis of cellulose in vitro and in vivo in cotton bolls," *Nature (London)*, 210:754–755.

28 Marx-figini, M. and G. V. Schulz. 1966. "On the kinetics and mechanism of cellulose biosynthesis in higher plants," *Biochim. Biophys. Acta*, 112:81–101.

29 Bacic, A., P. J. Harris, and B. A. Stone. 1988. "Structure and function of plant cell walls," in *The biochemistry of plants*, J. Preiss, ed. San Diego: Academic Press, Inc., pp. 297–371.

30 Timell, T. E. 1964. "Wood hemicelluloses: Part I," *Adv. Carbohydr. Chem.*, 19: 247–302.

31 Timell, T. E. 1965. "Wood hemicelluloses: Part II," *Adv. Carbohydr. Chem.*, 20: 409–483.

32 Wilkie, K. C. B. 1979. "The hemicelluloses of grasses and cereals," *Adv. Carbohydr. Chem. Biochem.*, 36:215–264.

33 Dekker, R. F. H. 1985. "Biodegradation of the hemicelluloses," in *Biosynthesis and Biodegradation of wood components*, T. Higuchi, ed. Orlando, FL: Academic Press, Inc., pp. 505–533.

34 Eda, S., A. Ohnishi, and K. Kato. 1976. "Xylan isolated from the stalk of *Nicotiana tabacum*," *Agric. Biol. Chem.*, 40:359–364.

35 Montgomery, R., F. Smith, and H. C. Srivastava. 1956. "Structure of corn hull hemicellulose. I. Partial hydrolysis and identification of 2-β-(α-D-glucopyrano-syluronic acid)-D-xylopyranose," *J. Am. Chem. Soc.*, 78:2837–2839.

36 Chanda, S. K., E. L. Hirst, J. K. N. Jones, and E. G. V. Percival. 1950. "The constitution of xylan from esparto grass (*Stipa tenacissima*, L.)," *J. Chem. Soc.*, 1950:1289–1297.

37 Lindberg, B., K.-G. Rosell, and S. Svensson. 1973. "Positions of the *O*-acetyl groups in birch xylan," *Sven. Papperstidn.*, 76:30–32.

38 Reicher, F., P. A. J. Gorin, M. R. Sierakowski, and J. B. C. Correa. 1989. "Highly uneven distribution of *O*-acetyl groups in the acidic D-xylan of *Mimosa scabrella* (Bracatinga)," *Carbohydr. Res.*, 193:23–31.

39 Johansson, M. H. and O. Samuelson. 1977. "Reducing end groups in birch xylan and their alkaline degradation," *Wood. Sci. Technol.*, 11:251–263.

40 Chanzy, H. D., A. Grosrenaud, J.-P. Joseleau, M. Dube, and R. H. Marchessault. 1982. "Crystallization behavior of glucomannan," *Biopolymers*, 21:301–319.

41 Puls, J. and J. Schuseil. 1993. "Chemistry of hemicelluloses: Relationship between hemicellulose structure and enzymes required for hydrolysis," in *Hemicellulose and hemicellulases*, M. P. Coughlan and G. P. Hazlewood, eds. London: Portland Press Ltd., pp. 1–27.

42 Darvill, J. E., M. McNeil, A. G. Darvill, and P. Albersheim. 1980. "Structure of plant cell walls. XI. Glucuronoarabinoxylan, a second hemicellulose in primary cell walls of suspension-cultured sycamore cells," *Plant Physiol.*, 66:1135–1139.

43 Mueller-Harvey, I., R. D. Hartley, P. J. Harris, and E. H. Curzon. 1986. "Linkage of *p*-coumaroyl and feruloyl groups to cell wall polysaccharides of barley straw," *Carbohydr. Res.*, 148:71–85.

44 Meier, H. 1961. "Isolation and characterization of an acetylated glucomannan from pine," *Acta Chem. Scand.*, 15:1381–1385.

45 Katz, G. 1965. "The location and significance of the *O*-acetyl groups in a glucomannan from Parana pine (*Araucaria angustifolia*)," *Tappi*, 48:34–41.

46 Clarke, A. E., R. L. Anderson, and B. A. Stone. 1979. "Form and function of arabinogalactans and arabinogalactan-proteins," *Phytochemistry*, 18:521–540.

47 Fincher, G. B., B. A. Stone, and A. E. Clarke. 1983. "Arabinogalactan-proteins. Structure, biosynthesis and function," *Annu. Rev. Plant Physiol.*, 34:47–70.

48 Schibeci, A., G. B. Fincher, B. A. Stone, and A. B. Wardrop. 1982. "Isolation of plasma membranes from protoplasts of *Lolium multiforum* (ryegrass) endosperm cells," *Biochem. J.*, 205:511–519.

49 Wood, T. M. and S. I. McCrae. 1986. "The effect of acetyl groups on the hydrolysis of ryegrass cell walls by xylanase and cellulase from *Trichoderma koningii*," *Phytochemistry*, 25:1053–1055.

50 Aspinall, G.O. 1982. "Chemical characterization and structure determination of polysaccharides," in *The polysaccharides. Vol. 1*, G.O. Aspinall, ed. San Diego: Academic Press, pp. 35–131.

51 Comtat, J. and J. P. Joseleau. 1981. "Mode of action of a xylanase and its significance for the structural investigation of the branched L-arabino-D-glucurono-D-xylan from redwood (*Sequoia sempervirens*)," *Carbohydr. Res.*, 95:101–112.

52 Debiere, P., B. Priem, G. Strecker, and M. Vignon. 1990. "Purification and properties of an endo-1,4-xylanase excreted by a hydrolytic thermophilic anaerobe, *Clostridium thermolacticum*: A proposal for its action mechanism on larchwood 4-*O*-methylglucuronoxylan," *Eur. J. Biochem.*, 187:573–580.

53 Kovac, P., J. Alfoldi, P. Kocis, E. Petrakova, and J. Hirsch. 1982. "Carbon-13 NMR spectra of a series of oligoglycuronic acid derivatives and a structurally related (4-*O*-methylglucurono)xylan," *Cell. Chem. Technol.*, 16:261–269.

54 Vleigenthart, J. F. G., R. A. Hoffman, and J. P. Kamerling. 1992. "A ^{1}H-NMR spectroscopic study on oligosaccharides obtained from wheat arabinoxylans," in *Xylans and xylanases*, J. Visser, G. Beldman, M. A. Kusters-van Someren, and A. G. J. Voragen, eds. Amsterdam: Elsevier Science Publishers B.V., pp. 17–37.

55 Shimizu, K., M. Hashi, and K. Sakurai. 1978. "Isolation from a softwood xylan of oligosaccharides containing two 4-*O*-methyl-D-glucuronic acid residues," *Carbohydr. Res.*, 62:117–126.

56 Dutton, G. G. S. and J.-P. Joseleau. 1977. "Hemicellulose of redwood (*Sequoia sempervirens*)," *Cellul. Chem. Technol.*, 11:313–319.

57 Atkins, E. D. T. 1992. "Three-dimensional structure, interactions and properties of xylans," in *Xylans and xylanases*, J. Visser, G. Beldman, M. A. Kusters-van Someren, and A. G. J. Voragen, eds. Amsterdam: Elsevier Scientific Publishers B.V., pp. 39–50.

58 Joseleau, J.-P., J. Comtat, and K. Ruel. 1992. "Chemical structure of xylans and their interaction in the plant cell walls," in *Xylans and xylanases*, J. Visser, G. Beldman, M. A. Kusters-van Someren, and A. G. J. Voragen, eds. Amsterdam: Elsevier Science Publishers B.V., pp. 1–15.

59 Watanabe, T., J. Ohnishi, Y. Yamasaki, S. Kaitzu, and T. Koshijima. 1989. "Binding-site analysis of the ether linkages between lignin and hemicelluloses in lignin-carbohydrate complexes by DDQ-oxidation," *Agric. Biol. Chem.*, 53:2233–2252.

60 Watanabe, T. and T. Koshijima. 1989. "Structural features of 2,3-dichloro-5,6-dicyano-1,4-benzoquinone-oxidized lignin acetate revealed by carbon-13 nuclear magnetic resonance spectroscopy," *Mokuzai Gakkaishi*, 31:130–134.

61 Cornu, A., J. M. Besle, P. Mosoni, and E. Grenet. 1994. "Lignin-carbohydrate complexes in forages: Structure and consequences in the ruminal degradation of cell wall carbohydrates," *Reprod. Nutr. Dev.*, 34:385–398.

62 Comtat, J., J.-P. Joseleau, C. Bosso, and F. Barnoud. 1974. "Characterization of structurally similar neutral and acidic tetrasaccharides obtained from the enzymic hydrolyzate of a 4-O-methyl-D-glucurono-D-xylan," Carbohydr. Res., 38:217–224.

63 Das, N. N., S. C. Das, A. K. Sarkar, and A. K. Mukherjee. 1984. "Lignin-xylan ester linkage in mesta fiber (*Hibiscus cannabinus*)," *Carbohydr. Res.*, 129:197–207.

64 Takahashi, N. and K. Koshijima. 1988. "Ester linkages between lignin and glucuronoxylan in a lignin-carbohydrate complex from beech (*Fagus crenata*) wood," *Wood Sci. Technol.*, 22:231–241.

65 Tanaka, K., F. Nakatsubo, and T. Higuchi. 1976. "Reactions of guaicylglycerol-β-guaiacyl ether with several sugars. I. Reaction of quinonemethide with D-glucuronic acid," *Mokuzai Gakkaishi*, 22:589–590.

66 Azuma, J. 1989. "Analysis of lignin-carbohydrate complexes of plant cell walls," in *Plant Fibers. Modern methods for plant analysis. Vol. 10*, H. F. Linskens and J. F. Jackson, eds. Berlin: Springer-Verlag, pp. 100–126.

67 Joseleau, J.-P. and R. Kesraoui. 1986. "Glycosidic bonds between lignin and carbohydrates," *Holzforschung*, 40:163–168.

68 Neilson, M. J. and G. N. Richards. 1982. "Chemical structures in a lignin-carbohydrate complex isolated from the bovine rumen," *Carbohydr. Res.*, 104:121–128.

69 Holmberg, B. and S. Runius. 1925. "Alcoholate digestion of wood," *Svensk. Kem. Tid.*, 37:189–197.

70 Bolker, H. I. 1963. "A lignin carbohydrate bond as revealed by infra-red spectroscopy," *Nature (London)*, 197:489–490.

71 Nakamura, Y. and T. Higuchi. 1976. "Ester linkage of *p*-coumaric acid in bamboo lignin," *Holzforschung*, 30:187–191.

72 Fry, S. 1986. "Crosslinking of matrix polymers in the growing cell walls of angiosperms," *Annu. Rev. Plant Physiol.*, 37:165–186.

73 Scalbert, A., B. Monties, C. Rolando and A. Sierra-Escudo. 1986. "Formation of ether linkage between phenolic acids and gramineae lignin: A possible mechanism involving quinone methides," *Holzforschung*, 40:191–195.

74 Hartley, R. D., W. H. Morrison, D. S. Himmelsbach, and W. S. Borneman. 1990. "Cross-linking of cell wall phenolic arabinoxylans in graminaceous plants," *Phytochemistry*, 29:3705–3709.

75 Geissman, T. and H. Neukom. 1971. "Vernetzung von Phenolearbonsaureestern von Polysaccharides durch oxydative phenolische Kupplung," *Helv. Chem. Acta*, 54:1108–1112.

76 Nishitani, K. and D. J. Nevins. 1989. "Enzymic analysis of feruloylated arabinoxylans (Feraxan) derived from *Zea mays* cell walls," *Plant Physiol.*, 91:242–248.

77 Selvendran, R. R. 1985. "Developments in the chemistry and biochemistry of pectic acid and hemicellulosic polymers," *J. Cell Sci.*, Suppl. 2:51–88.

78 Ruel, K., J. Comtat, and F. Barnoud. 1977. "Ultrastructural and histological localization of xylans in the primary walls of *Arundo donax* tissues," *C. R. Hebd. Seances Acad. Sci.*, Ser. D 284:1421–1424.

79 Vian, B., D. Reis, M. Mosiniak, and J. C. Roland. 1986. "The glucuronoxylans and the helicoidal shift in cellulose microfibrils in linden wood: Cytochemistry *in muro* and on isolated molecules," *Protoplasma*, 131:185–199.

80 Reis, D., B. Vian, H. Chanzy, and J. C. Roland. 1991. "Liquid crystal-type assembly of native cellulose-glucuronoxylans extracted from plant cell wall," *Biol. Cell*, 73:173–178.

81 Ruel, K. and J.-P. Joseleau. 1984. "Use of enzyme-gold complexes for the ultrastructural localization of hemicelluloses in the plant cell wall," *Histochemistry*, 81:573–580.

82 Mora, F., K. Ruel, J. Comtat, and J.-P. Joseleau. 1986. "Aspect of native and redeposited xylans at the surface of cellulose microfibrils," *Holzforschung*, 40:85–91.

83 Northcote, D. H., D. Davey, and J. Lay. 1989. "Use of antisera to localize callose, xylan and arabinogalactan in the cell-plate, primary and secondary walls of plant cells," *Planta*, 178:353–366.

84 Meikle, P. J., N. J. Hoogenraad, I. Bonig, A. E. Clarke, and B. A. Stone. 1994. "A (1-3,1-4)-β-glucan-specific monoclonal antibody and its use in the quantitation and immunocytochemical location of (1-3,1-4)-β-glucans," *Plant J.*, 5:1–9.

85 Meikle, P. J., I. Bonig, N. J. Hoogenraad, A. E. Clarke, and B. A. Stone. 1991. "The location of (1-3)-β-glucans in the walls of pollen tubes of *Nicotiana alata* using a (1-3)-β-glucan-specific monoclonal antibody," *Planta*, 185:1–8.

86 Barry, P., G. Prensier, and E. Grenet. 1991. "Immunogold labelling of arabinoxylans in the plant cell walls of normal and bn3 mutant maize," *Biol. Cell*, 71:307–312.

87 Martel, P. and I. E. P. Taylor. 1993. "Neutron diffraction and proton nuclear magnetic resonance: Complementary probes of *in situ* cellulose dimensions and primary plant cell wall structure," *Can. J. Botany*, 71:1375.

88 Dekker, R. F. H. and R. N. Richard. 1975. "Structures of the oligosaccharides from the enzymic hydrolysis of hemicellulose by a hemicellulase of *Ceratocystis paradoxa*," *Carbohydr. Res.*, 43:335–344.

89 Wilkie, K. C. B. and S. L. Woo. 1977. "A heteroxylan and hemicellulosic materials from bamboo leaves, and a reconsideration of the general nature of commonly occurring xylans and other hemicelluloses," *Carbohydr. Res.*, 57:145–162.

90 Buchala, A. J. and H. Meier. 1972. "Hemicelluloses from the stalk of *Cyperus papyrus*," *Phytochemistry*, 11:3275–3278.

91 Brillouet, J. M. and J. P. Joseleau. 1987. "Investigation of the structure of a heteroxylan from the outer pericarp (Beeseing bran) of wheat kernel," *Carbohydr. Res.*, 159:109–126.

92 Karacsonyi, S., J. Alfoldi, M. Kubavkova, and L. Stupka. 1983. "Chemical and carbon-13 NMR studies in the distribution of the *O*-acetyl groups in the *O*-acetyl(4-*O*-methyl-D-glucurono)-D-xylan from white willow (*Salix alba* L.)," *Cellul. Chem. Technol.*, 17:637–645.

93 Saavedra, F., S. Karacsonyi, and J. Alfoldi. 1988. "Studies of the polysaccharides of sugar cane (*Saccharum officinarum*): Structural features of the water insoluble D-xylans," *Carbohydr. Res.*, 180:61–71.

94 Kato, Y. and D. J. Nevins. 1985. "Isolation and identification of *O*-(5-*O*-feruloyl-α-L-arabinosyl)-(1→3)-D-xylopyranosyl-(1→4)-D-xylopyranose as component of *Zea* shoot cell walls," *Carbohydr. Res.*, 137:139–150.

95 Whistler, R. L. and E. L. Richards. 1970. "Hemicelluloses," in *The carbohydrates*, W. Pigman and D. Horton, eds. New York: Academic Press, pp. 447–469.

96 Klyosov, A. A., A. P. Sinitsyn, and M. L. Rabinovitch. 1980. "The comparative role of exoglucosidase and cellobiosidase in glucose formation from cellulose," *Enzyme Eng.*, 5:153.

97 Wood, T. M. and S. I. McCrae. 1982. "Purification and some properties of a (1→4)β-D-glucan glucohydrolase associated with the cellulase from the fungus *Penicillium funiculosum*," *Carbohydr. Res.*, 110:291–303.

98 Beguin, P. and J. P. Aubert. 1994. "The biological degradation of cellulose," *FEMS Microbiol. Rev.*, 13:25–58.

99 Dekker, R. F. H. and G. N. Richards. 1976. "Hemicellulases: Their occurrence, purification, properties and mode of action," *Adv. Carbohydr. Chem. Biochem.*, 32:277–352.

100 Coughlan, M. P. 1985. "The properties of fungal and bacterial cellulases with comment on their production and application," in *Biotechnology & genetic engineering reviews, Vol. 3*, G. E. Russel, ed. pp. 39–109.

101 Coughlan, M. P. and L. G. Ljungdahl. 1988. "Comparative biochemistry of fungal and bacterial cellulolytic enzyme systems," in *Biochemistry of genetics of cellulose degradation*, pp. 11–30.

102 Eriksson, K.-E. L., R. A. Blanchette, and P. Ander. 1990. *Microbial and enzymatic degradation of wood and wood components*. Berlin: Springer-Verlag.

103 Berghem, L. E. R., L. G. Pettersson, and U. B. Axio-Fredriksson. 1976. "The mechanism of enzymatic cellulose degradation. Purification and some properties of two different 1,4-β-glucan glucanohydrolases from *Trichoderma viride*," *Eur. J. Biochem.*, 61:621–630.

104 Berghem, L. E. R., L. G. Pettersson, and U. B. Axio-Fredriksson. 1975. "The mechanism of enzymatic cellulose degradation. Characterization and enzymatic properties of a β-1,4-glucan cellobiohydrolase from *Trichoderma viride*," *Eur. J. Biochem.*, 53:55–62.

105 Wood, T. M. and S. I. McCrae. 1978. "The cellulase of *Trichoderma koningii*. Purification and properties of some endoglucanase components with special reference to their action on cellulose when acting alone and in synergism with the cellobiohydrolase," *Biochem. J.*, 171:61–72.

106 Kanda, T., K. Wakabayashi, and K. Nisizawa. 1976. "Xylanase activity of an endo-cellulase of carboxymethyl-cellulase type from *Irpex lacteus* (*Polyporus tulipiferae*)," *J. Biochem. (Tokyo)*, 79:989–995.

107 Hall, J., G. P. Hazlewood, P. J. Barker, and H. J. Gilbert. 1988. "Conserved reiterated domains in *Clostridium thermocellum* endoglucanases are not essential for catalytic activity," *Gene*, 69:29–38.

108 Gerwig, G. J., J. P. Kamerling, J. F. G. Vliegenthart, E. Morag, R. Lamed and E. A. Bayer. 1993. "The nature of the carbohydrate-peptide linkage region in glycoproteins from the cellulosomes of *Clostridium thermocellum* and *Bacteroides cellulosolvens*," *J. Biol. Chem.*, 268:26956–26960.

109 Fierobe, H.-P., C. Bagnara-Tardif, C. Gaudin, F. Guerlesquin, P. Sauve, A. Belaich and J.-P. Belaich. 1993. "Purification and characterization of endoglucanase C from *Clostridium cellulolyticum*. Catalytic comparison with endoglucanase A," *Eur. J. Biochem.*, 217:557–565.

110 Ong, E., D. G. Kilburn, R. C. Miller, and R. A. J. Warren. 1994. "*Streptomyces lividans* glycosylates the linker region of a beta-1,4-glycanase from *Cellulomonas fimi*," *J. Bacteriol.*, 176:999–1008.

111 Langsford, M. L., G. B. Singh, B. Moser, R. C. Miller, Jr., R. A. J. Warren, and Killburn. 1987. "Glycosylation of bacterial cellulases prevents proteolytic cleavage between functional domains," *FEBS Lett.*, 225:163–167.

112 Merivuori, H., J. A. Sands, and B. S. Montenecourt. 1985. "Effects of tunicamycin on secretion and enzymatic activities of cellulase from *Trichoderma reesei*," *Appl. Microbiol. Biotechnol.*, 23:60–66.

113 Olden, K., B. A. Bernard, M. J. Humphries, T.-K. Yeo, S. J. White, S. A. Newton, H. C. Bauer and J. B. Parent. 1985. "Function of glycoprotein glycans," *Trends Biochem. Sci.*, 10:78–81.

114 Chanzy, H., B. Henrissat, and R. Vuong. 1984. "Colloidal gold labelling of 1,4-b-D-glucan cellobiohydrolase adsorbed on cellulose substrates," *FEBS Lett.*, 172:193–197.

115 Coughlan, M. P. and A. P. Moloney. 1988. "Isolation of 1,4-β-D-glucan 4-glucanohydrolase of *Talaromyces emersonii*," *Methods Enzymol.*, 160:363–368.

116 Voragen, A. G. J., G. Beldman, and F. M. Rombouts. 1988. "Cellulases of a mutant strain of *Trichoderma viride* QM 9414," *Methods Enzymol.*, 160:243–251.

117 Bhikhabhai, R., G. Johansson, and G. Pettersson. 1984. "Isolation of cellulolytic enzymes from *Trichoderma reesei* QM 9414," *J. Appl. Biochem.*, 6:336–345.

118 Hakansson, U., L. Fagerstam, G. Pettersson, and L. Andersson. 1978. "Purification and characterization of a low molecular weight 1,4-b-glucan glucanohydrolase from the cellulolytic fungus *Trichoderma viride* QM 9414," *Biochim. Biophys. Acta*, 524:385–392.

119 Niku-Paavola, M.-L., A. Lappalainen, T.-M. Enari, and M. Nummi. 1985. "A new appraisal of the endoglucanases of the fungus *Trichoderma reesei*," *Biochem. J.*, 231:75–81.

120 Goksoyr, J. 1988. "Cellulases from *Sporocytophaga myxococcoides*," *Methods Enzymol.*, 160:338–342.

121 Stewart, J. C. and J. Heptinstall. 1988. "Cellulase and hemicellulase from *Aspergillus fumigatus Fresenius*," *Methods Enzymol.*, 160:264–274.

122 Durbin, M. L. and L. N. Lewis. 1988. "Cellulases in *Phaseolus vulgaris*," *Methods Enzymol.*, 160:342–351.

123 McGavin, M. and C. W. Forsberg. 1988. "Isolation and characterization of endoglucanases 1 and 2 from *Bacteroides succinogenes* S85," *J. Bacteriol.*, 170: 2914–2922.

124 Horikoshi, K., M. Nakao, Y. Kurono, and N. Sashihara. 1984. "Cellulases of an alkalophilic *Bacillus* strain isolated from soil," *Can. J. Microbiol.*, 30:774–779.

125 Sashihara, N., T. Kudo, and K. Horikoshi. 1984. "Molecular cloning and expression of cellulase genes of alkalophilic *Bacillus* sp. strain N-4 in *Escherichia coli*," *J. Bacteriol.*, 158:503–506.

126 Park, J. S., J. Hitomi, S. Horinouchi, and T. Beppu. 1993. "Identification of two amino acids contributing the high enzyme activity in the alkaline pH range of an alkaline endoglucanase from a *Bacillus* sp.," *Prot. Eng.*, 6:921–926.

127 Hitomi, J., J. S. Park, M. Nishiyama, S. Horinouchi, and T. Beppu. 1994. "Substrate-dependent change in the pH-activity profile of alkaline endo-1,4-beta-glucanase from an alkaline *Bacillus* sp.," *J. Biochem. Tokyo.*, 116:554–559.

128 Wood, T. M. and S. I. McCrae. 1977. "Cellulase from *Fusarium solani*. Purification and properties of the C1 component," *Carbohydr. Res.*, 57:117–133.

129 Eriksson, K.-E., B. Pettersson, and U. Westermark. 1974. "Oxidation: An important enzyme reaction in fungal degradation of cellulose," *FEBS Lett.*, 49:282.

130 Wood, T. M., S. I. McCrae, and C. C. MacFarlane. 1980. "The isolation, purification and properties of the cellobiohydrolase component of *Penicillium funiculosum* cellulase," *Biochem. J.*, 189:51–65.

131 Enari, T.-M. and M.-L. Niku-Paavola. 1987. "Enzymatic hydrolysis of cellulose: Is the current theory of the mechanisms of hydrolysis valid?" *CRC Crit. Rev. Biotechnol.*, 5:67–87.

132 Enoki, A., H. Tanaka, and G. Fuse. 1988. "Degradation on lignin-related compounds, pure cellulose and wood components by white- and brown-rot fungi," *Holzforschung*, 42:85–93.

133 Koenigs, J. W. 1972. "Production of extracellular hydrogen peroxide by wood-rotting fungi," *Phytopathology*, 62:100–110.

134 Messner, R., E. M. Kubicek-Pranz, A. Gsur, and C. P. Kubicek. 1991. "Cellobiohydrolase II is the main condial-bound cellulase in *Trichoderma reesei* and other *Trichoderma* strains," *Arch. Microbiol.*, 155:601–606.

135 MacKenzie, C. R., D. Bilous, and K. G. Johnson. 1984. "Purification and characterization of an exoglucanase from *Streptomyces flavogriseus*," *Can. J. Microbiol.*, 30:1171–1178.

136 Creuzet, N., J.-F. Berenger, and C. Frixon. 1983. "Characterization of exoglucanase and synergistic hydrolysis of cellulose in *Clostridium stercorarium*," *FEMS Microbiol. Lett.*, 20:347–350.

137 Gardner, R. M., K. C. Doerner, and B. A. White. 1987. "Purification and characterization of an exo-β-1,4-glucanase from *Ruminococcus flavefaciens* FD-1," *J. Bacteriol.*, 169:4581–4588.

138 Meinke, A., N. R. Gilkes, E. Kwan, D. G. Kilburn, R. A. J. Warren, and R. C. Miller, Jr. 1994. "Cellobiohydrolase A (CbhA) from the cellulolytic bacterium *Cellulo-*

monas fimi is a β-1,4-exocellobiohydrolase analogous to *Trichoderma reesei* CBH II," *Mol. Microbiol.*, 12:413–422.

139 Coughlan, M. P. and L. G. Ljungdahl. 1988. "Comparative biochemistry of fungal and bacterial cellulolytic enzyme systems," in *Biochemistry and genetics of cellulose degradation. FEMS symposium 43*, J.-P. Aubert, P. Beguin, and J. Millet, eds. London: Academic Press, pp.11–30.

140 Hungate, R. E. 1966. *The rumen and its microbes*. New York: Academic Press, Inc.

141 Huang, L. and C. W. Forsberg. 1987. "Isolation of a cellodextrinase from *Bacteroides succinogenes*," *Appl. Environ. Microbiol.*, 53:1034–1041.

142 Huang, L. and C. W. Forsberg. 1988. "Purification and comparison of the periplasmic and extracellular forms of the cellodextrinase from *Bacteroides succinogenes*," *Appl. Environ. Microbiol.*, 54:1488–1493.

143 Huang, L., C. W. Forsberg, and D. Y. Thomas. 1988. "Purification and characterization of a chloride-stimulated cellobiosidase from *Bacteroides succinogenes* S85," *J. Bacteriol.*, 170:2923–2932.

144 Kim, C.-H. and D.-S. Kim. 1995. "Purification and specificity of a specific endo-β-1,4-D-glucanase (Avicelase II) resembling exo-cellobiohydrolase from *Bacillus circulans*," *Enzyme Microb. Technol.*, 17:248–254.

145 Kim, C.-H. 1995. "Characterization and substrate specificity of an endo-β-1,4-D-glucanase I (Avicelase I) from an extracellular multienzyme complex of *Bacillus circulans*," *Appl. Environ. Microbiol.*, 61:959–965.

146 Howell, J. A. and J. D. Stuck. 1975. "Kinetics of Solka Floc cellulose hydrolysis by *Trichoderma viride* cellulase," *Biotechnol. Bioeng.*, 17:873–893.

147 Selby, K. and C. C. Maitland. 1965. "The fractionation of *Myrothecium verrucaria* cellulase by gel filtration," *Biochem. J.*, 94:578–583.

148 Chirico, W. J. and Brown, R. D. Jr., 1987. "Purification and characterization of a β-glucosidase from *Trichoderma reesei*," *Eur. J. Biochem.*, 165:333–341.

149 Umile, C. and C. P. Kubicek. 1986. "A constitutive, plasma-membrane bound β-glucosidase in *Trichoderma reesei*," *FEMS Microbiol. Lett.*, 34:291–295.

150 Coughlan, M. P. and A. McHale. 1988. "Purification of β-D-glucoside glucohydrolases of *Talaromyces emersonii*," *Methods Enzymol.*, 160:437–443.

151 Wilson, R. W. and D. J. Niederpruem. 1967. "Control of β-glucosidases in *Schizophyllum commune*," *Can. J. Microbiol.*, 13:1009–1020.

152 Groleau, D. and C. W. Forsberg. 1981. "Cellulolytic activity of the rumen bacterium *Bacteroides succinogenes*," *Can. J. Microbiol.*, 27:517–530.

153 Ohmiya, K. and S. Shimizu. 1988. "β-Glucosidase from *Ruminococcus albus*," *Methods Enzymol.*, 160:408–414.

154 Pettipher, G. L. and M. J. Latham. 1979. "Production of enzymes degrading plant cell walls and fermentation of cellobiose by *Ruminococcus flavefaciens* in batch and continuous culture," *J. Gen. Microbiol.*, 110:29–38.

155 Ait, N., N. Creuzet, and J. Cattaneo. 1982. "Properties of β-glucosidase purified from *Clostridium thermocellum*," *J. Gen. Microbiol.*, 128:569–577.

156 Forano, E., V. Broussolle, G. Gaudet, and J. A. Bryant. 1994. "Molecular cloning, expression, and characterization of a new endoglucanase gene from *Fibrobacter-succinogenes* S85," *Curr. Microbiol.*, 28:7–14.

157 Fournier, R. A., M. M. Frederick, J. R. Frederick, and P. J. Reilly. 1985. "Purification and characterization of endo-xylanases from *Aspergillus niger* III: An enzyme of pI 3.65," *Biotechnol. Bioeng.*, 27:539–546.

158 Tan, L. U. L., P. Mayers, and J. N. Saddler. 1987. "Purification and characterization of a thermostable xylanase from a thermophilic fungus *Thermoascus aurantiacus*," *Can. J. Microbiol.*, 33:689–692.

159 Grepinet, O., M.-C. Chebrou, and P. Beguin. 1988. "Nucleotide sequence and deletion analysis of the xylanase gene *xynZ* of *Clostridium thermocellum*," *J. Bacteriol.*, 170:4582–4588.

160 Hall, J., G. P. Hazlewood, N. S. Huskisson, A. J. Durrant, and H. J. Gilbert. 1989. "Conserved serine-rich sequences in xylanase and cellulase from *Pseudomonas fluorescens* subspecies *cellulosa*: Internal signal sequence and unusual protein processing," *Mol. Microbiol.*, 3:1211–1219.

161 Gilbert, H. J., D. A. Sullivan, G. Jenkins, L. E. Kellet, N. P. Minton, and J. Hall. 1988. "Molecular cloning of multiple xylanase genes from *Pseudomonas fluorescens* subsp. *cellulosa*," *J. Gen. Microbiol.*, 134:3239–3247.

162 Gilkes, N. R., M. L. Langsford, D. G. Kilburn, R. C. Miller, Jr., and R. A. J. Warren. 1984. "Mode of action and substrate specificities of cellulases from cloned bacterial genes," *J. Biol. Chem.*, 259:10455–10459.

163 Frederick, M. M., C. H. Kiang, J. R. Frederick, and P. J. Reilly. 1985. "Purification and characterization of endo-xylanases from *Aspergillus niger* I: Two isozymes active on xylan near branch points," *Biotechnol. Bioeng.*, 27:525–532.

164 Nishitani, K. and D. J. Nevins. 1991. "Glucuronoxylan xylanohydrolase. A unique xylanase with the requirement for appendant glucuronosyl units," *J. Biol. Chem.*, 266:6539–6543.

165 Ritschkoff, A. C., J. Buchert, and L. Viikari. 1994. "Purification and characterization of a thermophilic xylanase from the brown-rot fungus *Gloeophyllum trabeum*," *J. Biotechnol.*, 32:67–74.

166 Coughlan, M. P., M. G. Tuohy, E. X. F. Filho, J. Puls, M. Claeyssens, M. Vrsanska, and M. M. Hughes. 1993. "Enzymological aspects of microbial hemicellulases with emphasis on fungal systems," in *Hemicellulose and hemicellulases*, M. P. Coughlan and G. P. Hazlewood, eds. London: Portland Press Ltd., pp. 53–84.

167 Biely, P. and M. Vrsanska. 1983. "Synthesis and hydrolysis of 1,3-xylosidic linkages by endo-1,4-β-xylanase of *Cryptococcus albidus*," *Eur. J. Biochem.*, 129:645–651.

168 Filho, E. X. F., M. G. Tuohy, J. Puls, and M. P. Coughlan. 1991. "The xylan-degrading enzyme systems of *Penicillium capsulatum* and *Talaromyces emersonii*," *Biochem. Soc. Trans.*, 19:25S.

169 Reilly, P. J. 1981. "Xylanases—Structure and function," *Basic Life Sci.*, 18: 111–129.

170 Matte, A. and C. W. Forsberg. 1992. "Purification, characterization, and mode of action of endoxylanase 1 and 2 from *Fibrobacter succinogenes* S85," *Appl. Environ. Microbiol.*, 58:157–168.

171 Coughlan, M. P. 1992. "Towards an understanding of the mechanism of action of main-chain hydrolysing xylanases," in *Xylans and xylanases*, J. Visser, G. Beldman, M. A. Kusters-van Someren, and A. G. J. Voragen, eds. Elsevier Science Publishers B.V., pp. 111–139.

172 Wong, K. K. Y., L. V. L. Tan, and J. N. Saddler. 1988. "Multiplicity of β-1,4–xylanase in microorganisms: Functions and applications," *Microbiol. Rev.*, 52:305–317.

173 Honda, H., T. Kudo, Y. Ikura, and K. Horikoshi. 1985. "Two types of xylanases of alkophilic *Bacillus* sp. no. C-125," *Can. J. Microbiol.*, 31:538–542.

174 Okazaki, W., T. Akiba, K. Horikoshi, and R. Akahoshi. 1985. "Purification and characterization of xylanases from alkalophilic thermophilic *Bacillus* spp.," *Agric. Biol. Chem.*, 49:2033–2039.

175 Horikoshi, K. and Y. Atsukawa. 1973. "Xylanase produced by alkophilic *Bacillus* no. C-59-2," *Agric. Biol. Chem.*, 37:2097–2103.

176 Uchino, F. and T. Nakane. 1981. "A thermostable xylanase from a thermophilic acidophilic *Bacillus* sp.," *Agric. Biol. Chem.*, 45:1121–1127.

177 Ratto, M., I. M. Mathrani, B. Ahring, and L. Viikari. 1994. "Application of Thermostable xylanase of *Dictyoglomus* sp. in enzymatic treatment of kraft pulps," *Appl. Microbiol. Biotechnol.*, 41:130–133.

178 Simpson, H. D., U. R. Haufler, and R. M. Daniel. 1991. "An extremely thermostable xylanase from the thermophilic eubacterium *Thermotoga*," *Biochem. J.*, 277: 413–417.

179 Matsuo, M. and T. Yasui. 1984. "Purification and some properties of β-xylosidase from *Emericella nidulans*," *Agric. Biol. Chem.*, 48:1853–1869.

180 Rodionova, N. A., I. M. Tavobilov, and A. M. Bezborodov. 1983. "β-Xylosidase from *Aspergillus niger* 15: Purification and properties," *J. Appl. Biochem.*, 5:300–312.

181 Van Doorslaer, E., H. Kersters-Hilderson, and C. K. De Bruyne. 1985. "Hydrolysis of β-D-xylo-oligosaccharides by β-D-xylosidase from *Bacillus pumilus*," *Carbohydr. Res.*, 140:342–346.

182 Deshpande, V., A. Lachke, C. Mishra, S. Keskar, and M. Rao. 1986. "Mode of action and properties of xylanase and beta-xylosidase from *Neurospora crassa*," *Biotechnol. Bioeng.*, 28:1832–1837.

183 Deleyn, F., M. Claeyssens, and C. K. De Bruyne. 1982. "β-D-Xylosidase from *Penicillium wortmanni*," *Methods Enzymol.*, 83:639–644.

184 Matsuo, M. and T. Yasui. 1984. "Purification and some properties of β-xylosidase from *Trichoderma viride*," *Agric. Biol. Chem.*, 48:1845–1852.

185 Gum, E. and R. Brown. 1976. "Structural characteristics of a glycoprotein cellulase 1,4-β-D-glucan cellobiohydrolase C from *Trichoderma viride*," *Biochim. Biophys. Acta*, 446:371–386.

186 Salovuori, I., M. Makarow, H. Rauvala, J. Knoweles, and L. Kaariainen. 1987. "Low molecular weight high-mannose type glycans in a secreted protein of the filamentous fungus *Trichoderma reesei*," *Bio-Technology*, 5:152–156.

187 Fagerstam, L. G., G. Pettersson, and J. A. Engstrom. 1984. "The primary structure of a 1,4-β-glucan cellobiohydrolase from the fungus *Trichoderma reesei* QM9414," *FEBS Lett.*, 167:309–315.

188 Gerwig, G., J. P. Kamerling, J. F. G. Vliegenhart, E. Morag, R. Lamed, and E. A. Bayer. 1992. "Novel oligosaccharide constituents of the cellulase complex of *Bacteroides cellulosolvens*," *Eur. J. Biochem.*, 205:799–808.

189 Gerwig, G. J., J. P. Kamerling, J. F. G. Vliegenhart, E. Morag (Morgenstern), R. Lamed, and E. A. Bayer. 1991. "Primary structure of *O*-linked carbohydrate chains in the cellulosome of different *Clostridium thermocellum* strains," *Eur. J. Biochem.*, 196:115–122.

190 Gilbert, H. J., J. Hall, G. P. Hazlewood, and L. M. A. Ferreira. 1990. "The N-terminal region of an endoglucanase from *Pseudomonas fluorescens* subspecies *cellulosa* constitutes a cellulose-binding domain that is distinct from the catalytic centre," *Mol. Microbiol.*, 4:759–767.

191 Gilkes, N. R., R. A. J. Warren, R. C. Miller, Jr. and D. G. Kilburn. 1988. "Precise excision of the cellulose binding domains from two *Cellulomonas fimi* cellulases by a homologous protease and the effect on catalysis," *J. Biol. Chem.*, 263: 10401–10407.

192 Ghangas, G. S. and D. B. Wilson. 1989. "Cloning of the *Thermomonospora fusca* endoglucanase E2 gene in *Streptomyces lividans*: Affinity purification and functional domains of the cloned gene product," *Appl. Environ. Microbiol.*, 54: 2521–2526.

193 McGavin, M. and C. W. Forsberg. 1989. "Catalytic and substrate-binding domains of endoglucanase 2 from *Bacteroides succinogenes*," *J. Bacteriol.*, 171:3310–3315.

194 Warren, R. A. J., C. F. Beck, N. R. Gilkes, D. G. Kilburn, M. L. Langsford, R. C. Miller, Jr., G. P. O'Neil, M. Scheufens, and W. K. K. Wong. 1986. "Sequence conservation and region shuffling in an endoglucanase and an exoglucanase from *Cellulomonas fimi*," *Proteins*, 1:335–341.

195 O'Neil, G. P., S. H. Goh, R. A. J. Warren, D. G. Kilburn, and R. C. Miller, Jr. 1986. "Structure of the gene encoding the exoglucanase of *Cellulomonas fimi*," *Gene*, 44:325–341.

196 Willick, G. E. and V. L. Seligy. 1985. "Multiplicity in cellulases of *Schizophyllum commune*: derivation partly from heterogeneity in transcription and glycosylation," *Eur. J. Biochem.*, 151:89–96.

197 Kubicek, C. P., T. Panda, G. Schreferl-Kunar, F. Gruber, and R. Messner. 1987. "*O*-linked but not *N*-linked glycosylation is necessary for the secretion of endoglucanase I and II by *Trichoderma reesei*," *Can. J. Microbiol.*, 33:698–703.

198 Morosoli, R., P. Lecher, and S. Durand. 1988. "Effect of tunicamycin on xylanase secretion in the yeast *Cryptococcus albidus*," *Arch. Biochem. Biophys.*, 265: 183–189.

199 Rabinovitch, M. L., N. Van Viet, and A. A. Klesov. 1982. "Adsorption of cellulolytic enzymes on cellulose and kinetics of action of adsorbed enzymes. Two types of interaction of the enzymes with an insoluble substrate," *Biokihimiya*, 47: 465–477.

200 Beguin, P. 1990. "Molecular biology of cellulose degradation," *Ann. Rev. Microbiol.*, 44:219–248.

201 Henrissat, B. 1992. "Analysis of hemicellulase sequences. Relationship to other glycanases," in *Xylans and xylanases*, J. Visser, G. Beldman, M. A. Kusters-van Someren, and A. G. J. Voragen, eds. Elsevier Science Publishers B.V., pp. 97–110.

202 Henrissat, B., M. Claeyssens, P. Tomme, L. Lemesle, and J. P. Mornon. 1989. "Cellulase families revealed by hydrophobic cluster analysis," *Gene*, 81:83–95.

203 Gilkes, N. R., B. Henrissat, D. G. Kilburn, R. C. Miller, and R. A. J. Warren. 1991. "Domains in microbial β-1,4-glycanases: Sequence conservation, function, and enzyme families," *Microbiol. Rev.*, 55:303–315.

204 Henrissat, B. 1991. "A classification of glycosyl hydrolases based on amino acid sequence similarities," *Biochem. J.*, 280:309–316.

205 Henrissat, B. and A. Bairoch. 1993. "New families in the classification of glycosyl hydrolases based on amino acid sequence similarities," *Biochem. J.*, 293:781–788.

206 Shen, H., P. Tomme, A. Meinke, N. R. Gilkes, D. G. Kilburn, R. A. J. Warren, and R. C. Miller, Jr. 1994. "Stereochemical course of hydrolysis catalysed by *Cellulomonas fimi* CenE, a member of a new family of β-1,4-glucanases," *Biochem. Biophys. Res. Commun.*, 199:1223–1228.

207 Meinke, A., N. R. Gilkes, D. G. Kilburn, R. C. Miller, Jr., and R. A. J. Warren. 1991. "Multiple domains in endoglucanase B (CenB) from *Cellulomonas fimi*: Functions and relatedness to domains in other polypeptides," *J. Bacteriol.*, 173: 7126–7135.

208 Rouvinen, J., T. Bergfors, T. Teeri, J. K. C. Knowles, and T. A. Jones. 1990. "Three-dimensional structure of cellobiohydrolase II from *Trichoderma reesei*," *Science*, 249:380–386.

209 Gilkes, N. R., M. Claeyssens, R. Aebersold, B. Henrissat, A. Meinke, H. D. Morrison, D. G. Kilburn, R. A. J. Warren, and R. C. Miller, Jr.. 1991. "Structural and functional relationships in two families of β-1,4-glycanases," *Eur. J. Biochem.*, 202:367–377.

210 Spezio, M., D. B. Wilson, and P. A. Karplus. 1993. "Crystal structure of the catalytic domain of a thermophilic endocellulase," *Biochemistry*, 32:9906–9916.

211 Bhikhabhai, R. and G. Pettersson. 1984. "The disulfide bridges in a cellobiohydrolase and an endoglucanase from *Trichoderma reesei*," *Biochem. J.*, 222:729–736.

212 Van Tilbeurgh, H., P. Tomme, M. Claeyssens, R. Bhikhabhai, and G. Pettersson. 1986. "Limited proteolysis of the cellobiohydrolase I from *Trichoderma reesei*. Separation of functional domains," *FEBS Lett.*, 204:223–227.

213 Tomme, P., H. Van Tilbeurgh, G. Pettersson, J. Van Damme, J. Vandekerckhove, J. Knowles, T. Teeri, and M. Claeyssens. 1988. "Studies of the cellulolytic systems of *Trichoderma reesei* QM 9414: Analysis of domain function in two cellobiohydrolases by limited proteolysis," *Eur. J. Biochem.*, 170:575–581.

214 Millwardsadler, S. J., D. M. Poole, B. Henrissat, G. P. Hazlewood, J. H. Clarke, and H. J. Gilbert. 1994. "Evidence for a general role for high-affinity non-catalytic cellulose binding domains in microbial plant cell wall hydrolases," *Mol. Microbiol.*, 11:375–382.

215 Hefford, M. A., K. Laderoute, G. E. Willick, M. Yaguchi and V. Seligy. 1992. "Bipartite organization of the *Bacillus subtilis* endo-β-1,4-glucanase revealed by C-terminal mutations," *Prot. Eng.*, 5:433–439.

216 Maglione, G., O. Matsushita, J. B. Russell, and D. B. Wilson. 1992. "Properties of a genetically reconstructed *Prevotella ruminicola* endoglucanase," *Appl. Environ. Microbiol.*, 58:3593–3597.

217 Poole, D. M., A. J. Durrant, G. P. Hazlewood, and H. J. Gilbert. 1991. "Characterization of hybrid proteins consisting of the catalytic domains of *Clostridium* and *Ruminococcus* endoglucanases, fused to *Pseudomonas* non-catalytic cellulose-binding domains," *Biochem. J.*, 279:787–792.

218 Srisodsuk, M., T. Reinikainen, M. Penttila, and T. T. Teeri. 1993. "Role of the interchain linker peptide of *Trichoderma reesei* cellobiohydrolase I in its interaction with crystalline cellulose," *J. Biol. Chem.*, 268:20756–20761.

219 Shen, H., M. Schmuck, I. Pilz, N. R. Gilkes, D. G. Kilburn, R. C. Miller, Jr., and R. A. J. Warren. 1991. "Deletion of the linker connecting the catalytic and cellulose-binding domains of endoglucanase A (CenA) of *Cellulomonas fimi* alters its conformation and catalytic activity," *J. Biol. Chem.*, 266:11335–11340.

220 Donner, T. R., B. R. Evans, K. A. Affholter, and J. Woodward. 1994. "Role of cellulose-binding domain of cellobiohydrolase I in cellulose hydrolysis," *ACS Symp. Ser.*, 566:75–83.

221 Linder, M., M.-L. Mattinen, M. Kontteli, G. Lindeberg, J. Shahlberg, T. Drakenberg, T. Reinikainen, G. Pettersson, and A. Annila. 1995. "Identification of functionally important amino acids in the cellulose-binding domain of *Trichoderma reesei* cellobiohydrolase I," *Protein Sci.*, 4:1056–1064.

222 Kraulis, P. J., G. M. Clore, M. J. Nilges, T. A. Jones, G. Pettersson, J. Knowles, and A. M. Gronenborn. 1989. "Determination of the three-dimensional structure of the C-terminal domain of cellobiohydrolase I from *Trichoderma reesei.* A study using nuclear magnetic resonance and hybrid distance geometry-dynamical simulated annealing," *Biochemistry*, 28:7241–7257.

223 Reinikainen, T., L. Ruohonen, T. Nevanen, L. Laaksonen, P. Kraulis, T. A. Jones, J. K. C. Knowles, and T. T. Teeri. 1992. "Investigation of the function of mutated cellulose-binding domains of *Trichoderma reesei* cellobiohydrolase I," *Proteins*, 14:475–482.

224 Din, N., I. J. Forsythe, L. D. Burtnick, N. R. Gilkes, R. C. Miller, R. A. J. Warren, and D. G. Kilburn. 1994. "The cellulose-binding domain of endoglucanase a (Cena) from *Cellulomonas fimi*—evidence for the involvement of tryptophan residues in binding," *Mol. Microbiol.*, 11:747–755.

225 Xu, G.-Y., E. Ong, N. R. Gilkes, D. G. Kilburn, D. R. Muhandiram, M. Harris-Brandt, J. P. Carver, L. E. Kay, and T. S. Harvey. 1995. "Solution structure of a cellulose-binding domain from *Cellulomonas fimi* by nuclear magnetic resonance spectroscopy," *Biochemistry*, 34:6993–7009.

226 Poole, D. B., G. P. Hazlewood, N. S. Huskisson, R. Virden, and H. J. Gilbert. 1993. "The role of conserved tryptophan residues in the interaction of a bacterial cellulose binding domain with its ligand," *FEMS Microbiol. Lett.*, 106:77–84.

227 Ferreira, L. M. A., A. J. Durrant, J. Hall, G. P. Hazlewood, and H. J. Gilbert. 1990. "Spatial separation of protein domains is not necessary for catalytic activity or substrate binding in a xylanase," *Biochem. J.*, 269:261–264.

228 Tomme, P., D. P. Driver, E. A. Amandoron, R. C. Miller, Jr., R. A. J. Warren, and D. G. Kilburn. 1995. "Comparison of a fungal (Family I) and bacterial (Family II) cellulose-binding domain," *J. Bacteriol.*, 177:4356–4363.

229 Zhang, J. -X. and H. J. Flint. 1992. "A bifunctional xylanase encoded by the xynA gene of the rumen cellulolytic bacterium *Ruminococcus flavefaciens* 17 comprises two dissimilar domains linked by an asparagine/glutamine-rich sequence," *Mol. Microbiol.*, 6:1013–1023.

230 Black, G. W., G. P. Hazlewood, and G.-P. Xue. 1994. "Xylanase B from *Neocallimastix patriciarum* contains a non-catalytic 455-residue linker sequence comprised of 57 repeats of an octapeptide," *Biochem. J.*, 299:381.

231 Din, N., N. R. Gilkes, B. Tekant, R. C. Miller, Jr., R. A. J. Warren, and D. G. Kilburn. 1991. "Non-hydrolytic disruption of cellulose fibers by the binding domain of a bacterial cellulase," *Bio-Technology*, 9:1096–1099.

232 Nidetzky, B., W. Steiner, and M. Claeyssens. 1994. "Cellulose hydrolysis by the cellulases from *Trichoderma reesei*: Adsorptions of two cellobiohydrolases, two endocellulases and their core proteins on filter paper and their relation to hydrolysis," *Biochem. J.*, 303:817–823.

233 Tokatlidis, K., P. Dhurjati, and P. Beguin. 1993. "Properties conferred on *Clostridium-thermocellum* endoglucanase CelC by grafting the duplicated segment of endoglucanase CelD," *Prot. Eng.*, 6:947–952.

234 Tokatlidis, K., S. Salamitou, P. Beguin, P. Dhurhati, and J.-P. Aubert. 1991. "Interaction of the duplicated segment carried by *Clostridium thermocellum* cellulases with cellulosome components," *FEBS Lett.*, 291:185–188.

235 Fujino, T., P. Beguin, and J. -P. Aubert. 1992. "Cloning a *Clostridium thermocellum*

DNA fragment encoding polypeptides that bind the catalytic components of the cellulosome," *FEMS Microbiol. Lett.*, 94:165–170.

236 Salamitou, S., O. Raynaud, M. Lemaire, M. P. Coughlan, P. Beguin, and J.-P. Aubert. 1994. "Recognition specificity of the duplicated segements present in *Clostridium thermocellum* endoglucanase CelD and in the cellulosome-integrating protein CipA," *J. Bacteriol.*, 176:2822–2827.

237 Gilbert, H. J., G. P. Hazlewood, J. L. Laurie, C. G. Orpin, and G. P. Xue. 1992. "Homologous catalytic domains in a rumen fungal xylanase: Evidence for gene duplication and prokaryotic origin," *Mol. Microbiol.*, 6:2065–2072.

238 Garciacampayo, V., S. I. McCrae, J. X. Zhang, H. J. Flint, and T. M. Wood. 1993. "Mode of action, kinetic properties and physicochemical characterization of two different domains of a bifunctional (1→4)-β-D-xylanase from *Ruminococcus clavefaciens* expressed separately in *Escherichia coli*," *Biochem. J.*, 296:235–243.

239 Paradis, F. W., H. Zhu, P. J. Krell, J. P. Phillips, and C. W. Forsberg. 1993. "The *xynC* gene from *Fibrobacter succinogenes* S85 codes for a xylanase with two similar catalytic domains," *J. Bacteriol.*, 175:7666–7672.

240 Zhu, H., F. W. Paradis, P. J. Krell, J. P. Phillips, and C. W. Forsberg. 1994. "Enzymatic specificities and modes of action of the two catalytic domains of the XynC xylanase from *Fibrobacter succinogenes* S85," *J. Bacteriol.*, 176: 3885–3894.

241 Warren, R. A. J., B. Gerhard, N. R. Gilkes, J. B. Owolabi, D. G. Kilburn, and R. C. Miller, Jr. 1987. "A bifunctional exoglucanase-endoglucanase fusion protein," *Gene*, 61:421–427.

242 Nomenclature Committee, I. U. B. M. B. 1992. *Enzyme nomenclature 1992*. London: Academic Press, Inc.

243 Takenishi, S. and T. Tsujisaka. 1973. "Hemicellulases. V. Structures of the oligosaccharides from the enzyme hydrolysate of rice-straw arabinoxylan by a xylanase of *Aspergillus niger*," *Agric. Biol. Chem.*, 37:1385–1391.

244 Biely, P., J. Puls and H. Schneider. 1985. "Acetyl xylan esterases in fungal cellulolytic systems," *FEBS Lett.*, 186:80–84.

245 Kormelink, F. J. M., M. J. F. Searle-van Leeuwen, T. M. Wood and A. G. J. Voragen. 1991. "(1,4)-β-D-Arabinoxylan arabinofuranohydrolase: A novel enzyme in the bioconversion of arabinoxylan," *Appl. Microbiol. Biotechnol.*, 35:231–232.

246 MacKenzie, C. R. and D. Bilous. 1988. "Ferulic acid esterase activity from *Schizophyllum commune*," *Appl. Environ. Microbiol.*, 54:1170–1173.

247 MacKenzie, C. R., D. Bilous, H. Schneider, and K. G. Johnson. 1987. "Induction of cellulolytic and xylanolytic enzyme systems in *Streptomyces* spp.," *Appl. Environ. Microbiol.*, 53:2835–2839.

248 Puls, J. and K. Poutanen. 1989. "Mechanism of enzyme hydrolysis of hemicelluloses (xylans) and procedures for determination of the enzyme activities involved," in *Enzyme systems for lignocellulose degradation*, M. P. Coughlan, ed. London: Elsevier Applied Science, pp. 151–165.

249 Murao, S., R. Sakamoto, and M. Arai. 1988. "Cellulases of *Aspergillus aculeatus*," *Methods Enzymol.*, 160:274–299.

250 Svensson, B. 1978. "Purification of a fungal endo-β-glucanase with high activity on barley β-glucan," *Carlsberg. Res. Commun.*, 43:103–115.

251 Hurst, P. L., J. Nielsen, P. A. Sullivan, and M. G. Shepherd. 1977. "Purification and properties of a cellulase from *Aspergillus niger*," *Biochem. J.*, 165:33–41.

252 Okada, G. 1988. "Cellulase of *Aspergillus niger*," *Methods Enzymol.*, 160:259–264.

253 Ikeda, R., T. Yamamoto, and M. Funatsu. 1973. "Purification and some physical properties of acid-cellulase from *Aspergillus niger*," *Agric. Biol. Chem.*, 37: 1153–1159.

254 Idogaki, H. and Y. Kitamoto. 1992. "Purification and some properties of a carboxymethyl cellulase from *Coriolus versicolor*," *Biosci. Biotech. Biochem.*, 56: 970–971.

255 Rouau, X. and M.-J. Foglietti. 1985. "Purification and partial characterization of three endo-glucanases from *Dichomitus squalens*," Carbohydr. Res., 142: 299–314.

256 Tanaka, M., M. Taniguchi, R. Matsuno, and T. Kamikubo. 1988. "Cellulases from *Eupenicillium javanicum*," *Methods Enzymol.*, 160:251–259.

257 Wood, T. M. 1971. "The cellulase of *Fusarium solani*. Purification and specificity of the β(1-4)-glucanase and the β-D-glucosidase components," *Biochem. J.*, 121: 353–362.

258 Kozlovskaya, L. N., N. A. Rodionova, V. K. Akparov, V. A. Paseshnichenko and A. M. Bezborodov. 1980. "Isolation and characterization of β-glucosidase from the fungus *Geotrichum condidum* 3C," *Prikl. Biokhim. Mikrobio.*, 16:46–55.

259 Hayashida, S., K. Ohta, and K. Mo. 1988. "Cellulases of *Humicola insolens* and *Humicola grisea*," *Methods Enzymol.*, 160:323–332.

260 Kanda, T., K. Wakabayashi, and K. Nisizawa. 1970. "Purification and properties of two cellulase components obtained from *Irpex lacteus* (*Polyporus tulipiferae*)," *J. Ferment. Technol.*, 48:607–615.

261 Kleman-Leyer, K. M. and T. K. Kirk. 1994. "Three native cellulose-depolymerizing endoglucanases from solid-substrate cultures of the brown rot fungus *Meruliporia (Serpula) incrassata*," *Appl. Environ. Microbiol.*, 60:2839–2845.

262 Eriksson, K.-E. 1975. "Enzyme mechanisms inovolved in the degradation of wood components," in *Symposium on enzymatic hydrolysis of cellulose*, M. Bailey, T.-M. Enari, and M. Linko, eds. Helsinki: SITRA, pp. 263–280.

263 Pettersson, G., E. B. Cowling, and J. Porath. 1963. "Studies on cellulolytic enzymes. I. Isolation of a low-molecular weight cellulase from *Polysporus versicolor*," *Biochim. Biophys. Acta*, 67:1–8.

264 Clarke, A. J. and M. Yaguchi. 1985. "The role of carboxyl groups in the function of endo-β-1,4-glucanase from *Schizophyllum commune*," *Eur. J. Biochem.*, 149: 233–238.

265 Paice, M. G., M. Desrochers, D. Rho, L. Jurasek, C. Roy, C. F. Rollin, E. De Miguel, and M. Yaguchi. 1984. "Two forms of endoglucanase from the basidomyte *Schizophyllum commune* and their relationship to other β-1,4-glycoside hydrolases," *Bio-Technology*, 2:535–539.

266 Sadana, J. C., H. Lachke, and R. V. Patil. 1984. "Endo-(1-4)-β-D-glucanases from *Sclerotium rolfsii*. Purification, substrate specificity, and mode of action," *Carbohydr. Res.*, 133:297–312.

267 Streamer, M., K. -E. Eriksson, and L. G. Pettersson. 1975. "Extracellular enzyme system utilized by the fungus *Sporotrichum pulverulentum* for the breakdown of cellulose," *Eur. J. Biochem.*, 59:607–613.

268 Tong, C. C., A. L. Cole, and M. G. Shepherd. 1980. "Purification and properties of the cellulases from the thermophilic fungus *Thermoascus aurantiacus*," *Biochem. J.*, 191:83–94.

269 Shoemaker, S. P. and R. D. Brown, Jr. 1978. "Characterization of endo-1,4-β-D-glucanases purified from *Trichoderma viride*," *Biochim. Biophys. Acta*, 523: 147–161.

270 Bray, M. R. and A. J. Clarke. 1992. "Pattern of xylo-oligosaccharide hydrolysis and subsite structure of *Schizophyllum commune* xylanase A," in *Xylans and xylanases*, J. Visser, G. Beldman, M. A. Kusters-van Someren, and A. G. J. Voragen, eds. Elsevier Science Publishers B.V., pp. 423–428.

271 Sharma, P., J. K. Gupta, D. V. Vadehra, and D. K. Dube. 1990. "Purification and properties of an endoglucanase from a *Bacillus* isolate," *Enzyme Microb. Technol.*, 12:132–137.

272 Nakamuar, K. and K. Kitamura. 1988. "Cellulases of *Cellulomonas uda*," *Methods Enzymol.*, 160:211–216.

273 Petre, D., J. Millet, R. Longin, P. Beguin, H. Girard, and J. -P. Aubert. 1986. "Purification and properties of the endoglucanase C of *Clostridium thermocellum* produced in *Escherichia coli*," *Biochimie*, 68:687–695.

274 Petre, J., R. Longin, and J. Millet. 1981. "Purification and properties of an endo-β-1,4-glucanase from *Clostridium thermocellum*," *Biochimie*, 63:629–639.

275 Halliwell, G. and N. Halliwell. 1989. "Cellulolytic enzyme components of the cellulase complex of *Clostridium thermocellum*," *Biochim. Biophys. Acta*, 992:223–229.

276 Creuzet, N. and C. Frixon. 1983. "Purification and characterization of an endoglucanase from a newly isolated thermophilic anaerobic bacterium," *Biochimie*, 65: 149–156.

277 Ng, T. and J. G. Zeikus. 1988. "Endoglucanase from *Clostridium thermocellum*," *Methods Enzymol.*, 160:351–355.

278 Ng, T. K. and J. G. Zeikus. 1981. "Purification and characterization of an endoglucanase (1,4-β-D-glucan glucanohydrolase) from *Clostridium thermocellum*," *Biochem. J.*, 199:341–350.

279 McGavin, M. J., C. W. Forsberg, B. Crosby, A. W. Bell, D. Dignard and D. Y. Thomas. 1989. "Structure of the *cel3* gene from *Fibrobacter succinogenes* S85 and characteristics of the encoded gene product, endoglucanase 3," *J. Bacteriol.*, 171: 5587–5595.

280 Yamane, K. and H. Suzuki. 1988. "Cellulases of *Pseudomonas fluorescens* var. *cellulosa*," *Methods Enzymol.*, 160:200–210.

281 Wood, T. M. 1988. "Cellulase of *Ruminococcus albus*," *Methods Enzymol.*, 160: 216–221.

282 Wittmann, S., F. Shareck, D. Kluepfel, and R. Morosoli. 1994. "Purification and characterization of the CelB endoglucanase from *Streptomyces lividans* 66 and DNA sequence of the encoding gene," *Appl. Environ. Microbiol.*, 60:1701–1703.

283 Wilson, D. B. 1988. "Cellulases of *Thermomonospora fusca*," *Methods Enzymol.*, 160:314–323.

284 McGinnis, K., C. Kroupis, and D. B. Wilson. 1993. "Dimerization of *Thermomonospora fusca* β-1,4-endoglucanase E2," *Biochemistry*, 32:8146–8152.

285 Bronnonmeier, K., A. Kern, W. Liebl, and W. L. Staudenbauer. 1995. "Purification of *Thermotoga maritima* enzymes for the degradation of cellulosic materials," *Appl. Environ. Microbiol.*, 61:1399–1407.

286 Schmidhalter, D. R. and G. Canevascini. 1993. "Purification and characterization of two exo-cellobiohydrolases from the brown-rot fungus *Coniphora puteana* (Schum ex Fr) Karst," *Arch. Biochem. Biophys.*, 300:551–558.

287 Kanda, T., K. Wakabayashi, and K. Nisizawa. 1976. "Purification and properties of an endo-cellulase of avicelase type from *Irpex lacteus (Polyporus tulipiferae)*," *J. Biochem. (Tokyo)*, 79:977–988.

288 Kanda, T. and K. Nisizawa. 1988. "Exocellulase of *Irpex lacteus (Polyporus tulipiferae)*," *Methods Enzymol.*, 160:403–408.

289 Wood, T. M. 1988. "Cellobiohydrolases of *Penicillium pinophilum*," *Methods Enzymol.*, 160:398–403.

290 Limam, F., S. E. Chaabouni, R. Ghrir, and N. Marzouki. 1995. "Two cellobiohydrolases of *Penicillium occitanis* mutant Pol 6: Purification and properties," *Enzyme Microb. Technol.*, 17:340–346.

291 Sadana, J. C. and R. V. Patil. 1988. "1,4-β-D-glucan cellobiohydrolase from *Sclerotium rolfsii*," *Methods Enzymol.*, 160:307–314.

292 Fracheboud, D. and G. Canevascini. 1989. "Isolation, purification, and properties of the exocellulase from *Sporotrichum (Chrysosporium) thermophile*," *Enzyme Microb. Technol.*, 11:220–229.

293 Wood, T. M. and S. I. McCrae. 1972. "The purification and properties of the CI component of *Trichoderma koningii* cellulase," *Biochem. J.*, 128:1183–1192.

294 Nummi, M., M.-L. Niku-Paavola, A. Lappalainen, T.-M. Enari, and V. Raunio. 1983. "Cellobiohydrolase from *Trichoderma reesei*," *Biochem. J.*, 215:677–683.

295 Gum, E. K., Jr. and R. D. Brown, Jr. 1976. "Comparison of four purified extracellular 1,4-β-D-glucan cellobiohydrolase enzymes from *Trichoderma viride*," *Biochim. Biophys. Acta*, 446:371–386.

296 Morag, E., I. Halevy, E. A. Bayer, and R. Lamed. 1991. "Isolation and properties of a major cellobiohydrolase from the cellulosome of *Clostridium thermocellum*," *J. Bacteriol.*, 173:4155–4162.

297 Bartley, T. D., K. Murphy-Holland, and D. E. Eveleigh. 1984. "A method for the detection and differentiation of cellulase components in polyacrylamide gels," *Anal. Biochem.*, 140:157–161.

298 Ohmiya, K. and S. Shimizu. 1988. "Cellobiosidase from *Ruminococcus albus*," *Methods Enzymol.*, 160:391–398.

299 Rudick, M. J. and A. D. Elbein. 1975. "Glycoprotein enzymes secreted by *Aspergillus fumigatus*: Purification and properties of a second β-glucosidase," *J. Bacteriol.*, 124:534–541.

300 Kitpreechavanich, V., M. Hayashi, and S. Nagai. 1986. "Purification and characterization of extracellular β-xylosidase and β-glucosidase from *Aspergillus fumigatus*," *Agric. Biol. Chem.*, 50:1703–1711.

301 Rudick, M. J. and A. D. Elbein. 1973. "Glycoprotein enzymes secreted by *Aspergillus fumigatus*: Purification and properties of a β-glucosidase," *J. Biol. Chem.*, 248:6506–6513.

302 Unno, T., K. Ide, T. Yazaki, Y. Tanaka, T. Nakakuki, and G. Okada. 1993. "High recovery purification and some properties of a β-glucosidase from *Aspergillus niger*," *Biosci. Biotechnol. Biochem.*, 57:2172–2173.

303 McCleary, B. V. and J. Harrington. 1988. "Purification of β-D-glucosidase from *Aspergillus niger*," *Methods Enzymol.*, 169:575–583.

304 Mega, T. and Y. Matsushima. 1979. "Comparative studies of three exo-glycosidases of *Aspergillus oryzae*," *J. Biochem. (Tokyo)*, 85:335–341.

305 Sternberg, D., P. Vijayakumar, and E. T. Reese. 1977. "β-Glucosidase: Microbial production and effect on enzymatic hydrolysis of cellulose," *Can. J. Microbiol.*, 23:139–147.

306 Legler, G., M. Von Radloff, and M. Kempfle. 1972. "Composition, N-terminal amino acids, and chain length of a β-glucosidase from *Aspergillus wentii*," *Biochim. Biophys. Acta*, 257:40–48.

307 Saha, B. C., S. N. Freer, and R. J. Bothast. 1994. "Production, purification, and properties of a thermostable β-glucosidase from a color variant strain of *Aureobasidium pullulans*," *Appl. Environ. Microbiol.*, 60:3774–3780.

308 Umezurike, G. M. 1971. "Kinetic properties of β-glucosidase from *Botryodiplodia theobromae*," *Biochim. Biophys. Acta*, 250:182–191.

309 Umezurike, G. M. 1975. "The subunit structure of β-glucosidase from *Botryodiplodia theobromae* Pat," *Biochem. J.*, 145:361–368.

310 Roth, W. W. and V. R. G. Srinivasan. 1978. "Affinity chromatographic purification of β-glucosidase of *Candida guilliermondii*," *Prep. Biochem.*, 8:57–71.

311 Wilson, C. A., S. I. McCrae, and T. M. Wood. 1994. "Characterization of a β-glucosidase from the anaerobic rumen fungus *Neocallimastix frontalis* with particular reference to attack on cello-oligosaccharides," *J. Biotechnol.*, 37: 217–227.

312 Inamdar, A. N. and J. G. Kaplan. 1966. "Purification and properties of an inducible β-glucosidase of baker's yeast," *Can. J. Biochem.*, 44:1099–1108.

313 Duerkson, J. D. and H. Halvorson. 1958. "Purification and properties of an inducible β-glucosidase of yeast," *J. Biol. Chem.*, 233:1113–1120.

314 Fleming, L. W. and J. D. Duerksen. 1967. "Purification and characterization of yeast β-glucosidase," *J. Bacteriol.*, 93:135–141.

315 Clarke, A. J. 1990. "Chemical modification of a β-glucosidase from *Schizophyllum commune*: Evidence for essential carboxyl groups," *Biochim. Biophys. Acta*, 1040: 145–152.

316 Lo, A. C., G. Willick, R. Bernier, Jr., and M. Desrochers. 1988. "Purification and assay of β-glucosidase from *Schizophyllum commune*," *Methods Enzymol.*, 160:432–437.

317 Sadana, J. C., R. V. Patil, and J. G. Shewale. 1988. "β-D-Glucosidases from *Sclerotium rolfsii*," *Methods Enzymol.*, 160:424–431.

318 Deshpande, V., K. E. Eriksson, and B. Pettersson. 1978. "Production, purification and partial characterization of 1,4-β-D-glucosidase enzymes from *Sporotrichum pulverulentum*," *Eur. J. Biochem.*, 90:191–198.

319 Himmel, M. E., M. P. Tucker, S. M. Lastick, K. K. Oh, J. W. Fox, D. D. Spindler and K. Grohmann. 1986. "Isolation and characterization of a 1,4-β-D-glucan glucohydrolase from the yeast, *Torulopsis wickerhamii*," *J. Biol. Chem.*, 261: 12948–12955.

320 Wood, T. M. and S. I. McCrae. 1982. "Purification and some properties of the extracellular β-D-glucosidase of the cellulolytic fungus *Trichoderma koningii*," *J. Gen. Microbiol.*, 128:2973.

321 Enari, T.-M., M.-L. Niku-Paavola, L. Harju, A. Lappalainen, and M. Nummi. 1981. "Purification of *Trichoderma reesei* and *Aspergillus niger* β-glucosidase," *J. Appl. Biochem.*, 3:157.

322 Inglin, M., B. A. Feinberg, and J. R. Loewenberg. 1980. "Partial purification and characterization of a new intracellular β-glucosidase of *Trichoderma reesei*," *Biochem. J.*, 185:515–519.

323 Berghem, L. E. R. and L. G. Pettersson. 1974. "The mechanism of enzymatic cellulose degradation. Isolation and some properties of a β-glucosidase from *Trichoderma viride*," *Eur. J. Biochem.*, 46:295–305.

324 Gong, C. S., M. R. Ladisch and G. T. Tsao. 1977. "Cellobiase from *Trichoderma viride*: Purification, properties, kinetics, and mechanism," *Biotechnol. Bioeng.*, 19:959–981.

325 Saddler, J. N. and A. W. Khan. 1980. "Cellulolytic enzyme system of *Acetivibrio cellulolyticus*," *Can. J. Microbiol.*, 27:288–294.

326 Day, A. G. and S. G. Withers. 1986. "The purification and characterization of a β-glucosidase from *Alcaligenes faecalis*," *Biochem. Cell. Biol.*, 64:914–922.

327 Han, Y. W. and V. R. Srinivasan. 1969. "Purification and characterization of β-glucosidase of *Alcaligenes faecalis*," *J. Bacteriol.*, 100:1355–1363.

328 Ait, N., N. Cruezet, and J. Cattaneo. 1979. "Characterization and purification of thermostable β-glucosidase from *Clostridium thermocellum*," *Biochem. Biophys. Res. Commun.*, 90:537–546.

329 Garibaldi, A. and L. N. Gibbins. 1975. "Partial purification and properties of a β-glucosidase from *Erwinia herbicola* Y46," *Can. J. Microbiol.*, 21:513–520.

330 Perezpons, J. A., A. Cayetano, X. Rebordosa, J. Lloberas, A. Guasch, and E. Querol. 1994. "A beta-glucosidase gene (*bgl3*) from *Streptomyces* sp. strain QM-B814—Molecular cloning, nucleotide sequence, purification and characterization of the encoded enzyme, a new member of family 1 glycosyl hydrolases," *Eur. J. Biochem.*, 223:557–565.

331 Biswas, S. R., S. C. Jana, A. K. Mishra, and G. Nanda. 1990. "Production, purification, and characterization of xylanase from a hyperxylanolytic mutant of *Aspergillus ochraceus*," *Biotechnol. Bioeng.*, 35:244–251.

332 Shei, J. C., A. R. Fratzke, M. M. Frederick, J. R. Frederick, and P. J. Reilly. 1985. "Purification and characterization of endo-xylanases from *Aspergillus niger* II. An enzyme of pI 4.5," *Biotechnol. Bioeng.*, 27:533–538.

333 John, M., B. Schmidt, and J. Schmidt. 1979. "Purification and some properties of five endo-1,4-β-D-xylanases and a β-xylosidase produced by *Aspergillus niger*," *Can. J. Biochem.*, 57:125–134.

334 Gorbacheva, I. V. and N. A. Rodionova. 1977. "Studies on the xylan degrading enzymes. I. Purification and characterization of endo-1,4-β-xylanase from *Aspergillus niger* str. 14," *Biochim. Biophys. Acta*, 487:79–93.

335 Ghosh, M. 1994. "Purification and some properties of a xylanase from *Aspergillus sydowii*," *Appl. Environ. Microbiol.*, 60:4620–4623.

336 Okada, H. and A. Shinmyo. 1988. "Xylanase of *Bacillus pumilus*," *Methods Enzymol.*, 160:632–673.

337 Luthi, E., N. B. Jasmat, and P. L. Bergquist. 1990. "Xylanase from the extremely thermophilic bacterium *Caldocellum saccharolyticum*: Overexpression of the gene in *Escherichia coli* and characterization of the gene product," *Appl. Environ. Microbiol.*, 56:2677–2683.

338 Dekker, R. F. H. and G. N. Richards. 1975. "Purification, properties, and mode of action of hemicellulase II produced by *Ceratocystis paradoxa*," *Carbohydr. Res.*, 42:107–123.

339 Morosoli, R., C. Roy, and M. Yaguchi. 1986. "Isolation and primary sequence of a xylanase from the yeast *Cryptococcus albidus*," *Biochim. Biophys. Acta*, 870: 473–478.

340 Ganju, R. K., P. J. Vithayathil, and S. K. Murthy. 1989. "Purification and characterization of two xylanases from *Chaetomium thermophile* var. *coprophile*," *Can. J. Microbiol.*, 35:836–842.

341 Kitpreechavanich, V., M. Hayashi, and S. Nagai. 1984. "Purification and properties of endo-1,4-β-xylanase from *Humicola lanuginosa*," *J. Ferment. Technol.*, 62: 415–420.

342 Anand, L., S. Krishnamurthy, and P. J. Vithayathil. 1990. "Purification and properties of xylanase from the thermophilic fungus *Humicola lanuginosa* (Griffen and Maublanc) Bunce," *Arch. Biochem. Biophys.*, 246:546–553.

343 Hoebler, C. and J. M. Brillouet. 1984. "Purification and properties of an endo-(1,4)-β-D-xylanase from *Irpex lacteus (Polysporus tulipiferae)*," *Carbohydr. Res.*, 128: 141–155.

344 Mishra, C., I. T. Forrester, B. D. Kelley, R. R. Burgess, and G. F. Leatham. 1990. "Characterization of a major xylanase purified from *Lentinula edodes* cultures grown on a commercial solid lignocellulosic substrate," *Appl. Microbiol. Biotechnol.*, 33: 226–232.

345 Matsuo, M. and T. Yasui. 1988. "Xylanase of *Malbranchea pulchella* var. *sulfurea*," *Methods Enzymol.*, 160:671–674.

346 Filho, E. X. F., J. Puls, and M. P. Coughlan. 1993. "Physicochemical and catalytic properties of a low-molecular-weight endo-1,4-β-D-xylanase from *Myrothecium verrucaria*," *Enzyme Microb. Technol.*, 15:535–540.

347 Eyzaguirre, J., J. Scarpa, A. Belancic, and J. Steiner. 1992. "The xylanase system of *Penicillium purpurogenum*," in *Xylans and xylanases*, J. Visser, G. Beldman, M. A. Kusters-van Someren, and A. G. J. Voragen, eds. Amsterdam: Elsevier Science Publishers B.V., pp. 505–510.

348 Green, F. III, C. A. Clausen, J. A. Micales, T. L. Highley, and K. E. Wolter. 1989. "Carbohydrate-degrading complex of the brown-rot fungus *Postia placenta*: Purification of β-1,4-xylanase," *Holzforschung*, 43:25–31.

349 Paice, M. G., L. Jurásek, M. R. Carpenter, and L. B. Smillie. 1978. "Production, characterization, and partial amino acid sequence of xylanase A from *Schizophyllum commune*," *Appl. Environ. Microbiol.*, 36:802–808.

350 Varadi, J., V. Necesany, and P. Kovacs. 1971. "Cellulase and xylanase of fungus *Schizophyllum commune*. III. Purification and properties of xylanase," *Drevarsky vyskum*, 14:147–159.

351 Cavazzoni, V., M. Manzoni, C. Pavini, and M. C. Bonferoni. 1989. "β-D-Xylanase produced by *Schizophyllum radiatum*," *Appl. Microbiol. Biotech.*, 30: 247–251.

352 Comtat, J. 1983. "Isolation, properties, and postulated role of some of the xylanases from the Basidiomycete *Sporotrichum dimorphosporum*," *Carbohydr. Res.*, 118: 215–231.

353 Hayashida, S., K. Ohta, and K. Mo. 1988. "Xylanase of *Talaromyces byssochlamydoides*," *Methods Enzymol.*, 160:675–678.

354 Ghosh, V. K. and J. K. Deb. 1988. "Production and characterization of xylanase from *Thielaviopsis basicola*," *Appl. Microbiol. Biotechnol.*, 29:44–47.

355 Kubackova, M., S. Karacsonyi, L. Bilisics, and R. Toman. 1978. "Some properties of an endo-1,4-β-D-xylanase from the ligniperdous fungus *Trametes hirsuta*," *Folia Microbiol.*, 23:202–209.

356 Tan, L. U. L., K. K. Y Wong, E. K. C. Yu, and J. N. Saddler. 1985. "Purification and characterization of two β-xylanases from *Trichoderma harzianum*," *Enzyme Microb. Technol.*, 7:425–430.

357 Wong, K. K. Y., L. V. L. Tan, J. N. Saddler, and M. Yaguchi. 1986. "Purification of a third distinct xylanase from the xylanolytic system of *Trichoderma harzianum*," *Can. J. Microbiol.*, 32:570–576.

358 Wood, T. M. and S. I. McCrae. 1986. "Studies of two low-molecular-weight endo-(1,4)-β-D-xylanases constitutively synthesized by the cellulolytic fungus *Trichoderma koningii*," *Carbohydr. Res.*, 148:321–330.

359 Huang, L., T. H. Hseu, and T.-T. Wey. 1991. "Purification and characterization of an endoxylanase from *Trichoderma koningii* G-39," *Biochem. J.*, 278:329–333.

360 John, M. and J. Schmidt. 1988. "Xylanases and β-xylosidases of *Trichoderma lignorum*," *Meth. Enzymol.*, 160:662–671.

361 Baker, C. J., C. H. Whalen, and D. F. Bateman. 1977. "Xylanase from *Trichoderma pseudokoningii:* Purification, characterization and effects on isolated plant cell walls," *Phytopathology*, 67:1250–1258.

362 Poutanen, K. and J. Puls. 1989. "The xylanolytic enzyme system of *Trichoderma reesei*," *ACS Symp. Ser.*, 399:630–640.

363 Biely, P., M. Vrsanska, and M. Claeyssens. 1991. "The endo-1,4-β-glucanase I from *Trichoderma reesei*. Action on β-1,4-oligomers and polymers derived from D-glucose and D-xylose," *Eur. J. Biochem.*, 200:157–163.

364 Lappalainen, A. 1986. "Purification and characterization of xylanolytic enzymes from *Trichoderma reesei*," *Biotechnol. Appl. Biochem.*, 8:437–448.

365 Toda, S., H. Suzuki, and K. Nisizawa. 1971. "Enzymic properties and the substrate specificities of *Trichoderma* cellulases with special reference to their activity toward xylan," *J. Ferment. Technol.*, 49:499–521.

366 Sinner, M. and H. H. Dietrichs. 1975. "Enzymic hyrolysis of hardwood xylans. II. Isolation of five β-1-4-xylanases from three commercial enzyme products," *Holzforschung*, 29:168–177.

367 Sinner, M. and H. H. Dietrichs. 1975. "Enzymatic hydrolysis of hardwood xylans. III. Characterization of five isolated β-1,4-xylanases," *Holzforschung*, 29:207–214.

368 Sinner, M. and H. H. Dietrichs. 1976. "Enzymatic hydrolysis of hardwood xylans. IV. Degradation of isolated xylans," *Holzforschung*, 30:50–59.

369 Shikata, S. and K. Nisizawa. 1975. "Purification and properties of an exo-cellulase component of novel type from *Trichoderma viride*," *J. Biochem. (Tokyo)*, 78: 499–512.

370 Irie, T., T. Konishi, Y. Tagoyama, and A. Ogiso. 1990. "Purification of xylanase from *Trichoderma viride* (*reesei*) mutant K-10-34 and its xylobiose-forming properties," *Hakko kogaku Kaishi*, 68:457–463.

371 Gibson, T. S. and B. V. McCleary. 1987. "A simple procedure for the large-scale production of β-D-xylanase from *Trichoderma viride*," *Carbohydr. Polym.*, 7: 225–240.

372 Dean, J. F. D. and J. D. Anderson. 1991. "Ethylene biosynthesis-inducing xylanase II. Purification and characterization of the enzyme produced by *Trichoderma viride*," *Plant Physiol.*, 95:316–323.

373 Dean, F. J. D., K. C. Gross, and J. D. Anderson. 1991. "Ethylene biosynthesis-inducing xylanase. III. Product characterization," *Plant Physiol.*, 96:571–576.

374 Ujiie, M., C. Roy, and M. Yaguchi. 1991. "Low molecular-weight xylanase from *Trichoderma viride*," *Appl. Environ. Microbiol.*, 57:1860–1862.

375 Beldman, G., M. F. Searle-van Leeuwen, F. M. Rombouts, and A. G. J. Voragen. 1985. "The cellulase of *Trichoderma viride*. Purification, characterization, and comparison of all detectable endoglucanases, exoglucanases and β-glucosidases," *Eur. J. Biochem.*, 146:301–308.

376 Beldman, G., A. G. J. Voragen, F. M. Rombouts, and F. G. J. Voragen. 1988. "Specific and non-specific glucanases from *Trichoderma viride*," *Biotechnol. Bioeng.*, 31:160–167.

377 Hashimoto, S., T. Muramatsu, and M. Funatsu. 1971. "Xylanase from *Trichoderma viride*. I. Isolation and some properties of crystalline cellulose," *Agric. Biol. Chem.*, 35:501–508.

378 Ishihara, M., K. Shimizu, and T. Ishihara. 1978. "Hemicellulases of brown-rot fungus *Tyromyces palustris* III. Partial purification and mode of action of an extracellular xylanase," *Mokuzai Gakkaishi*, 24:108–115.

379 Kubata, B. K., K. Takamizawa, K. Kawai, T. Suzuki, and H. Horitsu. 1995. "Xylanase IV, an exoxylanase of *Aeromonas caviae* ME-1 which produces xylotetraose as the only low-molecular-weight oligosaccharide from xylan," *Appl. Environ. Microbiol.*, 61:1666–1668.

380 Kubata, B. K., T. Suzuki, H. Horitsu, K. Kawai, and K. Takamizawa. 1994. "Purification and characterization of *Aeromonas caviae* Me-1 xylanase V, which produces exclusively xylobiose from xylan," *Appl. Environ. Microbiol.*, 60: 531–535.

381 Debeire-Gosselin, M., M. Loonis, E. Samain, and P. Debeire. 1992. "Purification and properties of a 22kDa endoxylanase excreted by a new strain of thermophilic *Bacillus*," in *Xylans and xylanases*, J. Visser, G. Beldman, M. A. Kusters-van Someren, and A. G. J. Voragen, eds. Amsterdam: Elsevier Science Publishers B.V., pp. 463–466.

382 Nakamura, S., R. Aono, K. Wakabayashi, and K. Horikoshi. 1992. "Alkaline xylanase produced by newly isolated alkaliphilic *Bacillus* sp," in *Xylans and xylanases*, J. Visser, G. Beldman, M. A. Kusters-van Someren, and A. G. J. Voragen, eds. Amsterdam: Elsevier Science Publishers B.V., pp. 443–446.

383 Esteban, R., J. R. Villanueva, and T. G. Villa. 1982. "β-D-Xylanases of *Bacillus circulans* WL-12," *Can. J. Microbiol.*, 28:733–739.

384 Esteban, R., A. Chordi, and T. G. Villa. 1983. "Some aspects of a 1,4-β-D-xylanase and a β-D-xylosidase secreted by *Bacillus coagulans* strain 26," *FEMS Microbiol. Lett.*, 17:163–166.

385 Morales, P., A. Madarro, A. Flors, J. M. Sendra, and J. A. Perez-Gonzalez. 1995. "Purification and characterization of a xylanase and an arabinofuranosidase from *Bacillus polymyxa*," *Enzyme Microb. Technol.*, 17:424–429.

386 Panbangred, W., A. Shinmyo, S. Kimoshita, and H. Okada. 1983. "Purification and properties of endoxylanase produced by *Bacillus pumilus*," *Agric. Biol. Chem.*, 47: 957–963.

387 Bernier, R., Jr., M. Desrochers, L. Jurasek, and M. G. Paice. 1983. "Isolation and characterization of a xylanase from *Bacillus subtilis*," *Appl. Environ. Microbiol.*, 46:511–514.

388 Kato, Y. and D. J. Nevins. 1984. "Enzymic dissociation of *Zea* shoot cell wall polysaccharides. III. Purification and partial characterization of an endo-(1-4)-β-D-xylanase from a *Bacillus subtilis* enzyme preparation," *Plant Physiol.*, 75:753–758.

389 Karjalainen, R., S. Peltonen, O. Kajander, and M.-L. Niku-Paavola. 1992. "Production and characterization of a xylanase from a phytopathogenic fungus, *Bipolaris sorokiniana*," in *Xylans and xylanases*, J. Visser, G. Beldman, M. A. Kusters-van

Someren, and A. G. J. Voragen, eds. Amsterdam: Elsevier Science Publishers B.V., pp. 529–533.

390 Lee, S. F., C. W. Forsberg, and J. B. Rattray. 1987. "Purification and characterization of two endoxylanases from *Clostridium acetobutylicum* ATCC 824," *Appl. Environ. Microbiol.*, 53:644–650.

391 Zappe, H., D. T. Jones, and D. R. Woods. 1987. "Cloning and expression of a xylanase gene from *Clostridium acetobutylicum* P262 in *Escherichia coli*," *Appl. Microbiol. Biotechnol.*, 27:57–63.

392 Berenger, J.-F., C. Frixon, J. Bigliardi, and N. Creuzet. 1985. "Production, purification and properties of thermostable xylanase from *Clostridium stercorarium*," *Can. J. Microbiol.*, 31:635–643.

393 Debeire-Gosselin, M., J. P. Touzel, and P. Debeire. 1992. "Isoxylanases from the thermophile *Clostridium thermolacticum*," in *Xylans and xylanases*, J. Visser, G. Beldman, M. A. Kusters-van Someren, and A. G. J. Voragen, eds. Amsterdam: Elsevier Science Publishers B.V., pp. 471–474.

394 Elegir, G., G. Szakacs, and T. W. Jeffries. 1994. "Purification, characterization, and substrate specificities of multiple xylanases from *Streptomyces* sp. strain B-12-2," *Appl. Environ. Microbiol.*, 60:2609–2615.

395 Nakajima, T., K.-I. Tsukamoto, T. Watanabe, K. Kainuma, and K. Matsuda. 1984. "Purification and some properties of an endo-1,4-β-D-xylanase from *Streptomyces* sp.," *J. Ferment. Technol.*, 62:269–276.

396 Kusakabe, I., M. Kawaguchi, T. Yasui, and T. Kabayashi. 1977. "Purification and some properties of extracellular xylanase from *Streptomyces* sp. E-86. Studies on xylanase system of *Streptomyces*, part VII," *Nippon Nogeikagaku Kaishi*, 51:429–437.

397 Marui, M., K. Nakanishi and T. Yasui. 1985. "Purification and properties of three types of xylanases induced by methyl β-xyloside from *Streptomyces* sp.," *Agric. Biol. Chem.*, 49:3399–3407.

398 Sreenath, H. K. and R. Joseph. 1982. "Purification and properties of extracellular xylan hydrolases of *Streptomyces exfoliatus*," *Folia Microbiol.*, 27:107–115.

399 Johnson, K. G., D. Brener, R. Vidoli, and D. W. Griffith. 1984. "Xylan-degrading enzymes from *Streptomyces flavogriseus* CD-45," Abstract K152, Annu. Meet. Am. Soc. Microbiol.

400 Ruiz-Arribas, A., J. M. Fernandez-Abalos, P. Sanchez, A. L. Garda, and R. I. Santamaria. 1995. "Overproduction, purification, and biochemical characterization of a xylanase (Xys1) from *Streptomyces halstedii* JM8," *Appl. Environ. Microbiol.*, 61:2414–2419.

401 Morosoli, R., J.-L. Bertrand, F. Mondou, F. Shareck, and D. Kluepfel. 1986. "Purification and properties of a xylanase from *Streptomyces lividans*," *Biochem. J.*, 239:587–592.

402 Tsujibo, H., K. Miyamoto, T. Kuda, K. Minami, T. Sakamoto, T. Hasegawa and Y. Inamori. 1992. "Purification, properties, and partial amino acid sequences of thermostable xylanases from *Streptomyces thermoviolaceus* OPC-520," *Appl. Environ. Microbiol.*, 58:371–375.

403 Iizuka, H. and T. Kawaminami. 1965. "Studies on the xylanase from Steptomyces. Part 1. Purification and some properties of xylanase from *Streptomyces xylophagus* nov. sp.," *Agric. Biol. Chem.*, 29:520–524.

404 Shao, W., S. DeBlois, and J. Wiegel. 1995. "A high-molecular-weight, cell-associated xylanase isolated from exponentially growing *Thermoanaerobacterium* sp. strain JW/SL-YS485," *Appl. Environ. Microbiol.*, 61:937–940.

405 Irwin, D., E. D. Jung, and D. B. Wilson. 1994. "Characterization and sequence of a *Thermomonospora fusca* xylanase," *Appl. Environ. Microbiol.*, 60:763–770.

406 Claeyssens, M., F. G. Loontiens, H. Kersters-Hilderson, and C. K. DeBruyne. 1971. "Partial purification and properties of an *Aspergillus niger* β-D-xylosidase," *Enzymologia*, 40:177–198.

407 Yasui, T. and M. Matsuo. 1988. "Beta-xylosidase/beta-glucosidase of *Chaetomium trilaterale*," *Methods Enzymol.*, 160:696–700.

408 Uziie, M., M. Matsuo, and T. Yasui. 1985. "Purification and some properties of *Chaetomium trilaterale* β-xylosidase," *Agric. Biol. Chem.*, 49:1159–1166.

409 Peciarova, A. and P. Biely. 1982. "β-Xylosidases and a nonspecific wall-bound β-glucosidase of the yeast *Crytococcus albidus*," *Biochim. Biophys. Acta*, 716: 391–399.

410 Matsuo, M. and T. Yasui. 1988. "β-Xylosidases of several fungii," *Methods Enzymol.*, 160:684–695.

411 Lachke, A. H. 1988. "1,4-β-D-Xylan xylohydrolase of *Sclerotium rolfsii*," *Methods Enzymol.*, 160:679–684.

412 Poutanen, K. and J. Puls. 1988. "Characterization of *Trichoderma reesei* β-xylosidase and its use in the hydrolysis of solubilized xylans," *Appl. Microbiol. Biotechnol.*, 28:425–432.

413 Kersters-Hilderson, H., F. G. Loontiens, M. Claeyssens, and C. K. DeBruyne. 1969. "Partial purification and properties of induced β-D-xylosidase of *Bacillus* pumilus," *Eur. J. Biochem.*, 7:434.

414 Kersters-Hilderson, H., M. Claeyssens, E. Van Doorslaer, and C. K. DeBruyne. 1982. "β-D-Xylosidase from *Bacillus pumilus*," *Methods Enzymol.*, 83:631–639.

415 Berger, E., W. A. Jones, D. T. Jones, and D. R. Woods. 1989. "Cloning and sequencing of an endoglucanase (*end1*) gene from *Butyrivibrio fibrisolvens* H17c," *Mol. Gen. Genet.*, 219:193–198.

416 Wong, W. K. K., B. Gerhard, Z. M. Guo, D. G. Kilburn, R. A. J. Warren, and R. C. Miller, Jr. 1986. "Characterization and structure of an endoglucanase gene *cenA* of *Cellulomonas fimi*," *Gene*, 44:315–324.

417 Meinke, A., C. Braun, N. R. Gilkes, D. G. Kilburn, R. C. Miller, Jr., and R. A. J. Warren. 1991. "Unusual sequence organization in CenB, an inverting endoglucanase from *Cellulomonas fimi*," *J. Bacteriol.*, 171:308–314.

418 Al-Tawheed, A. R. 1988. "Molecular characterization of cellulase genes from *Cellulomonas flavigena*," M.Sc. thesis. Dublin, Ireland: Trinity College.

419 Yaglonsky, M. D., T. Bartley, K. O. Elliston, S. K. Kahrs, Z. P. Shalita, and D. E. Eveleigh. 1988. "Characterization and cloning of the cellulase complex of *Microbispora bispora*," *FEMS Symp.*, 43:249–266.

420 Hall, J. and H. J. Gilbert. 1988. "The nucleotide sequence of a carboxymethylcellulase gene from *Pseudomonas fluorescens* subsp. *cellulosa*," *Mol. Gen. Genet.*, 213:112–117.

421 Kellett, L. E., D. M. Poole, L. M. A. Ferreira, A. J. Durrant, G. P. Hazlewood, and H. J. Gilbert. 1990. "Xylanase B and an arabinofuranosidase from *Pseudomonas fluorescens* subsp. *cellulosa* contain identical cellulose-binding domains are encoded by adjacent genes," *Biochem. J.*, 272:369–376.

422 Teeri, T. T., P. Lehtovaara, S. Kauppinen, I. Salovuori, and J. Knowles. 1987. "Homologous domains in *Trichoderma reesei* cellulolytic enzymes: Gene sequence and expression of cellobiohydrolase II," *Gene*, 51:43–52.

423 Penttila, M., P. Lehtovaara, H. Nevalainen, R. Bhikhabbai, and J. Knowles. 1986. "Homology between cellulase genes of *Trichoderma reesei*: Complete nucleotide sequence of the endoglucanase I gene," *Gene*, 45:253–263.

424 Saloheimo, M., P. Lechtovaara, M. Penttila, T. T. Teeri, J. Stahlberg, G. Johansson, G. Pettersson, M. Claeyssens, P. Tomme, and J. K. C. Knowles. 1988. "EG III, a new endoglucanase from *Trichoderma reesei*: The characterization of both gene and enzyme," *Gene*, 63:11–21.

425 Saloheimo, A., B. Henrissat, A. M. Hoffren, O. Teleman, and M. Penttila. 1994. "A novel, small endoglucanase gene, *egl5*, from *Trichoderma reesei* isolated by expression in yeast," *Mol. Microbiol.*, 13:219–228.

426 Hansen, C. K. 1992. Ph.D. diss. Technical University of Denmark, Copenhagen.

427 Gibbs, M. D., D. J. Saul, E. Luthi, and P. L. Berquist. 1992. "The β-mannanase from *Caldocellum saccharlolyticum*," *Appl. Environ. Microbiol.*, 58:3864–3867.

428 Gibbs, M. D., D. J. Saul, E. Luthi, and P. L. Bergquist. 1992. "The β-mannanase from *Caldocellum saccharolyticum* is part of a multichain enzyme," *Appl. Environ. Microbiol.*, 58:3864–3867.

429 Xue, G.-P., K. S. Gobius, and C. G. Orpin. 1992. "A novel polysaccharide hydrolase cDNA (*celD*) from *Neocallimastix patriciarum* encoding three multi-functional catalytic domains with high endoglucanase, cellobiohydrolase and xylanase activities," *J. Gen. Microbiol.*, 138:2397–2403.

430 Flint, H. J., J. Martin, C. A. McPherson, A. S. Daniel and J. -X. Zhang. 1993. "A bifunctional enzyme, with separate xylanase and β(1,3-1,4)-glucanase domains, encoded by the *xynD* gene of *Ruminococcus flavefaciens*," *J. Bacteriol.*, 175: 2943–2951.

431 Hazlewood, G. P. and H. J. Gilbert. 1993. "Molecular biology of hemicellulases," in *Hemicellulose and hemicellulases*, M. P. Coughlan and G. P. Hazlewood, eds. London: Portland Press Ltd., pp. 103–126.

432 Shoemaker, S., V. Schweickart, M. Lander, D. Gelgand, S. Kwok, K. Myambo, and M. Innis. 1983. "Molecular cloning of exo-cellobiohydrolase I derived from *Trichoderma reesei* strain L27," *Biotechnology*, 1:691–696.

433 Cheng, C., N. Tsukagoshi, and S. Udaka. 1990. "Nucleotide sequence of the cellobiohydrolase gene from *Trichoderma viride*," *Nucleic Acid Res.*, 18:55–59.

434 Koch, A., C. T. O. Weigel, and G. Schulz. 1993. "Cloning, sequencing, and heterologous expression of a cellulase encoding cDNA (*cbh1*) from *Penicillium janitellum*," *Gene*, 124:57–65.

435 Sims, P., C. James, and P. Broda. 1988. "The identification, molecular cloning and characterisation of a gene from *Phanerochaete chrysosporium* that shows strong homology to the exo-cellobiohydrolase I gene from *Trichoderma reesei*," *Gene*, 74: 411–422.

436 Azevedo, M., M. S. S. Felipe, S. Astolfi-Filho, and A. Radford. 1990. "Cloning, sequencing and homologies of the cbh-I (exoglucanase) gene of *Humicola grisea*," *J. Gen. Microbiol.*, 136:2569–2579.

437 Raguz, S., E. Yague, D. A. Wood, and C. F. Thurston. 1992. "Isolation and characterization of a cellulose-growth-specific gene from *Agaricus bisporus*," *Gene*, 119:183–190.

438 Chow, C. M., E. Yague, S. Raguz, D. A. Wood, and C. F. Thurston. 1994. "The *cel3* gene of *Agaricus bisporus* codes for a modular cellulase and is transcriptionally regulated by the carbon source," *Appl. Environ. Microbiol.*, 60:2779–2785.

439 Ramussen, G., J. Mikkelsen, M. Schulein, F. Hagen, C. Hjort, and S. Hastrup. 1991.

"A cellulase preparation comprising an endoglucanase enzyme," World Pat WO 91/17243, 50–51.

440 Meinke, A., N. R. Gilkes, D. G. Kilburn, R. C. Miller, Jr., and R. A. J. Warren. 1991. "Cellulose-binding polypeptides from *Cellulomonas fimi*: Endoglucanse D (CenD), a family A β-1,4-glucanase," *J. Bacteriol.*, 175:1910–1918.

441 Hamamoto, T., F. Foong, O. Shoseyov, and R. H. Doi. 1992. "Analysis of functional domains of endoglucanases from *Clostridium cellulovorans* by gene cloning, nucleotide sequencing and chimeric protein construction," *Mol. Gen. Genet.*, 231:472–479.

442 Ferreira, L. M. A., G. P. Hazlewood, P. J. Barker, and H. J. Gilbert. 1993. "The cellodextrinase from *Pseudomonas fluorescens* subsp. *cellulosa* consists of multiple functional domains," *Biochem. J.*, 279:793–799.

443 Hall, J., G. P. Hazlewood, N. S. Huskisson, A. J. Durrant, and H. J. Gilbert. 1989. "Homology of a xylanase and cellulase from *Pseudomonas fluorescens* subsp. *cellulosa*: Internal signal sequence and unusual protein processing," *Mol. Microbiol.*, 3:1211–1219.

444 Yablonsky, M. D., K.O. Elliston, and D. E. Eveleigh. 1989. "The relationship between the endoglucanase MbcelA of *Microbispora bispora* and the cellulases of *Cellulomonas fimi*," in *Enzyme systems for lignocellulose degradation*, M. Coughlan, ed. London: Elsevier Applied Science, pp. 73–83.

445 Lao, G., G. S. Ghangas, E. D. Jung, and D. B. Wilson. 1991. "DNA sequences of three β-1,4-endoglucanase genes from *Thermomonospora fusca*," *J. Bacteriol.*, 173:3397–3407.

446 Theberge, M., P. Lacaze, F. Shareck, R. Morosoli, and D. Kluepfel. 1992. "Purification and characterization of an endoglucanase from *Streptomyces lividans* 66 and DNA sequence of the gene," *Appl. Environ. Microbiol.*, 58:815–820.

447 Grabnitz, F. and W. L. Staudenbauer. 1988. "Characterization of two β-glucosidase genes from *Clostridium thermocellum*," *Biotechnol. Lett.*, 10:73–78.

448 Hazlewood, G. P., M. P. M. Romaniec, K. Davidson, O. Grepinet, and P. Beguin. 1988. "A catalogue of *Clostridium thermocellum* endoglucanase, β-glucosidase and xylanase genes cloned in *Escherichia coli*," *FEMS Microbiol. Lett.*, 51:231–236.

449 Wood, T. M. 1985. "Properties of cellulolytic enzyme systems," *Biochem. Soc. Trans.*, 13:407–410.

450 Fagerstam, L. G. and L. G. Pettersson. 1980. "The 1,4-β-glucan cellobiohydrolases of *Trichoderma reesei* QM 9414," *FEBS Lett.*, 119:97–110.

451 Knowles, J. K. C., T. T. Teeri, P. Lehtovaara, M. Penttila, and M. Saloheimo. 1988. "The use of gene technology to investigate fungal cellulolytic enzymes," in *Biochemistry and genetics of cellulose degradation*, J. P. Aubert, P. Beguin, and J. Millet, eds. New York: Academic Press, Inc., pp. 153–170.

452 Shoemaker, S. P. and R. D. Brown, Jr. 1978. "Enzymatic activities of endo-1,4-β-D-glucanases purified from *Trichoderma viride*," *Biochim. Biophys. Acta*, 523:133–146.

453 Gong, C. S., K. F. Chen, and G. T. Tsao. 1979. "Affinity chromatography of endoglucanase of *Trichoderma viride* by Concanavalin A-agarose," *Biotechnol. Bioeng.*, 21:167–171.

454 Chernoglazov, V., O. V. Ermolova, and A. A. Klesov. 1985. "Production of highly purified multiple forms of endo-1,4-glucanases from *Trichoderma viride*, differing in adsorption capacity by affinity chromatography on cellulose and highly effective chromatofocusing," *Biokihimiya (Eng. Transl.)*, 50:939–949.

455 Messner, R., F. Gruber and C. P. Kubicek. 1988. "Differential regulation of synthesis of multiple forms of specific endoglucanases by *Trichoderma reesei* QM 9414," *J. Bacteriol.*, 170:3689–3693.

456 Luderer, M. E. H., F. Hofer, K. A Hagspiel, D. Blaas, and C. P. Kubicek. 1991. "A re-appraisal of multiplicity of endoglucanase I from *Trichoderma reesei* using monoclonal antibodies and plasma desorption mass spectrometry," *Biochim. Biophys. Acta*, 1076:427–434.

457 Labudova, I. and V. Farkas. 1983. "Multiple enzyme forms in the cellulase system of *Trichoderma reesei* during its growth on cellulose," *Biochim. Biophys. Acta*, 744:135.

458 Sheir-Neiss, G. and B. S. Montenecourt. 1984. "Characterization of secreted cellulases of *Trichoderma reesei* wild type and mutants during controlled fermentations," *Appl. Microbiol. Biotechnol.*, 20:46.

459 Sprey, B. and C. Lambert. 1983. "Titration curves of cellulases from *Trichoderma reesei*: Demonstration of a cellulase-xylanase-beta-glucosidase-containing complex," *FEMS Microbiol. Lett.*, 18:217–222.

460 Hofer, F., E. Weissinger, H. Mischak, R. Messner, B. Meixner-Monori, D. Blaas, J. Visser and C. Kubicek. 1989. "A monoclonal antibody against the alkaline extracellular β-glucosidase from *Trichoderma reesei:* Reactivity with other *Trichoderma* β-glucosidases," *Biochim. Biophys. Acta*, 992:298–306.

461 Eriksson, K.-E. and B. Pettersson. 1982. "Purification and partial characterization of two acidic proteases from the white-rot fungus *Sporotrichum pulvervulentum*," *Eur. J. Biochem.*, 124:635–642.

462 Mischak. H., F. Hofer, R. Messner, E. Weissinger, M. Hayn, P. Tomme, H. Esterbauer, E. Kuchler, M. Claeyssens, and C. P. Kubicek. 1989. "Monoclonal antibodies against different domains of cellobiohydrolase I and II from *Trichoderma reesei*," *Biochim. Biophys. Acta*, 990:1–7.

463 Kammel, W. P. and C. P. Kubicek. 1985. "Absence of post-secretional modification in extracellular proteins of *Trichoderma reesei* during growth on cellulose," *J. Appl. Biochem.*, 7:138–144.

464 Stahlberg, J., G. Johansson, and G. Pettersson. 1988. "A binding-site deficient, catalytically active, core protein of endoglucanase III from the culture filtrate of *Trichoderma reesei*," *Eur. J. Biochem.*, 173:179–183.

465 Willick, G. E., R. Li, V. L. Seligy, M. Yaguchi, and M. Desrochers. 1984. "Extracellular proteins excreted by the Basidiomycete *Schizophyllum commune* in response to carbon source," *J. Bacteriol.*, 159:294–299.

466 Mo, K. and S. Hayashida. 1988. "Conversion of *Geotrichum candidum* endocellulase I to endocellulase II by limited proteolysis," *Agric. Biol. Chem.*, 52:1683–1688.

467 Malcolm, B. A., S. Rosenberg, M. J. Corey, J. S. Allen, A. de Baetselier, and J. F. Kirch. 1989. "Site-directed mutagenesis of the catalytic residues Asp-52 and Glu-35 of chicken egg white lysozyme," *Proc. Natl. Acad. Sci.*, 86:133–137.

468 Moloney, A. P., S. I. McCrae, T. M. Wood, and M. P. Coughlan. 1985. "Isolation and characterization of the 1,4-β-D-glucanhydrolases of *Talaromyces emersonii*," *Biochem. J.*, 225:365–374.

469 Reese, E.T. 1975. "Polysaccharases and the hydrolysis of insoluble substrates," in *Biological transformation of wood by microorganisms*, W. Leise, ed. New York: Springer-Verlag, pp. 165–181.

470 Wood, T. M. 1969. "The relationship between cellulolytic and pseudo-cellulolytic microorganisms," *Biochim. Biophys. Acta*, 192:531–534.

471 Henrissat, B., H. Driguez, C. Veit, and M. Schulein. 1985. "Synergism of cellulase from *Trichoderma reesei* in degradation of cellulose," *Biotechnology*, 3:722–726.

472 Woodward, J., M. K. Hayes, and N. E. Lee. 1988. "Hydrolysis of cellulose by saturating and non-saturating concentrations of cellulase: Implications for synergism," *Biotechnology*, 6:301–304.

473 Woodward, J. 1991. "Synergism in cellulase systems," *Bioresource Technol.*, 36: 67–75.

474 Wood, T. M. and S. I. McCrae. 1979. "Synergism between enzymes involved in the solubilization of native cellulose," *Adv. Chem. Ser.*, 181:181–209.

475 Fujii, M. and M. Shimizu. 1986. "Synergism of endoenzyme and exoenzyme on hydrolysis of soluble cellulose derivatives," *Biotechnol. Bioeng.*, 28:878–882.

476 Kyriacou, A., C. R. MacKenzie, and R. J. Neufeld. 1987. "Detection and the characterization of the specific and non-specific endoglucanases of *Trichoderma reesei:* Evidence demonstrating endoglucanase activity by cellobiohydrolase II," *Enzyme Microb. Technol.*, 9:25–31.

477 Ryu, D. D. Y., C. Kim, and M. Mandels. 1984. "Competition and sorption of cellulase components and its significance in a synergistic mechanism," *Biotechnol. Bioeng.*, 26:488–496.

478 Tomme, P., V. Heriban and M. Claeyssens. 1990. "Adsorption of two cellobiohydrolases from *Trichoderma reesei* to Avicel: Evidence for exo-exo synergism and possible 'loose complex' formation," *Biotechnol. Lett.*, 12:525–530.

479 Wood, W. A. and S. I. McCrae. 1986. "The cellulase of *Penicillium pinophilum.* Synergism between enzyme components in solubilizing cellulose with special reference to the involvement of two immunologically distinct cellobiohydrolases," *Biochem. J.*, 234:93–99.

480 Wood, T. M., S. I. McCrae, and K. M. Bhat. 1989. "The mechanism of fungal cellulase action. Synergism between enzyme components of *Penicillium pinophilum* cellulase in solubilizing hydrogen bond-ordered cellulose," *Biochem. J.*, 260: 37–43.

481 Nidetzky, B., W. Steiner, M. Hayn, and M. Claeyssens. 1994. "Cellulose hydrolysis by the cellulases from *Trichoderma reesei*—A new model for synergistic interaction," *Biochem. J.*, 298:705–710.

482 Hoshino, E., Y. Sasaki, K. Mori, M. Okazaki, K. Nisizawa, and T. Kanda. 1993. "Electron microscopic observation of cotton cellulose degradation by exo- and endo-type cellulases from *Irpex lacteus*," *J. Biochem. (Tokyo)*, 114:236–245.

483 Sprey, B. and H.-P. Bochem. 1992. "Effect of endoglucanase and cellobiohydrolase from *Trichoderma reesei* on cellulose microfibril structure," *FEMS Microbiol. Lett.*, 97:113.

484 Borneman, W.S. and D.E. Akin. 1990. "Lignocellulose degradation by rumen fungi and bacteria: Ultrastructure and cell wall degrading enzymes," in *Microbial and plant opportunities to improve lignocellulose utilization by ruminants*, D. E. Akin, L. G. Ljungdahl, J. R. Wilson, and P. J. Harris, eds. New York: Elsevier , pp. 325–339.

485 Castanares, A. and T. M. Wood. 1992. "Purification and characterization of a feruloyl/*p*-coumaroyl esterase from solid-state cultures of the aerobic fungus *Penicillium pinophilum*," *Biochem. Soc. Trans.*, 20:275S.

486 Faulds, C.B. and G. Williamson. 1992. "Ferulic acid release from plant polysaccharides by specific esterases," in *Xylans and xylanases*, J. Visser, G. Beldman, M. A.

Kusters-van Someren, and A. G. J. Voragen, eds. Amsterdam: Elsevier Science Publishers B.V., pp. 419–422.

487 Lee, S. F. and C. W. Forsberg. 1987. "Purification and characterization of an α-L-arabinofuranosidase from *Clostridium acetobutylicum* ATCC 824," *Can. J. Microbiol.*, 33:1011–1016.

488 Greve, L. C., J. M. Labavitch and R. E. Hungate. 1984. "α-L-Arabinofuranosidase from *Ruminococcus albus* 8: Purification and possible role in hydrolysis of alfalfa cell wall," *Appl. Environ. Microbiol.*, 47:1135–1140.

489 Bachman, S. L. and A. J. McCarthy. 1991. "Purification and cooperative activity of enzymes constituting the xylan-degrading system of *Thermomonospora fusca*," *Appl. Environ. Microbiol.*, 57:2121–2130.

490 Poutanen, K. 1988. "An α-L-arabinofuranosidase of *Trichoderma reesei*," *J. Biotechnol.*, 7:271–281.

491 Biely, P., C. R. MacKenzie, J. Puls, and H. Schneider. 1986. "Cooperativity of esterases and xylanases in the enzymatic degradation of acetyl xylan," *Biotechnology*, 4:731–733.

492 Poutanen, K. and J. Puls. 1989. "The xylanolytic enzyme system of *Trichoderma reesei*," in *ACS symposium series 399: Plant cell wall polymers: Biogenesis and biodegradation*, N. G. Lewis and M. G. Paice, eds. Washington D.C.: Am. Chem. Soc., pp. 630–639.

493 Lamed, R., E. Setter, R. Kenig, and E. A. Bayer. 1993. "The cellulosome—A discrete cell surface organelle of *Clostridium thermocellum* which exhibits separate antigenic, cellulose-binding and various cellulolytic activities," *Biotechnol. Bioeng. Symp.*, 13:163–181.

494 Lamed, R. and E. A. Bayer. 1988. "The cellulosome of *Clostridium thermocellum*," *Adv. Appl. Microbiol.*, 33:1–46.

495 Lamed, R., E. Setter, and E. A. Bayer. 1983. "Characterization of a cellulose-binding, cellulase-containing complex in *Clostridium thermocellum*," *J. Bacteriol.*, 156:828–836.

496 Bayer, E. A., E. Morag, and R. Lamed. 1994. "The cellulosome—A treasure trove for biotechnology," *Trends Biotechnol.*, 12:379–386.

497 Lamed, R. and E.A. Bayer. 1993. "The cellulosome concept—A decade later," in *Genetics, biochemistry and ecology of lignocellulose degradation*, K. Shimada, S. Hoshino, K. Ohmiya, K. Sakka, Y. Kobayashi, and S. Karita, eds. Tokyo, Japan: Uni Publishers Co., Ltd., pp. 1–12.

498 Gerngross, U. T., M. P. M. Romaniec, T. Kobayashi, N. S. Huskisson, and A. L. Demain. 1993. "Sequencing of a *Clostridium thermocellum* gene (*cipA*) encoding the cellulosomal SL-protein reveals an unusual degree of internal homology," *Mol. Microbiol.*, 8:325–334.

499 Lamed, E., J. Naimark, E. Morgenstern, and E. A. Bayer. 1987. "Specialized cell surface structures in cellulolytic bacteria," *J. Bacteriol.*, 169:3792–3800.

500 Fujino, T., P. Beguin and J.-P. Aubert. 1993. "Organization of a *Clostridium thermocellum* gene cluster encoding the cellulosomal scaffolding protein CipA and a protein possibly involved in attachment of the cellulosome to the cell surface," *J. Bacteriol.*, 175:1891–1899.

501 Shoseyov, O., M. Takagi, M. A. Goldstein, and R. H. Doi. 1992. "Primary sequence analysis of *Clostridium cellulovorans* cellulose binding protein A," *Proc. Natl. Acad. Sci. USA*, 89:3483–3487.

502 Morag, E., A. Lapidot, D. Govorko, R. Lamed, M. Wilchek, E. A. Bayer, and Y. Shoham. 1995. "Expression, purification, and characterization of the cellulose-binding domain of the scaffoldin subunit from the cellulosome of *Clostridium thermocellum*," *Appl. Environ. Microbiol.*, 61:1980–1986.

503 Goldstein, M. A., M. Takagi, S. Hashida, O. Shoseyov, R. H. Doi, and L. H. Segel. 1993. "Characterization of the cellulose binding domain of the *Clostridium cellulovorans* cellulose-binding protein A," *J. Bacteriol.*, 175:5762–5768.

504 Navarro, A., M.-C. Chebrou, P. Beguin, and J.-P. Aubert. 1991. "Nucleotide sequence of the cellulase gene *celF* of *Clostridium thermocellum*," *Res. Microbiol.*, 142:927–936.

505 Durrant, A. J., J. Hall, G. P. Hazlewood, and H. J. Gilbert. 1991. "The non-catalytic C-terminal region of endoglucanase I from *Clostridium thermocellum* contains a cellulose-binding domain," *Biochem. J.*, 273:289–293.

506 Morag, E., E. A. Bayer, and R. Lamed. 1993. "Affinity digestion for the near-total recovery of purified cellulosome from *Clostridium thermocellum*," *Enzyme Microb. Technol.*, 14:289–292.

507 Mayer, F., M. P. Coughlan, Y. Mori, and L. G. Ljungdahl. 1987. "Macromolecular organization of the cellulolytic enzyme complex of *Clostridium thermocellum* as revealed by electron microscopy," *Appl. Environ. Microbiol.*, 53:2785–2792.

508 Hon-nami, K., M. P. Coughlan, H. Hon-nami, and L. Ljungdahl. 1986. "Separation and characterization of the complexes constituting the cellulolytic enzyme system of *Clostridium thermocellum*," *Arch. Microbiol.*, 145:13–19.

509 Lamed, R., E. Setter, R. Kenig, and E. A. Bayer. 1983. "The cellulosome: A discrete cell surface organelle of *Clostridium thermocellum* which exhibits separate antigenic, cellulose-binding and various cellulolytic activities," *Bioeng. Symp.*, 13: 163–181.

510 Lamed, R. and E. A. Bayer. 1988. "The cellulosome concept: exocellular/extracellular enzyme reactor centers for efficient binding and cellulolysis," in *Biochemistry and genetics of cellulose degradation*, J. Millet, ed. London: Academic Press, Inc., pp. 101–116.

511 Beguin, P., P. Cornet, and J.-P. Aubert. 1985. "Sequence of a cellulase gene of the thermophilic bacterium *Clostridium thermocellum*," *J. Bacteriol.*, 162:102–105.

512 Beguin, P., P. Cornet, and J. Millet. 1983. "Identification of the endoglucanase encoded by the *celB* gene of *Clostridium thermocellum*," *Biochimie*, 65:495–500.

513 Grepinet, O. and P. Beguin. 1986. "Sequence of the cellulase gene of *Clostridium thermocellum* coding for endoglucanase B," *Nucleic Acids Res.*, 144: 1791–1799.

514 Schwarz, W. H., S. Schimming, K. P. Rucknagel, S. Burgschwaiger, G. Kreil and W. L. Staudenbauer. 1988. "Nucleotide sequence of the *celC* gene encoding endoglucanase C of *Clostridium thermocellum*," *Gene*, 63:23–30.

515 Joliff, G., P. Beguin, and J.-P. Aubert. 1986. "Nucleotide sequence of the cellulase gene *celD* encoding endoglucanase D of *Clostridium thermocellum*," *Nucleic Acid Res.*, 14:8605–8613.

516 Lemaire, M. and P. Beguin. 1993. "Nucleotide sequence of the *celG* gene of *Clostridium thermocellum* and characterization of its product, endoglucanase CelG," *J. Bacteriol.*, 156:828–836.

517 Yaguee, E., P. Beguin, and J.-P. Aubert. 1990. "Nucleotide sequence and deletion analysis of the cellulase-encoding gene *celH* of *Clostridium thermocellum*," *Gene*, 89:61–67.

518 Kobayashi, T., M. P. M. Romaniec, P. J. Barker, V. T. Gerngross, and A. L. Demain. 1993. "Nucleotide sequence of the gene *celM* encoding a new endoglucanase (CelM) of *Clostridium thermocellum* and the purification of the enzyme," *J. Ferment. Bioeng.*, 76:251–256.

519 Morag, E., E. A. Bayer, G. P. Hazlewood, H. J. Gilbert, and R. Lamed. 1993. "Cellulase Ss (CelS) is synonymous with the major cellobiohydrolase (subunit S8) from the cellulosome of *Clostridium thermocellum*," *Appl. Biochem. Biotechnol.*, 43:147–151.

520 Grabnitz, F., M. Seiss, K. P. Rucknagel and W. L. Staudenbauer. 1991. "Structure of the β-glucosidase gene *bglA* of *Clostridium thermocellum*. Sequence analysis reveals a superfamily of cellulases and β-glucosidases including lactase/phlorizin hydrolase," *Eur. J. Biochem.*, 200:301–309.

521 Grabnitz, F., K. P. Rucknagel, M. Seiss, and W. L. Staudenbauer. 1989. "Nucleotide sequence of the *Clostridium thermocellum bglB* gene encoding thermostable β-glucosidase B: Homology to fungal β-glucosidase," *Mol. Gen. Genet.*, 217:70–76.

522 Grepinet, O., M.-C. Cherbou, and P. Beguin. 1988. "Purification of *Clostridium thermocellum* xylanase Z expressed in *Escherichia coli* and identification of the corresponding product in the culture medium of *Clostridium thermocellum*," *J. Bacteriol.*, 170:4576–4581.

523 Tokatlidis, K., S. Salamitou, P. Beguin, P. Dhurjati, and J. P. Aubert. 1993. "Interaction of the duplicated segment carried by *Clostridium thermocellum* cellulases with cellulosome components," *FEBS Lett.*, 291:185–188.

524 Beattie, L., K. M. Bhat, and T. M. Wood. 1994. "The effect of cations on reassociation of the components of the cellulosome cellulase complex synthesized by the bacterium *Clostridium thermocellum*," *Appl. Microbiol. Biotechnol.*, 40:740–744.

525 Bayer, E. A. and R. Lamed. 1986. "Ultrastructure of the cell surface cellulosome of *C. thermocellum* and its interaction with cellulose," *J. Bacteriol.*, 167:828–836.

526 Salamitou, S., M. Lemaire, T. Fujino, H. Ohayon, P. Gounon, P. Beguin, and P. Aubert. 1994. "Subcellular localization of *Clostridium thermocellum* ORF3, a protein carrying a receptor for the docking sequence borne by the catalytic components of the cellulosome," *J. Bacteriol.*, 176:2828–2834.

527 Morag, E., E. A. Bayer and R. Lamed. 1992. "Unorthodox intrasubunit interactions in the cellulosome of *Clostridium thermocellum*," *Appl. Biochem. Biotechnol.*, 33: 205–217.

528 Morag, E., E. A. Bayer, and R. Lamed. 1990. "Relationship of cellulosomal and noncellulosomal xylanases of *Clostridium thermocellum* to cellulose-degrading enzymes," *J. Bacteriol.*, 172:6098–6105.

529 Kim, C.-H. and D.-S. Kim. 1993. "Extracellular cellulolytic enzymes of *Bacillus circulans* are present as two multiple-protein complexes—Cellulase complex characterization," *Appl. Biochem. Biotechnol.*, 32:83–94.

530 Lin, L.-L. and J. A. Thomson. 1991. "An analysis of the extracellular xylanases and cellulases of *Butyrivibrio fibrosolvens* H17c," *FEMS Microbiol. Lett.*, 84:197–204.

531 Fujino, T., S. Karita, and K. Ohmiya. 1993. "Nucleotide sequences of the *celB* gene encoding endo-1,4-β-glucanse-2, ORF1 and ORF2 forming a putative cellulase gene cluster of *Clostridium josui*," *J. Ferment. Bioeng.*, 76:243–250.

532 Pohlschroder, M., S. B. Leschine, and E. Canale-Parola. 1993. "Regulation of the multienzyme cellulase-xylanase system of *Clostridium papyrosolvens*," in *Genetics, biochemistry and ecology of lignocellulose degradation*, New York: Academic Press, pp. 86–94.

533 Gong, J. and C.W. Forsberg. 1993. "Cellulose binding proteins and their potential role in adhesion of *Fibrobacter succinogenes* to cellulose," in *Genetics, biochemistry and ecology of lignocellulose degradation*, New York: Academic Press, pp. 138–145.

534 Doerner, K. C. and B. A. White. 1990. "Assessment of the endo-1,4-β-glucanase components of *Ruminococcus flavefaciens* FD-1," *Appl. Environ. Microbiol.*, 56: 1844–1850.

535 Bond, K. and F. Stutzenberger. 1989. "A note on the localization of cellulosome formation in *Thermomonospora curvata*," *J. Appl. Bacteriol.*, 67:605–609.

536 Wilson, C. and C. M. Wood. 1992. "The anaerobic fungus *Neocallimastix frontalis*: Isolation and properties of a cellulosome-type enzyme fraction with the capacity to solubilize hydrogen-bond ordered cellulose," *Appl. Microbiol. Biotechnol.*, 37: 125–129.

537 Mandels, M. and E. T. Reese. 1960. "The induction of cellulase in fungi by cellobiose," *J. Bacteriol.*, 79:816–826.

538 Linden, J. C. and M. Shiang. 1991. "Bacterial cellulases. Regulation of synthesis," *ACS symp. ser.*, 460:331–348.

539 Bisaria, V. S. and S. Mishra. 1989. "Regulatory aspects of cellulase biosynthesis and secretion," *CRC Crit. Rev. Biotechnol.*, 9:61–103.

540 Shimada, K., S. Hoshimno, K. Ohmiya, K. Sakka, Y. Kobayashi, and S. Karita, eds. 1993. *Genetics, biochemistry and ecology of lignocellulose degradation*. Tokyo, Japan: Uni Publishers Co., Ltd.

541 Kubicek, C. P., G. Muhlbauer, M. Koltz, E. John, and E. M. Kubicek-Pranz. 1988. "Properties of a conidial-bound cellulase enzyme system from *Trichoderma reesei*," *J. Gen. Microbiol.*, 134:1215–1222.

542 Messner, R., E. M. Kubicek-Pranz, A. Bsur and C. P. Kubicek. 1991. "Cellobiohydrolase II is the main conidial-bound cellulase in *Trichoderma reesei* and other *Trichoderma* strains," *Arch. Microbiol.*, 155:601–605.

543 Chaudhary, K. and P. Tauro. 1982. "Sequential release of cellulose enzymes during germination of *Trichoderma reesei*," *J. Biosci.*, 4:281–286.

544 Kubicek, C. P. 1987. "Involvement of a conidial endoglucanase and a plasma-membrane-bound β-glucosidase in the induction of endoglucanase synthesis by cellulose in *Trichoderma reesei*," *J. Gen. Microbiol.*, 133:1481–1487.

545 Mandels, M., F. W. Parrish, and E. T. Reese. 1962. "Sophorose as an inducer of cellulase in *Trichoderma viride*," *J. Bacteriol.*, 83:400–408.

546 Zhu, Y. S., T. Q. Wu, W. Chen, C. Tan, J. H. Gao, J. X. Fei, and C. N. Shih. 1982. "Induction and regulation of cellulase synthesis in *Trichoderma pseudokoningii* mutants EA3-867 and N2-78," *Enzyme Microb. Technol.*, 4:3–12.

547 el Gregory, S., A. Leite, O. Crivellaro, D. E. Eveleigh, and H. el Dorry. 1989. "Mechanism by which cellulose triggers cellobiohydrolase I gene expression in *Trichoderma reesei*," *Proc. Natl. Acad. Sci. USA*, 86:6138–6141.

548 Seiboth, B., R. Messner, F. Gruber, and C. P. Kubicek. 1992. "Disruption of the *Trichoderma reesei cbh2* gene coding for cellobiohydrolase II leads to a delay in the triggering of cellulase formation by cellulose," *J. Gen. Microbiol.*, 138: 1259–1264.

549 Kubicek, C. P., R. Messner, F. Gruber, M. Mandels, and E. M. Kubicek-Pranz. 1993. "Triggering of cellulase biosynthesis by cellulose in *Trichoderma reesei*," *J. Biol. Chem.*, 268:19364–19368.

550 Fritscher, G., R. Messner, and D. P. Kubicek. 1990. "Cellobiose metabolism and cellobiohydrolase I biosynthesis by *Trichoderma reesei*," *Exp. Mycol.*, 14:405–415.

551 Rho, D., M. Desrochers, L. Jurásek, H. Driguez, and J. DeFaye. 1982. "Induction of cellulase in *Schizophyllum commune*: Thiocellobiose as a new inducer," *J. Bacteriol.*, 149:47–53.

552 Bruchmann, E.-E., H. Scharch, and H. Graf. 1987. "Role and properties of lactonase in a cellulase system," *Biotechnol. Appl. Biochem.*, 9:146–159.

553 Vaheri, M. P., M. E. O. Vaheri, and V. S. Kauppinen. 1979. "Formation and release of cellulolytic enzymes during growth of *Trichoderma reesei* on cellobiose and glycerol," *Eur. J. Appl. Microbiol. Biotechnol.*, 8:73–80.

554 Szakmary, K., A. Wotawa, and C. P. Kubicek. 1991. "Origin of oxidized cellulose degradation products and mechanism of their promotion of cellobiohydrolase I biosynthesis in *Trichoderma reesei*," *J. Gen. Microbiol.*, 137:2873–2878.

555 Nisizawa, T., H. Suzuki, and K. Nisizawa. 1971. "De novo synthesis of cellulase induced by sophorose in *Trichoderma viride* cells," *J. Biochem. (Tokyo)*, 70: 387–393.

556 Vaheri, M., M. Leisola, and V. Kaupinnen. 1979. "Transglycosylation products of cellulase system of *Trichoderma reesei*," *Biotechnol. Lett.*, 1:41–47.

557 Schmid, G. and C. Wandrey. 1987. "Purification and partial characterization of a cellodextrin glucohydrolase (β-glucosidase) from *Trichoderma reesei* Strain QM 9414," *Biotechnol. Bioeng.*, 30:571–585.

558 Claeyssens, M., H. Van Tilbeurgh, J. P. Kamerling, J. Berg, M. Vrsanska, and P. Biely. 1990. "Studies of the cellulolytic system of the filamentous fungus *Trichoderma reesei* QM 9414," *Biochem. J.*, 270:251–256.

559 Sternberg, D. and G. R. Mandels. 1979. "Induction of cellulolytic enzymes in *Trichoderma reesei* by sophorose," *J. Bacteriol.*, 139:1321–1326.

560 Eriksson, K. E. and S. G. Hamp. 1978. "Regulation of endo-1,4-β-glucanase production in *Sporotrichum pulverulentum*," *Eur. J. Biochem.*, 90:183–190.

561 Messner, R. and C. P. Kubicek. 1991. "Carbon source control of cellobiohydrolase I and II formation by *Trichoderma reesei*," *Appl. Environ. Microbiol.*, 57:630–635.

562 Wang, C.-H., T.-H. Hseu, and C.-M. Huang. 1988. "Induction of cellulase by cello-oligosaccharides in *Trichoderma koningii* G-39," *J. Biotechnol.*, 9:47–60.

563 Montenecourt, B. S., S. D. Nhlapo, H. Trimino-Vazquez, S. Cuskey, D. H. J. Schamhart, and D. E. Eveleigh. 1981. "Regulatory controls in relation to over-production of fungal cellulases," *Trends Biol. Ferment.*, 18:33–53.

564 Farkas, V., M. Gressik, N. Kolarova, Z. Sulova, and S. Sestak. 1990. "Biochemical and physiological changes during photo-induced conidation and depression of cellulase synthesis in *Trichoderma*," in *Trichoderma reesei cellulase: Biochemistry, genetics, physiology, and application*, C. P. Kubicek, D. E. Eveleigh, H. Esterbauer, W. Steiner, and E. M. Kubicek-Pranz, eds. Cambridge: Graham House, pp. 139–155.

565 Sestak, S. and V. Farkas. 1992. "Metabolic regulation of endoglucanase synthesis in *Trichoderma reesei*: Participation of cyclic AMP and glucose-6-phosphate," *Can. J. Microbiol.*, 39:342–347.

566 Kwon, K. S., Y. C. Hoh, and S. W. Hong. 1988. "Location and biosynthetic regulation of endo-1,4-β-glucanase in *Aspergillus nidulans*," *Microbios*, 54: 149–156.

567 Rao, M., S. Gaikwad, C. Mishra, and V. Deshpande. 1988. "Induction and catabolite repression of cellulase in *Penicillium funiculosum*," *Appl. Biochem. Biotechnol.*, 19:129–137.

568 Breuil, C. and D. J. Kushner. 1976. "Cellulase induction and the use of cellulose as a prepared growth substrate by *Cellvibrio gilvus*," *Can. J. Microbiol.*, 22: 1776–1781.

569 Lin, E. and D. B. Wilson. 1987. "Regulation of β-1,4-endoglucanase synthesis in *Thermomonospora fusca*," *Appl. Environ. Microbiol.*, 53:1352–1357.

570 Biely, P., Z. Kratky, M. Vrsanska, and D. Umrmanicova. 1980. "Induction and inducers of endo-1,4-β-xylanase in the yeast *Cryptococcus albidus*," *Eur. J. Biochem.*, 108:323–329.

571 Leathers, T. D., R. W. Detroy, and R. J. Bothast. 1986. "Induction and glucose repression of xylanase from a color variant strain of *Aureobasidium pullans*," *Biotechnol. Lett.*, 8:867–872.

572 Morosoli, R., S. Durand, and E. D. Letendre. 1987. "Induction of xylanase by β-methylxyloside in *Cryptococcus albidus*," *FEMS Microbiol. Lett.*, 48:261–266.

573 Morosoli, R., S. Durand, and F. Boucher. 1989. "Stimulation of xylanase synthesis in *Cryptococcus albidus* by cyclic AMP," *FEMS Microbiol. Lett.*, 57:57–60.

574 Kratky, Z. and P. Biely. 1980. "Inducible β-xyloside permease as a constituent of the xylan-degrading enzyme system of the yeast *Cryptococcus albidus*," *Eur. J. Biochem.*, 112:367–373.

575 Yasui, T., B. T. Nguyen, and K. Nakanishi. 1984. "Inducers for xylanase production by *Cryptococcus flavus*," *J. Ferment. Technol.*, 62:353–359.

576 Senior, D. J., P. R. Mayers, and J. N. Saddler. 1989. "Production and purification of xylanases," in *Plant cell wall polymers: Biogenesis and biodegradation*, N.G. Lewis and M.G. Paice, eds. *ACS symposium series no. 399*, pp. 641–654.

577 Hrmova, M., P. Biely, and M. Vrsanska. 1986. "Specificity of cellulase and β-xylanase induction in *Trichoderma reesei* QM 9414," *Arch. Microbiol.*, 144: 307–311.

578 Herzog, P., A. Torronen, A. Harkki, and C.P. Kubicek. 1992. "Mechanism by which xylan and cellulose trigger the biosynthesis of endo-xylanase I by *Trichoderma reesei*," in *Xylans and xylanases*, J. Visser, G. Beldman, M. A. Kusters-van Someren, and A. G. J. Voragen, eds. Amsterdam: Elsevier Science Publishers B.V., pp. 289–293.

579 Biely, P., C. R. MacKenzie, and H. Schneider. 1988. "Production of acetyl xylan esterase by *Trichoderma reesei* and *Schizophyllum commune*," *Can. J. Microbiol.*, 39:767–772.

580 Arhin, F. F., F. Shareck, D. Kluepfel, and R. Morosoli. 1994. "Effects of disruption of xylanase-encoding genes on the xylanolytic system of *Streptomyces lividans*," *J. Bacteriol.*, 176:4924–4930.

581 Kluepfel, D., N. Daigneault, R. Morosoli, and F. Shareck. 1992. "Purification and characterization of a new xylanase (xylanase C) produced by *Streptomyces lividans* 66," *Appl. Microbiol. Biotechnol.*, 36:626–631.

582 Kluepfel, D., S. Vats-Mehta, F. Aumont, F. Shareck, and R. Morosoli. 1990. "Purification and characterization of a new xylanase (xylanase B) produced by *Streptomyces lividans* 66," *Biochem. J.*, 267:45–50.

583 Khan, A. W., K. A. Lamb, and R. P. Overend. 1990. "Comparison of natural hemicellulose and chemically acetylated xylan as substrates for the determination of acetyl-xylan esterase activity in Aspergilli," *Enzyme Microb. Technol.*, 12: 127–131.

584 van der Veen, P., M. J. A. Flipphi, A. G. J. Voragen, and J. Visser. 1992. "Induction of arabinases from *Aspergillus niger* on monomeric substrates," in *Xylans and xylanases*, J. Visser, G. Beldman, M. A. Kusters-van Someren, and A. G. J. Voragen, eds. Amsterdam: Elsevier Science Publishers B.V., pp. 497–500.

585 Kubicek, C. P., R. Messner, F. Gruber, R. L. Mach, and E. M. Kubicek-Pranz. 1993. "The *Trichoderma* cellulase regulatory puzzle: From the interior life of a secretory fungus," *Enzyme Microb. Technol.*, 15:90–99.

586 Sternberg, D. and G. R. Mandels. 1980. "Regulation of the cellulolytic system in *Trichoderma reesei* by sophorose: Induction of cellulase and repression of β-glucosidase," *J. Bacteriol.*, 144:1197–1199.

587 Enari, T.-M. 1983. "Microbial Cellulases," in *Microbial enzymes and biotechnology*, Applied Science Publishers Ltd., pp. 183–223.

588 Ullmann, A. 1985. "Catabolite repression," *Biochimie*, 67:29–34.

589 Fennington, G., D. Neubauer and F. Stutzenberger. 1984. "Cellulase biosynthesis in a catabolite repression-resistant mutant of *Thermomonospora curvata*," *Appl. Environ. Microbiol.*, 47:201–204.

590 Wood, W. E., D. G. Neubauer, and F. J. Stutzenberger. 1984. "Cyclic AMP levels during induction and repression of cellulase biosynthesis in *Thermomonospora curvata*," *J. Bacteriol.*, 160:1047–1054.

591 Esteban, R., A. R. Nebreda, J. R. Villanueva, and T. G. Villa. 1984. "Possible role of cAMP in the synthesis of β-glucanases and β-xylanases of *Bacillus circulans* WL-12," *FEMS Microbiol. Lett.*, 23:91–94.

592 de Graaff, L. H., H. C. van den Broeck, and A. J. J. van Ooijen. 1994. "Regulation of the xylanase-encoding *xinA* gene of *Aspergillus tubigensis*," *Mol. Microbiol.*, 12: 479.

593 Fan, L. T. and Y. H. Lee. 1983. "Kinetic studies on enzymatic hydrolysis of insoluble cellulose: Derivation of a mechanistic kinetic model," *Biotechnol. Bioeng.*, 25:2707–2733.

594 Ladisch, M. R., C. S. Gong, and G. T. Tsao. 1980. "Cellobiose hydrolysis by endoglucanase (glucan glucanohydrolase) from *Trichoderma reesei:* Kinetics and mechanism," *Biotechnol. Bioeng.*, 22:1107–1126.

595 Ladisch, M. R., J. Hong, M. Voloch, and G. T. Tsao. 1981. "Cellulase kinetics," in *Trends in the biology of fermentation for fuels and chemicals*, A. Hollaender, R. Rabson, P. Rodgers, A. San Pietro, R. Valentine, and R. Wolfe, eds. New York: Plenum Publishing Co., pp. 55–83.

596 Mandels, M. and E. T. Reese. 1963. "Inhibition of enzymatic hydrolysis of cellulose and related chemicals," in *Advances in enzymatic hydrolysis of cellulose and related materials*, E. T. Reese, ed. Oxford: Pergamon Press, pp. 115–157.

597 Mandels, M. and E. T. Reese. 1965. "Inhibition of cellulases," *Annu. Rev. Phytopathol.*, 3:85–102.

598 Marsden, W. L. and P. P. Gray. 1986. "Enzymatic hydrolysis of cellulose in lignocellulosic materials," *CRC Crit. Rev. Biotechnol.*, 3:235–276.

599 Lee, Y. H. and L. T. Fan. 1983. "Kinetic studies of enzymatic hydrolysis of insoluble cellulose: (II) Analysis of extended hydrolysis times," *Biotechnol. Bioeng.*, 25: 939–966.

600 Huang, A. A. 1975. "Kinetic studies on insoluble cellulose-cellulase system," *Biotechnol. Bioeng.*, 17:1421–1433.

601 Yablonsky, M. D., T. Bartley, K. O. Elliston, S. K. Kahrs, Z. P. Shalita, and D. E. Eveleigh. 1988. "Characterization and cloning of the cellulase complex of *Microbispora bispora*," in *Proceedings of the FEMS Symposium on the biochemistry and genetics of cellulose degradation*, J. P. Aubert, P. Beguin, and J. Millet, eds. New York: Academic Press, p. 249.

602 Huang, X. and M. H. Penner. 1991. "Apparent substrate inhibition of the *Trichoderma reesei* cellulase system," *J. Agric. Food Chem.*, 39:2096–2100.

603 Wilson, D. B. 1995. "Cloning of *Thermomonospora fusca* cellulase genes in *Escherichia coli* and *Streptomyces lividans*," in *Microbial gene techniques*, K. W. Adolph, ed. San Diego: Academic Press, Inc., pp. 367–374.

604 Teather, R. M. and P. J. Wood. 1982. "Use of Congo Red-polysaccharide interactions in enumeration and characterization of cellulolytic bacteria from the bovine rumen," *Appl. Environ. Microbiol.*, 43:777–780.

605 Biely, P., D. Mislovicova, and R. Toman. 1985. "Soluble chromogenic substrates for the assay of endo-1,4-β-xylanases and endo-1,4-β-glucanases," *Anal. Biochem.*, 144:142–146.

606 Kluepfel, D. 1988. "Screening of prokaryotes for cellulose- and hemicellulose-degrading enzymes," *Methods Enzymol.*, 160:180–186.

607 Farkas, V., M. Liskova, and P. Biely. 1985. "Novel media for detection of microbial producers of cellulase and xylanase," *FEMS Microbiol. Lett.*, 28:137–140.

608 Knowles, J., P. Lehtovaara, and T. Teeri. 1987. "Cellulase families and their genes," *Trends Biotechnol.*, 5:255–261.

609 Schwarz, W. H., K. Bronnenmeler, F. Grabnitz, and W. L. Staudenbauer. 1987. "Activity staining of cellulases in polyacrylamide gels containing mixed linkage β-glucans," *Anal. Biochem.*, 164:72–77.

610 Beguin, P. 1983. "Detection of cellulase activity in polyacrylamide gels using Congo Red-stained replicas," *Anal. Biochem.*, 131:333–336.

611 Biely, P., O. Markovic, and D. Mislovicová. 1985. "Sensitive detection of endo-1,4-β-glucanases and endo-1,4-β-xylanases in gels," *Anal. Biochem.*, 144:147–151.

612 MacKenzie, C. R. and R. E. Williams. 1984. "Detection of cellulase and xylanase activity in isoelectric-focused gels using agar substrate gels supported on plastic film," *Can. J. Microbiol.*, 30:1522–1525.

613 Painbeni, E., S. Valles, J. Polaina, and A. Flors. 1992. "Purification and characterization of a *Bacillus polymyxa* β-glucosidase expressed in *Escherichia coli*," *J. Bacteriol.*, 174:3087–3091.

614 Biely, P., M. Vrsanska, and S. Kucar. 1992. "Identification and mode of action of endo-(1-4)-β-xylanases," in *Xylans and xylanases*, J. Visser, G. Beldman, M. A. Kusters-van Someren, and A. G. J. Voragen, eds. Amsterdam: Elsevier Science Publishers, B.V., pp. 81–95.

615 Miller, G. L. 1959. "Use of dinitrosalicylic acid reagent for determination of reducing sugar," *Anal. Chem.*, 31:426–428.

616 Nelson, N. 1944. "Photometric adaptation of the Somogyi method for the determination of glucose," *J. Biol. Chem.*, 153:375–380.

617 Somogyi, M. 1952. "Notes on sugar determination," *J. Biol. Chem.*, 195:19–23.

618 Bailey, M. J. 1988. "A note on the use of dinitrosalicylic acid for determining the products of enzymatic reactions," *Appl. Microbiol. Biotechnol.*, 29:494–496.

619 Johnson, E. A., M. Sakajoh, G. Halliwell, A. Madia, and A. L. Demain. 1982. "Saccharification of complex cellulosic substrates by the cellulase system from *Clostridium thermocellum*," *Appl. Environ. Microbiol.*, 43:1125–1132.

620 Tailliez, P., H. Girard, J. Millet, and P. Beguin. 1989. "Enhanced cellulose fermentation by an asporogenous and ethanol-tolerant mutant of *Clostridium thermocellum*," *Appl. Environ. Microbiol.*, 55:207–211.

621 Updegraff, D. M. 1969. "Semimicro determination of cellulose in biological materials," *Anal. Biochem.*, 32:420–424.

622 Berghem, L. E. R. and G. Pettersson. 1973. "The mechanism of enzymatic cellulose degradation. Purification of a cellulolytic enzyme from *Trichoderma viride* active on highly ordered cellulose," *Eur. J. Biochem.*, 37:21–30.

623 Wood, T. M. 1968. "Cellulolytic enzyme system of *Trichoderma koningii*. Separation of components attacking native cotton," *Biochem. J.*, 109:217–227.

624 Van Tilbeurgh, H. and M. Claeyssens. 1985. "Detection and differentiation of cellulase components using low molecular-mass fluorogenic substrates," *FEBS Lett.*, 187:283–288.

625 Van Tilbeurgh, H., M. Claeyssens, and C. K. De Bruyne. 1982. "The use of 4-methylumbelliferyl and other chromogenic glycosides in the study of cellulolytic enzymes," *FEBS Lett.*, 149:152–156.

626 Van Tilbeurgh, H., F. G. Loontiens, C. K. De Bruyne, and M. Claeyssens. 1988. "Fluorogenic and chromogenic glycosides as substrates and ligands of carbohydrases," *Methods Enzymol.*, 160:45–59.

627 Miller, G. L., J. Dean, and R. Blum. 1960. "A study of methods for preparing oligosaccharides from cellulose," *Arch. Biochem. Biophys.*, 91:21–26.

628 Pereira, A. N., M. Mobedshahi, and M. R. Ladisch. 1988. "Preparation of cellodextrins," *Methods Enzymol.*, 160:26–38.

629 Schmid, G. 1988. "Preparation of cellodextrins: Another perspective," *Methods Enzymol.*, 160:38–45.

630 Macarron, R., C. Acebal, M. P. Castillon, J. M. Dominguez, I. de la Mata, G. Pettersson, P. Tomme, and M. Claeyssens. 1993. "Mode of action of endoglucanase III from *Trichoderma reesei*," *Biochem. J.*, 289:867–873.

631 Bray, M. R. and A. J. Clarke. 1992. "Action pattern of xylo-oligosaccharide hydrolysis by *Schizophyllum commune* xylanase A," *Eur. J. Biochem.*, 204:191–196.

632 MacKenzie, C. R., R. C. A. Yang, G. B. Patel, D. Bilous, and S. A. Narang. 1989. "Identification of three distinct *Clostridium thermocellum* xylanase genes by molecular cloning," *Arch. Microbiol.*, 152:377–381.

633 Kremer, S. M. and P. M. Wood. 1992. "Continuous monitoring of cellulase action on microcrystalline cellulose," *Appl. Microbiol. Biotechnol.*, 37:750–755.

634 King, N. J. and D. B. Fuller. 1968. "The xylanase system of *Coniophora cerebella*," *Biochem. J.*, 108:571–576.

635 Sengupta, S., S. Khowla, and P. K. Goswami. 1987. "Assay of endo-β-D-xylanase activity with a soluble *O*-(carboxymethyl) derivative of larch wood D-xylan," *Carbohydr. Res.*, 168:156–161.

636 Biely, P., D. Kluepfel, R. Morosoli, and F. Shareck. 1993. "Mode of action of three endo-β-1,4-xylanses of *Streptomyces lividans*," *Biochim. Biophys. Acta*, 1162: 246–254.

637 Biely, P., D. Mislovicova, and R. Toman. 1988. "Remazol brilliant blue-xylan: A soluble chromogenic substrate for xylanases," *Methods Enzymol.*, 160:536–541.

638 Wakarchuk, W. W., W. L. Sung, R. L. Campbell, A. Cunningham, D. C. Watson, and M. Yaguchi. 1994. "Thermostabilization of the *Bacillus circulans* xylanase by the introduction of disulfide bonds," *Prot. Eng.*, 7:1379–1386.

639 Nystrom, J. M. and P. H. DiLuca. 1977. "Enhanced production of *Trichoderma* cellulase on high levels of cellulose in submerged culture," in *Proceedings of the bioconversion symposium*, New Delhi: Indian Institute of Technology, pp. 293–304.

640 Chahal, D. S. 1985. "Solid-state fermentation with *Trichoderma reesei* for cellulase production," *Appl. Environ. Microbiol.*, 49:205–210.

641 Strobel, G. A. 1963. "A xylanase system produced by *Diplodia viticola*," *Phytopathology*, 53:592–596.

642 Walker, D. J. 1967. "Some properties of xylanase and xylobiase from mixed rumen organisms," *Aust. J. Biol. Sci.*, 20:799–808.

643 Paice, M. G. and L. Jurásek. 1977. "Wood saccharifying enzymes from *Schizophyllum commune*," in *TAPPI forest biology wood chemistry conference proceedings*, Atlanta: TAPPI Publishing, pp. 113–117.

644 Bray, M. R. and A. J. Clarke. 1990. "Essential carboxy groups in xylanase A," *Biochem. J.*, 270:91–96.

645 Jurasek, L., J. Varadi, and R. Sopko. 1967. "Enzymatic properties of the lignolytically active supernatant of *Schizophyllum commune* shake culture," *Drevarsky Vyskum.*, 4:191–201.

646 Paice, M. G., R. Bernier, Jr., and L. Jurasek. 1988. "Viscosity-enhancing bleaching of hardwood kraft pulp with xylanase from a cloned gene," *Biotechnol. Bioeng.*, 32:235–239.

647 Bernier, R., Jr., H. Driguez, and M. Desrochers. 1983. "Molecular cloning of a *Bacillus subtilis* xylanase gene in *Escherichia coli*," *Gene*, 26:59–65.

648 Yang, R. C. A., R. MacKenzie, D. Bilous, V. L. Seligy, and S. Narang. 1987. "Expression and secretion of a *Cellulomonas fimi* exoglucanase in *Saccharomyces cerivisiae*," *Appl. Environ. Microbiol.*, 54:476–484.

649 Woodward, J. 1984. "Xylanases: Functions, properties, and applications," *Top. Enzyme Ferment. Biotechnol.*, 8:9–30.

650 Ghosh, B. K. and A. Ghosh. 1992. "Degradation of cellulose by fungal cellulase," in *Microbial degradation of natural products*, G. Winkelmann, ed. Weinheim: VCH Publishers Inc., pp. 83–126.

651 Wood, W. A. and S. T. Kellogg, eds. 1988. *Methods in enzymology. Vol. 160. Biomass. Part A. Cellulose and hemicellulose*. San Diego: Academic Press, Inc.

652 Halgasova, N., E. Kutejova, and J. Timko. 1994. "Purification and some characteristics of the acetylxylan esterase from *Schizophyllum commune*," *Biochem. J.*, 298:751–755.

653 Gama, F. M., C. J. Faro, J. A. Teixeira, and M. Mota. 1993. "New methodology for the characterization of endoglucanase activity and its application on the *Trichoderma longibrachiatum* cellulolytic complex," *Enzyme Microb. Technol.*, 15:57–61.

654 Baker, J. O., W. S. Adney, R. A. Nieves, S. R. Thomas, D. B. Wilson, and M. E. Himmel. 1994. "A new thermostable endoglucanase, *Acidothermus cellulolyticus* E1," *Appl. Biochem. Biotechnol.*, 45:245–256.

655 Leah, R., J. Kigel, I. Svendsen, and J. Mundy. 1995. "Biochemical and molecular characterization of a barley seed β-glucosidase," *J. Biol. Chem.*, 270:15789–15797.

656 Shao, W. and J. Wiegel. 1995. "Purification and characterization of two thermostable acetyl xylan esterases from *Thermoanaerobacterium* sp. strain JW/SL-YS485," *Appl. Environ. Microbiol.*, 61:729–733.

657 Fliess, A. and K. Schugerl. 1983. "Characterization of cellulases by HPLC separation," *Eur. J. Appl. Microbiol. Biotechnol.*, 17:314–318.

658 Schulein, M. 1988. "Cellulases of *Trichoderma reesei*," *Methods Enzymol.*, 160: 234–242.

659 Sakka, K., Y. Kojima, T. Kondo, S. Karita, K. Shimada, and K. Ohmiya. 1994. "Purification and characterization of xylanase a from *Clostridium stercorarium* F-9 and a recombinant *Escherichia coli*," *Biosci. Biotechnol. Biochem.*, 58:1496–1499.

660 Kormelink, F. J. M., B. Lefebvre, and F. Strozyk. 1993. "Purification and characterization of an acetyl xylan esterase from *Aspergillus niger*," *J. Biotechnol.*, 27:267.

661 Manin, C., F. Shareek, R. Morosoli, and D. Kluepfel. 1994. "Purification and characterization of an alpha-L-arabinofuranosidase from *Streptomyces lividans* 66 and DNA sequence of the gene (*abfA*)," *Biochem. J.*, 302:443–449.

662 Vaidya, M., R. Seeta, C. Mishra, V. Deshpande, and M. Rao. 1984. "A rapid and simplified procedure for purification of a cellulase from *Fusarium lini*," *Biotechnol. Bioeng.*, 24:41–45.

663 Rozie, H., F. M. Rombouts, and J. Visser. 1992. "One step purification of endo-xylanases with cross-linked xylan as an affinity adsorbent," in *Xylans and xylanases*, J. Visser, G. Beldman, M. A. Kusters-van Someren, and A. G. J. Voragen, eds. Amsterdam: Elsevier Science Publishers B.V., pp. 429–434.

664 Iwasaki, T., H. Susuki, and M. Funatsu. 1964. "Purification and characterization of two types of cellulases from *Trichoderma koningii*," *J. Biochem. (Tokyo)*, 55: 209–212.

665 Kitamidado, T. and N. Toyama. 1962. "Assay for cellulase based on the breakdown of filter paper," *J. Ferment. Technol.*, 40:85–88.

666 Mandels, M., R. Andreotti and C. Roche. 1976. "Measurement of saccharifying cellulase," *Biotechnol. Bioeng. Symp.*, 6:21–33.

667 Poincelot, R. P. and P. R. Day. 1972. "Simple dye release assay for determining cellulolytic activity of fungi," *Appl. Microbiol.*, 22:875–879.

668 McCleary, B. V. 1988. "Soluble, dye-labeled polysaccharides for the assay of endohydrolases," *Methods Enzymol.*, 160:74–86.

669 Ogawa, K. and N. Toyama. 1963. "Decomposition of native cellulose by commercial cellulase preparations," *J. Ferment. Technol.*, 41:282–288.

670 Wood, T. M. 1969. "Cellulase from *Fusarium solani*. Resolution of the enzyme complex," *Biochem. J.*, 115:457–464.

671 Li, L. H., R. M. Flora, and K. W. King. 1963. "Purification of β-glucosidases from *Aspergillus niger*, and initial observations on the C1 of *Trichoderma koningii*," *J. Ferment. Technol.*, 41:98–105.

672 Enari, T.-M. and M.-L. Niku-Paavola. 1988. "Nephelometric and turbidometric assay for cellulase," *Methods Enzymol.*, 160:117–126.

673 Halliwell, G. and M. Griffin. 1973. "The nature and mode of action of the cellulolytic component C1 of *Trichoderma koningii* on native cellulose," *Biochem. J.*, 135:587–594.

674 Sakamoto, R., M. Arai, and S. Murao. 1984. "Enzymatic properties of hydrocellulase from *Aspergillus aculeatus*," *J. Ferment. Technol.*, 62:561–567.

675 Deshpande, M. V., L. G. Pettersson, and K.-E. Eriksson. 1988. "Selective assay for exo-1,4-β-glucanases," *Methods Enzymol.*, 160:126–130.

676 Hirayama, T., S. Horie, H. Nagayama, and K. Matsuda. 1978. "Studies on cellulases of a phytopathogenic fungus, *Pyricularia oryzae cavara*," *J. Biochem. (Tokyo)*, 84: 27–37.

677 Hash, J. H. and K. W. King. 1958. "Some properties of an aryl-β-glucosidase from culture filtrates of *Myrothecium verrucaria*," *J. Biol. Chem.*, 232:395–402.

678 Sternberg, D. 1976. "Production of cellulase by *Trichoderma*," *Biotechnol. Bioeng. Symp.*, 6:35–53.

679 Gallo, B. J., R. Andreotti, C. Roche, D. Ryu, and M. Mandels. 1978. "Cellulase production by a new mutant strain of *Trichoderma reesei* MCG77," *Biotechnol. Bioeng. Symp.*, 8:89–101.

680 Tangnu, S. K., H. W. Blanch, and C. R. Wilke. 1981. "Enhanced production of cellulase, hemicellulase, and β-glucosidase by *Trichoderma reesei*," *Biotechnol. Bioeng.*, 23:1837–1849.

681 Ryu, D. D. and M. Mandels. 1980. "Cellulases: Biosynthesis and applications," *Enzyme Microb. Technol.*, 2:91–101.

682 King, K. W. 1965. "Enzymatic attack on highly crystalline hydrocellulose," *J. Ferment. Technol.*, 43:79–94.

683 Weber, M., M. J. Foglietti, and F. Percheron. 1980. "Affinity chromatography of a cellulase complex from *Trichoderma viride* on cross-linked cellulose," *J. Chromatogr.*, 188:377–382.

684 Reese, E. T. 1982. "Elution of cellulase from cellulose," *Process Biochem.*, 17: 2–10.

685 Halliwell, G. and M. Griffen. 1978. "Affinity chromatography of the cellulase system of *Trichoderma koningii*," *Biochem. J.*, 169:713–715.

686 Nummi, M., M.-L. Niku-Paavola, T.-M. Enari, and V. Raunio. 1981. "Isolation of cellulases by means of biospecific sorption on amorphous cellulose," *Anal. Biochem.*, 116:137.

687 Van Tilbeurgh, H., R. Bhikhabhai, L. G. Pettersson, and M. Claeyssens. 1984. "Separation of endo- and exo-type cellulases using a new affinity chromatography method," *FEBS Lett.*, 169:215–218.

688 Tomme, P., S. McCrae, T. M. Wood, and M. Claeyssens. 1988. "Chromatographic separation of cellulolytic enzymes," *Meth. Enzymol.*, 160:187–193.

689 Hakansson, U., L. Fagerstam, L. G. Pettersson, and L. Andersson. 1979. "A 1,4-β-glucan glucanohydrolase from the cellulolytic fungus *Trichoderma viride* QM9414. Purification, characterization and preparation of an immunoadsorbent for the enzyme," *Biochem. J.*, 179:141–149.

690 Fahnrich, P. and K. Irrgang. 1984. "Affinity chromatography of extracellular cellulase from *Chaetomium cellulolyticum*," *Biotechnol. Lett.*, 6:251–256.

691 Eriksson, K. E. and B. Pettersson. 1975. "Separation, purification and physico-chemical characterization of five endo-1,4-β-glucanases," *Eur. J. Biochem.*, 51:193.

692 Eriksson, K. E. and B. Pettersson. 1975. "Extracellular enzyme system utilized by the fungus *Sporotrichum pulverulentum (Chrysosporium lignorum)* for breakdown of cellulose. III. Purification and physicochemical characterization of an exo-1,4-β-glucanase," *Eur. J. Biochem.*, 51:213.

693 Slade, A., P. B. Hoj, N. A. Morrice, and G. B. Fincher. 1989. "Purification and characterization of three (1→4)-β-D-xylan endohydrolases from germinated barley," *Eur. J. Biochem.*, 185:533–539.

694 Woodward, J., J. P. Brown, B. R. Evans, and K. A. Affholter. 1994. "Papain digestion of crude *Trichoderma reesei* cellulase-purification and properties of cellobiohydrolase I and II core proteins," *Biotechnol. Appl. Biochem.*, 19:141–153.

695 Christakopoulos, P., P. W. Goodenough, D. Kekos, B. J. Macris, M. Claeyssens, and M. K. Bhat. 1994. "Purification and characterisation of an extracellular beta-glucosidase with transglycosylation and exo-glucosidase activities from *Fusarium oxysporum*," *Eur. J. Biochem.*, 224:379–385.

696 Gomez de Segura, B. and M. Fevre. 1993. "Purification and characterization of two 1,4-β-xylan endohydrolases from the rumen fungus *Neocallimastix frontalis*," *Appl. Environ. Microbiol.*, 59:3654.

697 Borneman, W. S., L. G. Ljungdahl, R. D. Hartley, and D. E. Akin. 1992. "Purification and partial characterization of two feruloyl esterases from the anaerobic fungus *Neocallimastix* strain MC-2," *Appl. Environ. Microbiol.*, 58:3762–3766.

698 Kelly, M. A., M. L. Sinnott, and M. Herrchen. 1987. "Purification and mechanistic properties of an extracellular α-L-arabinofuranosidase from *Monilinia fructigena*," *Biochem. J.*, 245:843–849.

699 Gilead, S. and Y. Shoham. 1995. "Purification and characterization of α-L-arabinofuranosidase from *Bacillus stearothermophilus* T-6," *Appl. Environ. Microbiol.*, 61: 170–174.

700 Kaneko, S., M. Sano, and I. Kusakabe. 1994. "Purification and some properties of alpha-L-arabinofuranosidase from *Bacillus subtilis* 3-6," *Appl. Environ. Microbiol.*, 60:3425–3428.

701 ben-Gershom, E. and J. Liebowitz. 1958. "Stereochemistry of enzymic maltose hydrolysis," *Enzymologia*, 20:148–152.

702 Li, L. H., R. M. Flora, and K. W. King. 1965. "Individual roles of cellulase components derived from *Trichoderma viride*," *Arch. Biochem. Biophys.*, 111: 439–447.

703 Parrish, F. W. and E. T. Reese. 1967. "Anomeric form of D-cellulose produced during enzymolysis," *Carbohydr. Res.*, 3:424–429.

704 Withers, S. G., D. Dombroski, L. A. Berven, D. G. Kilburn, R. C. Miller, R. A. J. Warren, and N. R. Gilkes. 1986. "Direct ^{1}H N.M.R. determination of the stereochemical course of hydrolases catalysed by glucanase components of the cellulase complex," *Biochem. Biophys. Res. Commun.*, 139:487–494.

705 Gebler, J., N. R. Gilkes, M. Claeyssens, D. R. Wilson, P. Beguin, W. W. Wakarchuk, D. G. Kilgurn, R. C. Miller, Jr., A. J. Warren, and S. Withers. 1992. "Stereoselective hydrolysis catalyzed by related β-1,4-glucanases and β-1,4-xylanases," *J. Biol. Chem.*, 18:12559–12561.

706 Withers, S. G. and I. P. Street. 1988. "Identification of a covalent α-D-glucopyranosyl enzyme intermediate formed on a β-glucosidase," *J. Am. Chem. Soc.*, 110: 8551–8553.

707 Koshland, D. E. 1953. "Stereochemistry and the mechanism of enzymatic reactions," *Biol. Rev.*, 28:416–436.

708 Sinnott, M. L. 1990. "Catalytic mechanisms of glycosyl group transfer," *Chem. Rev.*, 90:1171–1202.

709 Blake, C. C. F., L. N. Johnson, G. A. Mair, A. C. T. North, D. C. Phillips, and V. R. Sarma. 1967. "Crystallographic studies of the activity of hen egg-white lysozyme," *Proc. R. Soc. London B. Biol. Sci.*, 167:378–388.

710 Imoto, T., L. N. Johnson, A. C. T. North, D. C. Phillips, and J. A. Rupley. 1972. "Vertebrate lysozymes," in *The enzymes, Vol. 7*, P. D. Boyer, ed. New York: Academic Press, pp. 666–868.

711 Sinnott, M. 1987. "Glycosyl group transfer," in *Enzyme mechanisms*, M. I. Page and A. Williams, eds. London: Royal Society of Chemistry, pp. 259–297.

712 Withers, S. G. and R. Aebersold. 1995. "Approaches to labeling and identification of active site residues in glycosidases," *Protein Sci.*, 4:361–372.

713 Umezurike, G. M. 1979. "The β-glucosidase from *Botryodiplodia theobromae*," *Biochem. J.*, 179:503–507.

714 Umezurike, G. M. 1977. "The active-site of β-glucosidase from *Botryodiplodia theobromae*," *Biochem. J.*, 167:831–833.

715 Bedino, S., G. Testore, and F. Obert. 1986. "Kinetic properties and mechanism of action of an intracellular β-glucosidase from *Thermoascus aurantiacus* Miehe," *Ital. J. Biochem.*, 35:207–220.

716 Greenstein, J. P. and M. Winitz. 1961. *Chemistry of the amino acids. Vol. 1*, New York: John Wiley and Sons, pp. 498–500.

717 Davoodi, J., W. W. Wakarchuk, R. L. Campbell, P. R. Carey, and W. K. Surewicz. 1994. "An active site glutamic acid of *Bacillus circulans* xylanase has a pKa of 6.8: FTIR evidence," *Biophys. J.*, 68:A201.

718 Dale, M. P., H. E. Ensley, K. Kern, K. A. R. Sastry, and L. D. Byers. 1985. "Reversible inhibitors of β-glucosidase," *Biochemistry*, 24:3530–3539.

719 Dale, M. P., W. P. Kopfler, I. Chait, and L. D. Byers. 1986. "β-Glucosidase: Substrate, solvent, and viscosity variation as probes of the rate-limiting steps," *Biochemistry*, 25:2522–2529.

720 Kempton, J. B. and S. G. Withers. 1992. "Mechanism of *Agrobacterium* β-glucosidase: Kinetic studies," *Biochemistry*, 31:9961–9969.

721 Tull, D. and S. G. Withers. 1994. "Mechanisms of cellulases and xylanases: A detailed kinetic study of the exo-β-1,4-glycanase from *Cellulomonas fimi*," *Biochemistry*, 33:6363–6370.

722 Matsui, H., J. S. Blanchard, C. F. Brewer, and E. J. Hehre. 1989. "Alpha-secondary tri-fluoride kinetic isotope effects for the hydrolysis of alpha-D-glucopyranosyl fluoride by exo-alpha-glycanases," *J. Biol. Chem.*, 264:8714–8716.

723 Lui, W., N. B. Madsen, C. Braun, and S. G. Withers. 1991. "Reassessment of the catalytic mechanism of glycogen debranching enzyme," *Biochemistry*, 30: 1419–1424.

724 Van Doorslaer, E., O. van Opstal, H. Kersters-Hilderson, and C. K. De Bruyne. 1984. "Kinetic α-deuterium isotope effects for enzymatic and acid hydrolysis of aryl-β-D-glucopyranosides," *Bioorg. Chem.*, 12:158–163.

725 Lai, H.-Y. L. and B. Axelrod. 1973. "1-Aminoglycosides, a new class of specific inhibitors of glycosidases," *Biochem. Biophys. Res. Commun.*, 54:463–468.

726 Walker, D. E. and B. Axelrod. 1978. "Evidence for a single catalytic site on the β-glucosidase-β-D-galactosidase of almond emulsin," *Arch. Biochem. Biophys.*, 187:102–107.

727 Legler, G. 1978. "Inhibition of β-glucosidases from almonds by cationic and neutral β-glucosyl derivatives," *Biochim. Biophys. Acta*, 524:94–101.

728 Legler, G., M. L. Sinnott, and S. Withers. 1980. "Catalysis by β-glucosidase A3 of *Aspergillus wentii*," *J. Chem. Soc. (London), Perkins Trans. II*, 1980:1376–1383.

729 Clarke, A. J. and L. S. Adams. 1987. "Irreversible inhibition of *Schizophyllum commune* cellulase by divalent transition metal ions," *Biochim. Biophys. Acta*, 916:213–219.

730 Perkins, S. J., L. N. Johnson, P. A. Machin, and D. C. Phillips. 1979. "Crystal structures of hen egg-white lysozyme complexes with gadolinium (III) and gadolinium (III)-*N*-acetyl-D-glucosamine," *Biochem. J.*, 181:21–36.

731 Dobson, C. M. and R. J. P. Williams. 1977. "Nuclear magnetic resonance studies of the interaction of lanthanide cations with lysozyme," in *Metal-ligand interactions in organic chemistry and biochemistry. Part I*, B. Pullman and N. Goldblum, eds. Dordrecht, The Netherlands: Reidel Publishing Co., pp. 255–282.

732 Roeser, K.-R. and G. Legler. 1981. "Role of sugar hydroxyl groups in glucoside hydrolysis. Cleavage mechanism of deoxyglucosides and related substrates by β-glucosidase A3 from *Aspergillus wentii*," *Biochim. Biophys. Acta*, 657:321–333.

733 Legler, G., K.-R. Roeser and H.-K. Illig. 1979. "Reaction of β-D-glucosidase A3 from *Aspergillus wentii* with D-glucal," *Eur. J. Biochem.*, 101:85–92.

734 McGinnis, K. and D. B. Wilson. 1993. "Disulfide arrangement and chemical modification of β-1,4-endoglucanase E2 from *Thermomonospora fusca*," *Biochemistry*, 32:8151–8156.

735 Tomme, P., J. van Beeumen, and M. Claeyssens. 1992. "Modification of catalytically important carboxy residues in endoglucanase D from *Clostridium thermocellum*," *Biochem. J.*, 285:319–324.

736 Tomme, P., S. Chauvaux, P. Beguin, J. Millet, J.-P. Aubert, and M. Claeyssens. 1991. "Identification of a histidyl residue in the active center of endoglucanase D from *Clostridium thermocellum*," *J. Biol. Chem.*, 266:10313–10318.

737 Hurst, P. L., P. A. Sullivan, and M. G. Shepherd. 1977. "Chemical modification of a cellulase from *Aspergillus niger*," *Biochem. J.*, 167:549–556.

738 Tomme, P. and M. Claeyssens. 1989. "Identification of a functionally important carboxyl group in cellobiohydrolase I from *Trichoderma reesei*," *FEBS Lett.*, 243: 239–243.

739 de la Mata, I., M. P. Castillon, J. M. Dominguez, R. Macarron, and C. Acebal. 1993. "Chemical modification of β-glucosidase from *Trichoderma reesei* QM 9414," *J. Biochem. (Tokyo)*, 114:754–759.

740 Bray, M. R. and A. J. Clarke. 1994. "Identification of a glutamate residue at the active site of xylanase a from *Schizophyllum commune*," *Eur. J. Biochem.*, 219: 821–827.

741 Chauthaiwale, J. and M. Rao. 1994. "Chemical modification of xylanase from alkalothermophilic *Bacillus* species—evidence for essential carboxyl group," *Biochim. Biophys. Acta*, 1204:164–168.

742 Legler, G. and H. Gilles. 1970. "Zur Keetnis der katalytisch wirkenden Gruppen einer β-Glucosidase aus *Aspergillus wentii*," *Hoppe-Seyler's Z. Physiol. Chem.*, 351:741–748.

743 Legler, G. 1990. "Glycoside hydrolases: Mechanistic information from studies with reversible and irreversible inhibitors," *Adv. Carbohydr. Chem. Biochem.*, 48: 319–384.

744 Black, T. S., L. Kiss, D. Tull and S. G. Withers. 1993. "*N*-Bromoacetyl-glycopyranosylamines as affinity labels for a β-glucosidase and a cellulase," *Carbohydr. Res.*, 250: 195–202.

745 Shulman, M. L., S. D. Shiyan, and A. Y. Khorlin. 1976. "Specific irreversible inibition of sweet-almond β-glucosidase by some β-glucopyranosyl-epoxyalkanes and β-D-glucopyranosyl isothiocyanate," *Biochim. Biophys. Acta*, 445:169–181.

746 Caron, G. and S. G. Withers. 1989. "Conduritol aziridine: A new mechanism-based glucosidase inactivator," *Biochem. Biophys. Res. Commun.*, 163:495–499.

747 Withers, S. G. and K. Umezawa. 1991. "Cyclophellitol: A naturally occurring mechanism-based inactivator of β-glucosidases," *Biochem. Biophys. Res. Commun.*, 177:532–537.

748 Clarke, A. J. and H. Strating. 1989. "Affinity labelling of *Schizophyllum commune* cellulase with [1-3H]-4,5-epoxypentyl-β-cellobioside: Synthesis of inhibitor and stoichiometry of interaction," *Carbohydr. Res.*, 188:245–250.

749 Clarke, A. J. 1988. "Active-site-directed inactivation of *Schizophyllum commune* cellulase by 4′,5′-epoxypentyl-4-D-(β-D-glucopyranosyl)-223-D-glucopyranoside," *Biochem. Cell. Biol.*, 56:871–879.

750 Legler, G. and E. Bause. 1973. "Epoxy alkyl oligo-(1,4)-β-D-glucosides as active site-directed inhibitors of cellulases," *Carbohydr. Res.*, 28:45–52.

751 Hoj, P. B., E. B. Rodriguez, R. V. Stick, and B. A. Stone. 1989. "Differences in active site structure in a family of β-glucan endohydrolases deduced from the kinetics of inactivation by epoxyalkyl β-oligoglucosides," *J. Biol. Chem.*, 264: 4939–4947.

752 Macarron, R., J. van Beeumen, B. Henrissat, I. de la Mata, and M. Claeyssens. 1993. "Identification of an essential glutamate residue in the active site of endoglucanase III from *Trichoderma reesei*," *FEBS Lett.*, 316:137–140.

753 Hoj, P. B., E. B. Rodriquez, J. R. Iser, R. V. Stick, and B. A. Stone. 1991. "Active-site-directed inhibition by optically pure epoxyalkyl cellobiosides reveals differences in active site geometry of two 1,3-1,4-β-D-glucan 4-glucanohydrolases," *J. Biol. Chem.*, 266:11628–11631.

754 Bause, E. and G. Legler. 1974. "Isolation and amino acid sequence of a hexadecapeptide from the active site of β-glucosidase A3 from *Aspergillus wentii*," *Hoppe-Seyler's Z. Physiol. Chem.*, 355:438–442.

755 Bause, E. and G. Legler. 1980. "Isolation and structure of a tryptic glycopeptide from the active site of β-glucosidase A3 from *Aspergillus wentii*," *Biochim. Biophys. Acta*, 626:459–465.

756 Legler, G. 1968. "Untersuchungen zum Wirkungsmechanismus glykosidspaltender Enzyme, III," *Hoppe-Seyler's Z. Physiol. Chem.*, 349:767–774.

757 Legler, G. 1966. "Untersuchungen zum Wirkungsmechanismus glykosidspaltender Enzyme I. Darstellung und Eigenschaften spezifischer Inaktivatoren," *Hoppe-Seyler's Z. Physiol. Chem.*, 345:197–214.

758 Legler, G. and L. M. O. Osama. 1968. "Reinigung und Eigenschaften einer β-Glucosidase aus *Aspergillus oryzae*," *Hoppe-Seyler's Z. Physiol. Chem.*, 349: 1488–1492.

759 Umezurike, G. M. 1987. "The mechanism of action of β-glucosidase from *Botryodiplodia theobromae* Pat," *Biochem. J.*, 241:455–462.

760 Donsimoni, R., G. Legler, R. Bourbouze, and P. Lalegerie. 1988. "Study of β-glucosidase from *Helix pomatia* by active site-directed inhibitors," *Enzyme*, 39: 78–89.

761 Legler, G. and G. Harder. 1978. "Amino acid sequence at the active site of β-glucosidase A from bitter almonds," *Biochim. Biophys. Acta*, 524:102–108.

762 Legler, G. 1970. "Markierung des aktiven Zentrums der β-Glucosidasen A und B aus dem Submandel-Emulsin mit [^{3}H]6-Brom-6-desoxy-condurit-B-epoxid," *Hoppe-Seyler's Z. Physiol. Chem.*, 351:25–31.

763 Marshall, P. J., M. L. Sinnott, P. J. Smith, and D. Widdows. 1981. "Active-site directed irreversible inhibition of glycosidases by the corresponding glycosylmethyl-(*p*-nitrophenyl) triazenes," *J. Chem. Soc. Perkins Trans.*, 1:366–376.

764 Fowler, A. V., I. Zabin, M. L. Sinnott, and P. J. Smith. 1978. "Methionine 500, the site of covalent attachment of an active site-directed reagent of β-galactosidase," *J. Biol. Chem.*, 253:5283–5285.

765 Halazy, S., V. Berges, A. Ehrhard, and C. Danzin. 1990. "*Ortho*- and *para*-(difluoromethyl)aryl-β-D-glucosides: A new class of enzyme-activated irreversible inhibitors of β-glucosidases," *Bioorg. Chem.*, 18:330–344.

766 Withers, S. G., I. P. Street, P. Bird, and D. H. Dolphin. 1987. "2-Deoxy-2-fluoroglucosides: A novel class of mechanism based inhibitors," *J. Am. Chem. Soc.*, 109:7530–7531.

767 Withers, S. G., K. Rupitz, and I. P. Street. 1988. "2-Deoxy-2-fluoro-D-glycosyl fluorides: A new class of specific mechanism-based glycoside inhibitors," *J. Biol. Chem.*, 263:7929–7932.

768 Withers, S. G., R. A. J. Warren, I. P. Street, K. Rupitz, J. B. Kempton, and R. Aebersold. 1990. "Unequivocal demonstration of the involvement of a glutamate residue as a nucleophile in the mechanism of a 'retaining' glycosidase," *J. Am. Chem. Soc.*, 112:5887–5889.

769 Street, I. P., J. B. Kempton, and S. G. Withers. 1992. "Inactivation of a β-glucosidase through the accumulation of a stable 2-deoxy-2-fluoro-α-D-glucopyranosyl-enzyme intermediate: A detailed investigation," *Biochemistry*, 31:9970–9978.

770 Wang, Q., D. Tull, A. Meinke, N. R. Gilkes, R. A. J. Warren, R. Aebersold, and S. G. Withers. 1993. "Glu280 is the nucleophile in the active site of *Clostridium thermocellum* CelC, a family A endo-β-1,4-glucanase," *J. Biol. Chem.*, 268: 14096–14102.

771 McCarter, J. D., M. J. Adam, C. Braun, M. Namchuk, D. Tull, and S. G. Withers. 1993. "Synthesis of 2-deoxy-2-fluoro mono- and oligo-saccharide glycosides from glycals and evaluation as glycosidase inhibitors," *Carbohydr. Res.*, 249:77–90.

772 Tull, D., S. Wither, N. R. Gilkes, D. G. Kilburn, R. A. J. Warren, and R. Aebersold. 1991. "Glutamic acid 274 is the nucleophile in the active site of a 'retaining' exoglucanase from *Cellulomonas fimi*," *J. Biol. Chem.*, 266:15621–15625.

773 Miao, S. C., J. D. McCarter, M. E. Grace, G. A. Grabowski, R. Aebersold, and S. G. Withers. 1994. "Identification of Glu(340) as the active-site nucleophile in human glucocerebrosidase by use of electrospray tandem mass spectrometry," *J. Biol. Chem.*, 269:10975–10978.

774 Post, C. B. and M. J. Karplus. 1986. "Does lysozyme follow the lysozyme pathway? An alternative based on dynamic, structural, and stereoelectronic considerations," *J. Am. Chem. Soc.*, 108:1317–1319.

775 Franck, R. W. 1992. "The mechanism of beta-glycosidases: A reassessment of some seminal papers," *Bio-organic Chemistry*, 20:77–88.

776 Fleet, G. W. J. 1985. "An alternative proposal for the mode of inhibition of glycosidase activity by polyhydroxylated piperidines, pyrrolidines and indolizidines: Implications for the mechanism of action of some glycosidases," *Tetrahedron Lett.*, 26:5073–5076.

777 Strynadka, N. C. and M. N. James. 1991. "Lysozyme revisited: Crystallographic evidence for distortion of an *N*-acetylmuramic acid residue bound in site D," *J. Mol. Biol.*, 220:401–424.

778 Yaguchi, M., C. Roy, C. F. Rollin, M. G. Paice, and L. Jurásek. 1983. "A fungal cellulase shows sequence homology with the active site of hen egg-white lysozyme," *Biochem. Biophys. Res. Commun.*, 116:408–411.

779 Wakarchuk, W. W., N. M. Greenburg, D. G. Kilburn, R. C. Miller, Jr., and R. A. J. Warren. 1988. "Structure and transcription analysis of the gene encoding a cellobiase from *Agrobacterium* sp. strain ATCC 21400," *J. Bacteriol.*, 170:301–307.

780 Moranelli, F., J. R. Barbier, M. J. Dove, R. M. MacKay, V. L. Seligy, M. Yaguchi, and G. E. Willick. 1986. "A clone for *Schizophyllum commune* β-glucosidase: Homology with a yeast β-glucosidase," *Biochem. Intl.*, 12:905–912.

781 Malcolm, B. A., S. Rosenberg, M. J. Corey, J. S. Allen, A. Baetselier, and J. F. Kirsch. 1989. "Site-directed mutagenesis of the catalytic residues Asp-52 and Glu-35 of chicken egg white lysozyme," *Proc. Natl. Acad. Sci. USA*, 86:133–137.

782 Schimmel, P. 1990. "Hazards and their exploitation in the application of molecular biology to structure-function relationships," *Biochemistry*, 29:9495–9502.

783 Svensson, B. and M. Sogaard. 1993. "Mutational analysis of glycosylase function," *J. Biotechnol.*, 29:1–37.

784 Trimbur, D. E., R. A. J. Warren and S. G. Withers. 1992. "Region-directed mutagenesis of residues surrounding the active site nucleophile of β-glucosidase from *Agrobacterium faecalis*," *J. Biol. Chem.*, 267:10248–10251.

785 Withers, S. G., K. Rupitz, D. Trimbur, and R. A. J. Warren. 1992. "Mechanistic consequences of mutation of the active site nucleophile Glu 358 in *Agrobacterium* β-glucosidase," *Biochemistry*, 31:9979–9985.

786 Baird, S. D., M. A. Hefford, D. A. Johnson, W. L. Sung, M. Yaguchi, and V. L. Seligy. 1990. "The Glu residue in the conserved asn-pro-glu sequence of two highly divergent endo-β-1,4-glucanases is essential for enzymatic activity," *Biochem. Biophys. Res. Commun.*, 169:1035–1039.

787 Navas, J. and P. Beguin. 1992. "Site-directed mutagenesis of conserved residues of *Clostridium thermocellum* endoglucanase CelC," *Biochem. Biophys. Res. Commun.*, 189:807–812.

788 Py, B., I. Bortoli-German, J. Haiech, M. Chippaux, and F. Barras. 1991. "Cellulase EGZ of *Erwinia chrysanthemi*: Structural organization and importance of His98 and Glu133 residues for catalysis," *Prot. Bioeng.*, 4:325–333.

789 Belaich, A., H.-P. Fierobe, D. Baty, B. Busetta, C. Bagnara-Tardif, C. Gaudin, and J.-P. Belaich. 1992. "The catalytic domain of endoglucanase A from *Clostridium cellulolyticum*: Effects of arginine 79 and histidine 122 mutations on catalyis," *J. Bacteriol.*, 174:4677–4682.

790 Roig, V., H.-P. Fierobe, V. Ducros, M. Czhzek, A. Belaich, C. Gaudin, J.-P. Belaich, and R. Haser. 1993. "Crystallization and preliminary X-ray analysis of the catalytic domain of endoglucanase from *Clostridium cellulolyticum*," *J. Mol. Biol.*, 233: 325–327.

791 Dominguez, R., H. Souchon, and P. M. Alzari. 1994. "Characterization of two crystal forms of *Clostridium thermocellum* endoglucanase CelC," *Proteins*, 19: 158–160.

792 Rixon, J. E., L. M. A. Ferreira, A. J. Durrant, J. I. Laurie, G. P. Hazlewood, and H. J. Gilbert. 1992. "Characterization of the gene *celD* and its encoded product 1,4-β-D-glucan glucohydrolase D from *Pseudomonas fluorescens* subsp. *cellulosa*," *Biochem. J.*, 285:947–955.

793 Mitsuishi, Y., S. Nitisinprasert, M. Saloheimo, I. Biese, T. Reinikainen, M. Claeyssens, S. Keranen, J. K. C. Knowles, and T. T. Teeri. 1990. "Site-directed mutagenesis of the putative catalytic residues of *Trichoderma reesei* cellobiohydrolase I and endoglucanase I," *FEBS Lett.*, 275:135–138.

794 Covert, S. F., A. VanDen Wymelenberg, and D. Cullen. 1992. "Structure, organization, and transcription of a cellobiohydrolase gene cluster from *Phanerochaete chrysosporium*," *Appl. Environ. Microbiol.*, 58:2168–2175.

795 Lee, Y.-E., S. E. Lowe, B. Henrissat, and J. G. Zeikus. 1993. "Characterization of the active site and thermostability regions of endoxylanase from *Thermoanaerobacterium saccharolyticum* B6A-RI," *J. Bacteriol.*, 175:5890–5898.

796 Divne, C., J. Stahlberg, T. Reinikainen, L. Ruohonen, G. Pettersson, J. K. C. Knowles, T. T. Teeri, and T. A. Jones. 1994. "The three-dimensional crystal structure of the catalytic core of cellobiohydrolase I from *Trichoderma reesei*," *Science*, 265:524–528.

797 Derewenda, U., L. Swenson, R. Green, Y. Y. Wei, R. Morosoli, F. Shareck, D. Kluepfel, and Z. S. Derewenda. 1994. "Crystal structure, at 2.6-angstrom resolution, of the *Streptomyces lividans* xylanase A, a member of the F family of β-1,4-D-glycanases," *J. Biol. Chem.*, 269:20811–20814.

798 White, A., S. G. Withers, N. R. Gilkes, and D. R. Rose. 1994. "Crystal structure of the catalytic domain of the β-1,4-glycanase Cex from *Cellulomonas fimi*," *Biochemistry*, 33:12546–12552.

799 Moreau, A., M. Roberge, C. Manin, F. Shareck, D. Kluepfel, and R. Morosoli. 1994. "Identification of two acidic residues involved in the catalysis of xylanase A from *Streptomyces lividans*," *Biochem. J.*, 302:291–295.

800 MacLeod, A. M., T. Lindhorst, and S. G. Withers. 1994. "The acid/base catalyst in the exoglucanase/xylanase from *Cellulomonas fimi* is glutamic acid 127: Evidence from detailed kinetic studies on mutants," *Biochemistry*, 33:6371.

801 Wakarchuk, W. W., R. L. Campbell, W. L. Sung, J. Davoodi, and M. Yaguchi. 1994. "Mutational and crystallographic analyses of the active site residues of the *Bacillus circulans* xylanase," *Protein Sci.*, 3:467–475.

802 Moriyama, H., Y. Hata, H. Yamaguchi, M. Sato, A. Shinmyo, N. Tanaka, H. Okada, and Y. Katsube. 1987. "B Crystallization and preliminary X-ray studies of *Bacillus pumilus* IPO xylanase," *J. Mol. Biol.*, 193:237–238.

803 Torronen, A., J. Rouvinen and M. Ahlgren. 1993. "Crystallization and preliminary X-ray analysis of two major xylanases from *Trichoderma reesei*," *J. Mol. Biol.*, 233:313–316.

804 Rose, D. R., G. I. Birnbaum, L. V. L. Tan, and J. N. Saddler. 1987. "Crystallization and preliminary X-ray diffraction study of a xylanase from *Trichoderma harzianum*," *J. Mol. Biol.*, 194:755–756.

805 Golubev, A. M., A. Y. Kilimnik, K. N. Neustroev, and R. W. Pickersgill. 1993. "Crystals of β-xylanase from *Aspergillus oryzae*," *J. Mol. Biol.*, 230:661–663.

806 Pickersgill, R. W., P. Debeire, M. Debeire-Gosselin, and J. A. Jenkins. 1993. "Crystallization and preliminary X-ray analysis of a thermophilic *Bacillus* xylanase," *J. Mol. Biol.*, 230:664–666.

807 Viswamitra, M. A., P. Bhanumoorthy, S. Ramakumar, M. V. Manjula, P. J. Vithayathil, S. K. Murtjy, and A. P. Naren. 1993. "Crystallization and preliminary X-ray diffraction analysis of crystals of *Thermoascus aurantiacus* xylanase," *J. Mol. Biol.*, 232:987–988.

808 Katsube, Y., Y. Hata, H. Yamaguchi, H. Moriyama, A. Shinmyo, and H. Okada. 1989. "Estimation of xylanase active site from crystalline structure," in *Proceedings of the 2nd International Conference on Protein Engineering: Protein design in basic research, Medicine, and Industry*, M. Ikehara and K. Titani, eds. Tokyo: Japan Scientific Soc. Press, pp. 91–96.

809 Ko, E. P., H. Akatsuka, H. Moriyama, A. Shinmyo, Y. Hata, Y. Katsube, I. Urabe, and H. Okada. 1992. "Site-directed mutagenesis at aspartate and glutamate residues of xylanase from *Bacillus pumilus*," *Biochem. J.*, 288:117–121.

810 Torronen, A., A. Harkki, and J. Rouvinen. 1994. "Three-dimensional structure of endo-1,4-β-xylanase II from *Trichoderma reesei:* Two conformational states in the active site," *EMBO J.*, 13:2493–2501.

811 Torronen, A. and J. Rouvinen. 1995. "Structural comparison of two major endo-1,4-xylanases from *Trichoderma reesei*," *Biochemistry*, 34:847–856.

812 Bray, M. R. and A. J. Clarke. 1995. "Identification of an essential tyrosyl residue in the binding site of *Schizophyllum commune* xylanase A," *Biochemistry*, 34:2006–2014.

813 Hata, Y., K. Natori, Y. Katsube, T. Ooi, M. Arai, and H. Okada. 1994. "Crystallization and preliminary x-ray diffraction studies of an endoglucanase from *Aspergillus aculeatus*," *J. Mol. Biol.*, 241:278–280.

814 Abuja, P. M., I. Pilz, M. Claeyssens, and P. Tomme. 1988. "Domain structure of cellobiohydrolase II as studied by small angle X-ray scattering: Close resemblance to cellobiohydrolase I," *Biochem. Biophys. Res. Commun.*, 156:180–185.

815 Pilz, I., E. Schearz, D. G. Kilburn, R. C. Miller, Jr., R. A. J. Warren, and N. R. Gilkes. 1990. "The tertiary structure of a bacterial cellulase determined by small-angle X-ray scattering analysis," *Biochem. J.*, 271:277–280.

816 Damude, H. G., S. G. Withers, D. G. Kilburn, R. C. Miller, Jr., and R. A. J. Warren. 1995. "Site-directed mutation of the putative catalytic residues of endoglucanase CenA from *Cellulomonas fimi*," *Biochemistry*, 34:2220–2224.

817 Chauvaux, S., P. Beguin, and J.-P. Aubert. 1992. "Site-directed mutagenesis of essential carboxylic residues in *Clostridium thermocellum* endoglucanase CelD," *J. Biol. Chem.*, 267:4472–4478.

818 Joliff, G., P. Beguin, J. Millet, J.-P. Aubert, P. Alzari, M. Juy, and R. J. Poljak. 1986. "Crystallization and preliminary X-ray diffraction study of an endoglucanase from *Clostridium thermocellum*," *J. Mol. Biol.*, 189:249–250.

819 Juy, M., A. G. Amit, P. M. Alzari, R. J. Poljak, M. Claeyssens, P. Beguin and J.-P. Aubert. 1992. "Three-dimensional structure of a thermostable bacterial cellulase," *Nature (London)*, 357:89–91.

820 Davies, G., S. Tolley, K. Wilson, M. Schulein, H. F. Woldike, and G. Dodson. 1992. "Crystallization and preliminary X-ray analysis of a fungal endoglucanase I," *J. Mol. Biol.*, 228:970–972.

821 Davies, G. J., G. G. Dodson, R. E. Hubbard, S. P. Tolley, Z. Dauter, K. S. Wilson, C. Hjort, J. Moller Mikkelsen, G. Rasmussen, and M. Schulein. 1993. "Structure and function of endoglucanase V," *Nature (London)*, 365:362–364.

822 Ozaki, K., N. Sumitomo, Y. Hayashi, S. Kawai, and S. Ito. 1994. "Site-directed mutagenesis of the putative active site of endoglucanase K from *Bacillus* sp. KSM-330," *Biochim. Biophys. Acta*, 1207:159–164.

823 Legler, G. 1967. "Investigations on the mechanism of action of glycoside-hydrolysing enzymes. II. Isolation and enzymatic properties of two β-glucosidases from *Aspergillus wentii*," *Hoppe-Seyler's Z. Physiol. Chem.*, 348:1359–1366.

824 Maguire, R. J. 1977. "Kinetics of the hydrolysis of cellobiose and *p*-nitrophenyl-β-D-glucoside by cellobiase of *Trichoderma viride*," *Can. J. Biochem.*, 55:19–26.

825 Cubellis, M. V., C. Rozzo, P. Montecucchi, and M. Rossi. 1990. "Isolation and sequencing of a new β-galactosidase-encoding archaebacterial gene cloning and sequencing of the *celA* gene encoding endoglucanase A of *Butyrivibrio fibrisolvens* strain A46," *Gene*, 94:89–94.

826 Wakarchuk, W. W., N. M. Greenberg, D. G. Kilburn, R. C. Miller, Jr., and R. A. J. Warren. 1988. "Structure and transcription analysis of the gene encoding a cellobiase from *Agrobacterium* sp. strain ATCC 21400," *J. Bacteriol.*, 170: 301–307.

827 Mantei, N., M. Villa, T. Enzler, H. Wacker, W. Boll, P. James, W. Hunziker, and G. Semenza. 1988. "Complete primary structure of human and rabbit lactase phlorizin hydrolase: Implications for biosynthesis, membrane anchoring and evolution of the enzyme," *EMBO J.*, 7:2705–2713.

828 Porter, E. V. and B. M. Chassy. 1988. "Nucleotide sequence of the β-D-phosphogalactoside galactohydrolase gene of *Lactobacillus casei*: Comparison to analagous pbg genes of other gram-positive microorganisms," *Gene*, 62:263–276.

829 Boizet, B., D. Vileval, P. Slos, M. Novel, G. Novel, and A. Mercenier. 1988. "Isolation and structural analysis of the phospho-β-galactosidase gene from *Streptococcus lactis* Z268," *Gene*, 62:249–261.

830 Breidt, F. and G. C. Stewart. 1987. "Nucleotide and deduced amino sequences of the *Staphylococcus aureus* phospho-β-galactosidase gene," *Appl. Environ. Microbiol.*, 53:969–973.

831 Schnetz, K., C. Toloczyki, and B. Rak. 1987. "β-Glucoside (*bgl*) operon of *Escherichia coli* K-12: Nucleotide sequence, genetic organization, and possible evolutionary relationship to regulatory components of two *Bacillus subtilis* genes," *J. Bacteriol.*, 169:2579–2590.

832 Little, S., P. Cartwright, C. Campbell, A. Prenneta, J. McChesney, A. Mountain, and M. Robinson. 1989. "Nucleotide sequence of a thermostable β-galactosidase from *Sulfolobus slofataricus*," *Nucleic Acid Res.*, 17:7980.

833 Baird, S. D., D. A. Johnson, and V. Seligy. 1990. "Molecular cloning, expression and characterization of endo-β-1,4-glucanase genes from *Bacillus polymyxa* and *Bacillus circulans*," *J. Bacteriol.*, 172:1576–1586.

834 Meinke, A., N. R. Gilkes, D. G. Kilburn, R. C. Miller, Jr., and R. A. J. Warren. 1993. "Cellulose-binding polypeptides from *Cellulomonas fimi*: Endoglucanase D (CenD), a family A β-1,4-glucanase," *J. Bacteriol.*, 175:1910–1918.

835 Al-Tawheed, A. R. 1988. "Molecular characterization of cellulose genes from *Cellulomonas flavigena*," University of Dublin: Ph.D. diss.

836 Gough, C. L., J. M. Dow, J. Keen, B. Henrissat, and M. J. Daniels. 1990. "Nucleotide sequence of the *engXCA* gene encoding the major endoglucanase of *Xanthomonas campestris* pv. *campestris*," *Gene*, 89:53–59.

837 Hazlewood, G. P., K. Davidson, J. I. Laurie, M. P. M. Romaniec, and H. J. Gilbert. 1990. "Cloning and sequencing of the *celA* gene encoding endoglucanase A of *Butyrivibrio fibrisolvens* strain A46," *J. Gen. Microbiol.*, 136:2089–2097.

838 Sumitomo, N., K. Ozaki, S. Kawai, and S. Ito. 1990. "Nucleotide sequence of the gene for an alkaline endoglucanase from an alkalophilic *Bacillus* and its expression in *Escherichia coli* and *Bacillus subtilis*," *Biosci. Biotechnol. Biochem.*, 56: 872–877.

839 Osaki, K., S. Shikata, S. Kawai, S. Ito, and K. Okamoto. 1990. "Molecular cloning and nucleotide sequence of a gene for alkaline cellulase from *Bacillus* sp. KSM-635," *J. Gen. Microbiol.*, 136:1327–1334.

840 Park, S. H, H. K Kim, and M. Y. Pack. 1991. "Characterization and structure of the cellulase gene of *Bacillus subtilis* BSE616," *Agric. Biol. Chem.*, 55:441–448.

841 Nakamura, A., T. Uozumi, and T. Beppu. 1987. "Nucleotide sequence of a cellulase gene of *Bacillus subtilis*," *Eur. J. Biochem.*, 164:317–320.

842 MacKay, R. M., A. Lo, G. Willick, M. Zuker, S. Baird, M. Dove, F. Moranelli, and V. Seligy. 1986. "Structure of a *Bacillus subtilis* endo-β-1,4-glucanase gene," *Nucleic Acid Res.*, 14:9159–9170.

843 Robson, L. M. and G. H. Chambliss. 1986. "Cloning of the *Bacillus subtilis* DLG β-1,4-glucanase gene and its expression in *Escherichia coli* and *B. subtilis*," *J. Bacteriol.*, 165:612–619.

844 Fujino, T. and K. Ohmiya. 1992. "Nucleotide sequence of an endo-1,4-glucanase gene (*celA*) from *Clostridium josui*," *J. Ferment. Bioeng.*, 73:308–313.

845 Sakka, K., T. Shimanuki, and K. Shimada. 1991. "Nucleotide sequence of *celC307* encoding endoglucanase C307 of *Clostridium* sp. strain F1," *Agric. Biol. Chem.*, 55:347–350.

846 Wang, W. and J. A. Thomson. 1990. "Nucleotide sequence of the *celA* gene encoding a cellodextrinase of *Ruminococcus flavefaciens* FD-1," *Mol. Gen. Genet.*, 222:265–269.

847 Joorgesen, P. L. and C. K. Hansen. 1990. "Mutliple endo-β-1,4-glucanase-encoding genes from *Bacillus lautus* PL236 and characterization of the *celB* gene," *Gene*, 93: 55–60.

848 Matsushita, O., J. B. Russell, and D. B. Wilson. 1990. "Cloning and sequencing of a *Bacteroides ruminicola* B14 endoglucanase gene," J. Bacteriol., 172: 3620–3630.

849 Faure, E., A. Belaich, C. Bagnara, C. Gaudin, and J.-P. Belaich. 1989. "Sequence analysis of the *Clostridium cellulolyticum* endoglucanase-A-encoding gene, *celCCA*," *Gene*, 84:39–46.

850 Foong, F., T. Hamamoto, O. Shoseyov, and R. H. Doi. 1991. "Nucleotide sequence and characteristics of endoglucanase gene *engB* from *Clostridium cellulovorans*," *J. Gen. Microbiol.*, 137:1729–1736.

851 Hamamoto, T., F. Foong, O. Shoseyov, and R. H. Doi. 1992. "Analysis of functional domains of endoglucanases from *Clostridium cellulovorans* by gene cloning, nucleotide sequencing and chimeric protein construction," *Mol. Gen. Genet.*, 231: 472–479.

852 Luthi, E., N. B. Jasmat, R. A. Grayling, D. R. Love, and P. L. Bergquist. 1991. "Cloning, sequence analysis, and expression in *Escherichia coli* of a gene encoding for a β-mannanase from the extremely thermophilic bacterium *Caldocellum saccharolyticum*," *Appl. Environ. Microbiol.*, 57:694–700.

853 Poole, D. M., G. P. Hazlewood, J. I. Laurie, P. J. Barker, and H. J. Gilbert. 1990. "Nucleotide sequence of the *Ruminococcus albus* SY3 endoglucanase genes *celA* and *celB*," *Mol. Gen. Genet.*, 223:217–223.

854 Cunningham, C., C. A. McPherson, J. Martin, W. J. Harris, and H. J. Flint. 1991. "Sequence of a cellulase gene from the rumen anaerobe *Ruminococcus flavefaciens*," *Mol. Gen. Genet.*, 228:320–323.

855 Huang, J. and M. A. Schell. 1992. "Role of the two-component leader sequence and mature amino acid sequences in extracellular export of endoglucanase EGL from *Pseudomonas solanacearum*," *J. Bacteriol.*, 174:1314–1323.

856 Yoshigi, N., H. Taniguchi and T. Sasaki. 1990. "Cloning and sequencing of the endo-cellulase cDNA from *Robillarda* sp. Y-20," *J. Biochem (Tokyo)*, 108: 388–392.

857 Maat, J., M. Roza, J. Verbakel, H. Stam, M. J. Santos da Silva, M. Bosse, M. R. Egmond, M. L. D. Hagemans, R. F. M. van Gorcom, J. G. M. Hessing, C. A. M. J. J. van der Hondel, and C. van Rotterdam. 1992. "Xylanases and their application in bakery," in *Xylans and xylanases*, J. Visser, G. Beldman, M. A. Kusters-van Someren, and A. G. J. Voragen, eds. Amsterdam: Elsevier Science Publishers B.V., pp. 349–360.

858 de Graff, L. H., H. C. van den Broeck, A. J. J. van Ooijen, and J. Visser. 1992. "Structure and regulation of an *Aspergillus* xylanase gene," in *Xylans and xylanases*, J. Visser, G. Beldman, M. A. Kusters-van Someren, and A. G. J. Voragen, eds. Elsevier Science Publishers B.V., pp. 235–246.

859 Yang, R. C. A., C. R. MacKenzie, and S. A. Narang. 1988. "Nucleotide sequence of a *Bacillus circulans* xylanase gene," *Nucleic Acid Res.*, 16:7187.

860 Fukusaki, E., W. Panbangred, A. Shinmyo, and H. Okada. 1984. "The complete nucleotide sequence of the xylanase gene (*xynA*) of *Bacillus pumilus*," *FEBS Lett.*, 171:197–201.

861 Paice, M. G., R. Bourbonnais, M. Desrochers, L. Jurásek, and M. Yaguchi. 1986. "A xylanase gene from *Bacillus subtilis*: Nucleotide sequence and comparison with *B. pumilus* gene," *Arch. Microbiol.*, 144:201–206.

862 Yasui, T., M. Marui, I. Kusakabe, K. Nakanishi, H. Zappe, W. A. Jones, and D. R. Woods. 1990. "Nucleotide sequence of a *Clostridium acetobutylicum* P262 xylanase gene *(xynb)*," *Nucleic Acids Res.*, 18:2179–2655.

863 Zhang, J. and H. J. Flint. 1990. EMBL database accession number Z11127.

864 Shareck, F., C. Roy, M. Yaguchi, R. Morosoli, and D. Kluepfel. 1991. "Sequences of three genes specifying xylanases in *Streptomyces lividans*," *Gene*, 107:75–82.

865 Nagashima, M., Y. Okumoto, and M. Okanishi. 1989. "Nueleotide sequences of the gene of extracellular xylanase in *Streptomyces* species no. 36a and construction of secretion vectors using xylanase gene," in *Trends in actinomycetology in Japan*, Y. Koyama, ed. Tokyo: Society for Actinomycetes, pp. 91–96.

866 Yaguchi, M., C. Roy, D. C. Watson, C. F. Rollin, L. U. L. Tan, D. J. Senior, and J. N. Saddler. 1992. "The amino acid sequence of the 20 kD xylanase from *Trichoderma harzianum*," in *Xylans and xylanases*, J. Visser, G. Beldman, M. A. Kusters-van Someren, and A.G.J. Voragen, eds. Amsterdam: Elsevier Science Publishers B.V., pp. 435–438.

867 Torronnen, A., R. L. Mach, R. Messner, R. Gonzalez, N. Kalkkinen, A. Harkki, and C. Kubicek. 1992. "The two major xylanases from *Trichoderma reesei*: Characterization of both enzymes and genes," *Biotechnology*, 10:1461–1465.

868 Yaguchi, M., C. Roy, M. Ujiie, D. C. Watson, and W. Wakarchuk. 1992. "Amino acid sequence of the low-molecular xylanase from *Trichoderma viride*," in *Xylans and xylanases*, J. Visser, G. Beldman, M. A. Kusters-van Someren, and A. G. J. Voragen, eds. Amsterdam: Elsevier Science Publishers B.V., pp. 149–154.

869 Coutinho, J. B., B. Moser, D. G. Kilburn, R. A. J. Warren, and R. C. Miller, Jr. 1991. "Nucleotide sequence of the endoglucanase C gene (*cenC*) of *Cellulomonas fimi*, its high-level expression in *Escherichia coli*, and characterization of its products," *Mol. Microbiol.*, 5:1221–1235.

870 Berger, E., W. A. Jones, D. T. Jones, and D. R. Woods. 1990. "Sequencing and expression of a cellodextrinase (*cedl*) gene from *Butyrivibrio fibrisolvens* H17c cloned in *Escherichia coli*," *Mol. Gen. Genet.*, 223:310–318.

871 Tucker, M. L., M. L. Durbin, M. T. Clegg, and L. N. Lewis. 1987. "Avocado cellulase: Nucleotide sequence of a putative full-length cDNA clone and evidence for a small gene family," *Plant Mol. Biol.*, 9:197–203.

872 Giorda, R., T. Ohmachi, D. R. Shaw, and H. L. Ennis. 1990. "A shared internal threonine-glutamic acid-threonine-proline repeat defines a family of *Dictyostelium discodieum* spore germination specific proteins," *Biochemistry*, 29:7264–7269.

873 Jauris, S., K. P. Ruchnagel, W. H. Schwarz, P. Kratzsch, K. Bronnenmeier, and W. L. Staudenbauer. 1990. "Sequence analysis of the *Clostridium stercorarium celZ* gene encoding a thermoactive cellulase (Avicelase I): Identification of catalytic and cellulose-binding domains," *Mol. Gen. Genet.*, 223:258–267.

874 Bagnara-Tardif, C., C. Gaudin, A. Belaich, P. Hoest, T. Citard and J. P. Belaich. 1992. "Sequence analysis of a gene cluster encoding cellulases from *Clostridium cellulolyticum*," *Gene*, 119:17–28.

875 Nakamura, K., N. Misawa, and K. Kitamura. 1986. "Sequence of a cellulase gene of *Cellulomonas uda* CB4," *J. Biotechnol.*, 4:247–254.

876 Bueno, A., C. R. Vazquez de Aldana, J. Correa, T. G. Villa, and F. Del Rey. 1990. "Nucleotide sequence of a 1,3-1,4-β-glucanase-encoding gene of *Bacillus circulans* WL-12," *Nucleic Acids Res.*, 18:4248.

877 Chipman, D. M. and N. Sharon. 1969. "Mechanism of lysozyme action," *Science*, 165:454–465.

878 Hiromi, K. 1972. "Subsite affinities and action pattern of amylases," in *Proteins: Structure and function, Vol. 2*, M. Funatsu, K. Hiromi, K. Imahori, T. Murachi, and K. Norita, eds. New York: John Wiley and Sons, pp. 2–41.

879 Hiromi, K. 1972. "Analysis of action patterns of amylases with a special reference to their subsite affinities," in Molecular mechanisms of enzyme action, Y. Ogura, Y. Tonomura, and T. Nakamura, eds. University of Tokyo Press, Tokyo, pp. 241–262.

880 Suganuma, T., R. Matsuno, M. Ohnishi, and K. Hiromi. 1978. "A Study of the mechanism of action of taka-amylase A1 on linear oligosaccharides by product analysis and computer simulation," *J. Biochem. (Tokyo)*, 84:293–316.

881 Thoma, J. A. and J. D. Allen. 1976. "Subsite mapping of enzymes: Collecting and processing experimental data: A case study of an amylase-malto-oligosaccharide system," *Carbohydr. Res.*, 48:105–124.

882 Bhat, K. M., A. J. Hay, M. Claeyssens, and T. M. Wood. 1990. "Study of the mode of action and site-specificity of the endo-(1,4)-β-D-glucanases of the fungus *Penicillium pinophilum* with normal, 1-^{3}H-labelled, reduced and chromogenic cello-oligosaccharides," *Biochem. J.*, 266:371–378.

883 Van Tilbeurgh, H., G. Pettersson, R. Bhikabhai, H. De Boeck, and M. Claeyssens. 1985. "Studies of the cellulolytic system of *Trichoderma reesei* QM 9414. Reaction specificity and thermodynamics of interactions of small substrates and ligands with the 1,4-β-glucan cellobiohydrolase II," *Eur. J. Biochem.*, 148:329–334.

884 Chirico, W. J. and R. D. Brown, Jr. 1987. "β-Glucosidase from *Trichoderma reesei*. Substrate-binding region and mode of action on [1-^{3}H]cello-oligosaccharides," *Eur. J. Biochem.*, 165:343–351.

885 Biely, P., M. Vrsanka, and I. V. Gorbacheva. 1983. "The active site of an acidic endo-1,4-β-xylanase of *Aspergillus niger*," *Biochem. Biophys. Acta*, 743:155–161.

886 Biely, P., Z. Krátk, and M. Vrsanka. 1981. "A substrate-binding site of endo-1,4-β-xylanase of the yeast *Cryptococcus albidus*," *Eur. J. Biochem.*, 119:559–564.

887 Biely, P., M. Vrsanka, and Z. Krátk. 1981. "Mechanisms of substrate digestion by endo-1,4-β-xylanase of *Cryptococcus albidus*: Lysozyme-type pattern of action," *Eur. J. Biochem.*, 119:565–571.

888 Vrsanka, M., I. V. Gorbacheva, Z. Krátk, and P. Biely. 1982. "Reaction pathways of substrate degradation by an acidic endo-1,4-β-xylanase of *Aspergillus niger*," *Biochim. Biophys. Acta*, 704:114–122.

889 Nidetzky, B., W. Zachariae, G. Gercken, M. Hayn, and W. Steiner. 1994. "Hydrolysis of cellooligosaccharides by *Trichoderma reesei* cellobiohydrolases—Experimental data and kinetic modeling," *Enzyme Microb. Technol.*, 16:43–52.

890 Claeyssens, M., H. Van Tilbeurgh, P. Tomme, T. M. Wood, and S. I. McCrae. 1989. "Fungal cellulase systems. Comparison of the specificities of the cellobiohydrolases isolated from *Penicillium pinophilum* and *Trichoderma reesei*," *Biochem. J.*, 261: 819–825.

891 Thoma, J. A., G. V. K. Rao, C. Brothers, J. Spradlin, and L. H. Li. 1971. "Subsite mapping of enzymes: Correlation of product patterns with Michaelis parameters and substrate-induced strain," *J. Biol. Chem.*, 246:5621–5635.

892 Meagher, M. M., B. Y. Tao, J. M. Chow, and P. J. Reilly. 1988. "Kinetics and subsite mapping of a D-xylobiose- and D-xylose-producing *Aspergillus niger* endo-(1,4)-β-D-xylanase," *Carbohydr. Res.*, 173:273–283.

893 Mitsuishi, Y., T. Yamanobe, and M. Yagisawa. 1988. "The modes of action of three xylanases from mesophilic fungus strain Y-94 on xylo-oligosaccharides," *Agric. Biol. Chem.*, 52:921–927.

894 Robyt, J. F. and D. French. 1970. "The action pattern of porcine pancreatic α-amylase in relationship to the substrate-binding site of the enzyme," *J. Biol. Chem.*, 245:3917–3927.

895 Biely, P. and M. Vrsanka. 1988. "Xylanase of *Cryptococcus albidus*," *Methods Enzymol.*, 160:638–648.

896 Claeyssens, M. and B. Henrissat. 1992. "Specificity mapping of cellulolytic enzymes: Classification into families of structurally related proteins confirmed by biochemical analysis," *Protein Sci.*, 1:1293–1297.

897 Quiocho, F. A. 1988. "Molecular features and basic understanding of protein-carbohydrate interactions: The arabinose-binding protein-sugar complex," *Curr. Top. Microbiol. Immunol.*, 139:135–148.

898 Quiocho, F. A. 1986. "Carbohydrate binding proteins: Tertiary structures and protein-sugar interactions," *Annu. Rev. Biochem.*, 55:287–315.

899 Pettersson, G. 1968. "Structure and function of a cellulase from *Penicillium notatum* as studied by chemical modification and solvent accessibility," *Arch. Biochem. Biophys.*, 126:776–784.

900 Clarke, A. J. and M. Yaguchi. 1986. "Difference spectrophotometric study on the interaction of cellulase from *Schizophyllum commune* with substrate and inhibitors," *Biochim. Biophys. Acta*, 870:401–407.

901 Clarke, A. J. 1987. "Essential tryptophan residues in the function of cellulase from *Schizophyllum commune*," *Biochim. Biophys. Acta*, 912:424–431.

902 Baker, J. O., D. J. Mitchell, K. Grohmann, and M. E. Himmel. 1991. "Thermal unfolding of *Trichoderma reesei* CBHI," *ACS Symp. Ser.*, 460:313–330.

903 Deshpande, V., J. Hinge, and M. Rao. 1990. "Chemical modification of xylanases: Evidence for essential tryptophan and cysteine residues at the active site," *Biochim. Biophys. Acta*, 1041:172–177.

904 Bray, M. R., A. D. Carriere, and A. J. Clarke. 1994. "Quantitation of tryptophan and tyrosine resdiues in proteins by fourth derivative spectroscopy," *Anal. Biochem.*, 221:278–284.

905 Keskar, S. S., M. C. Srinivasan, and V. V. Deshpande. 1989. "Chemical modification of a xylanase from a thermotolerant *Streptomyces*: Evidence for essential tryptophan and cysteine residues at the active site," *Biochem. J.*, 261:49–55.

906 Kiss, L., I. Korodi, and P. Nanasi. 1981. "Study on the role of tyrosine side-chains at the active centre of emulsin β-D-glucosidase," *Biochim. Biophys. Acta*, 662:308–311.

907 Myers, B. II and A. N. Glazer. 1971. "Spectroscopic studies of the exposure of tyrosine residues in proteins with special reference to the subtilisins," *J. Biol. Chem.*, 246:412–419.

908 Lundblad, R. C. and C. M. Noyes. 1984. *Chemical reagents for protein modification*. Baton Rouge: CRC Press Inc., pp. 105–121.

909 Chanzy, H. and B. Henrissat. 1985. "Undirectional degradation of *Valonia* cellulose microcrystals subjected to cellulase action," *FEBS Lett.*, 184:285–288.

910 Moreau, A., F. Shareck, D. Kluepfel, and R. Morosoli. 1994. "Alteration of the cleavage mode and of the transglycosylation reactions of the xylanase a of *Streptomyces lividans* 1326 by site-directed mutagenesis of the Asn173 residue," *Eur. J. Biochem.*, 219:261–266.

911 Evans, B. R., R. Margalit, and J. Woodward. 1993. "Attachment of pentaamine ruthenium (III) to *Trichoderma reesei* cellobiohydrolase I increases its catalytic activity," *Biochem. Biophys. Res. Commun.*, 195:497–503.

912 Evans, B. R., R. Margalit, and J. Woodward. 1994. "Veratryl alcohol oxidase activity of a chemically modified cellulase protein," *Arch. Biochem. Biophys.*, 312:459–466.

913 Dominguez, J. M., C. Acebal, J. Jimenez, I. de la Mata, R. Macarron, and M. P. Castillon. 1992. "Mechanisms of thermoinactivation of endoglucanase I from *Trichoderma reesei* QM 9414," *Biochem. J.*, 287:583–588.

914 Tatu, U., S. K. Murthy, and P. J. Vithayathil. 1990. "Role of a disulfide cross-link in the conformational stability of a thermostable xylanase," *J. Protein Chem.*, 9:641–646.

915 Woodward, J., K. S. Whaley, G. S. Zachry, and D. L. Wohlpart. 1981. "Thermal stability of *Trichoderma reesei* C30 cellulase and *Aspergillus niger* β-glucosidase after pH and chemical modification," *Biotechnol. Bioeng. Symp.*, 11:619–629.

916 Reese, E. T. and M. Mandels. 1980. "Stability of the cellulase of *Trichoderma reesei* under use conditions," *Biotechnol. Bioeng.*, 22:323–335.

917 Reese, E. T., R. G. H. Siu, and H. S. Levinson. 1950. "The biological degradation of soluble cellulose derivatives and its relationship to the mechanism of cellulose hydrolysis," *J. Bacteriol.*, 59:485–497.

918 Oku, T., C. Roy, D. C. Watson, W. Wakarchuk, R. Campbell, M. Yaguchi, L. Jurasek, and M. G. Paice. 1993. "Amino acid sequence and thermostability of xylanase-A from *Schizophyllum-commune*," *FEBS Lett.*, 334:296–300.

919 Tempelaars, C. A. M., P. R. J. Birch, P. F. G. Sims, and P. Broda. 1994. "Isolation, characterization, and analysis of the expression of the CBHII gene of *Phanerochaete chrysosporium*," *Appl. Environ. Microbiol.*, 60:4387–4393.

920 Li, X. L. and L. G. Ljungdahl. 1994. "Cloning, sequencing, and regulation of a xylanase gene from the fungus *Aureobasidium pullulans* Y-2311-1," *Appl. Environ. Microbiol.*, 60:3160–3166.

921 Perito, B., E. Hanhart, T. Irdani, M. Iqbal, A. J. McCarthy, G. Mastromei. 1994. "Characterization and sequence analysis of a *Streptomyces rochei* A2 endoglucanase-encoding gene," *Gene*, 148:119–124.

922 Sinnott, M. L. 1993. "There is no experimental evidence for endocyclic cleavage in the action of e $\rightarrow$ e (retaining β) glycopyranosidases, and much against it," *Bioorg. Chem.*, 21:34–40.

Index